REVERSE ENGINEERING OF RUBBER PRODUCTS

Concepts, Tools, and Techniques

REVERSE ENGINEERING OF RUBBER PRODUCTS

Concepts, Tools, and Techniques

Saikat Das Gupta
Rabindra Mukhopadhyay
Krishna C. Baranwal
Anil K. Bhowmick

CRC Press is an imprint of the
Taylor & Francis Group, an **informa** business

CRC Press
Taylor & Francis Group
6000 Broken Sound Parkway NW, Suite 300
Boca Raton, FL 33487-2742

First issued in paperback 2017

Version Date: 20130806

ISBN 13: 978-0-8493-7316-9 (hbk)
ISBN 13: 978-1-138-07525-2 (pbk)

Visit the Taylor & Francis Web site at
http://www.taylorandfrancis.com

and the CRC Press Web site at
http://www.crcpress.com

To

Sreosree, Sanghamitra and Suchandra Das Gupta

Anshuman, Dhritiman and Nibha Mukhopadhyay

Arti Baranwal

Asmit and Kumkum Bhowmick

Contents

Preface

Reverse engineering is widely practiced in the rubber industry. Competitors' products are routinely analyzed in a company either to get an idea about specifications or to gain knowledge about compositions. There are scientific principles behind the analysis and reconstruction. The idea of this book is to highlight the scientific aspects of these areas. The book contains five chapters. Chapter 1 describes compounding ingredients and formulae construction. Principal chemical and analytical methods are discussed in Chapter 2. Routine physical test methods used in the rubber industry are presented in Chapter 3. The reverse engineering concept is introduced in Chapter 4. Chapter 5 is a compendium of case studies of formula reconstruction, covering large products such as tires, belts, etc., as well as small products like seals and hoses. These products represent a variety, and the presentation is deliberately different in each chapter so that readers become familiar with the methods followed in various organizations.

We hope that this book is useful to those in the rubber industry and academics interested in understanding the principles of various tests used in the industry. Our labor will be amply rewarded if this book satisfies its readers and bridges the gap that currently exists.

Our knowledge of this subject has been gained through a number of experiments done over the years by us and our colleagues. We appreciate the contributions of our colleagues in developing this knowledge and solving various problems. A number of people helped us in the preparation of this manuscript. To name just a few, we gratefully acknowledge the contributions of Sugata Chakraborty, Rohit Ameta, Samar Bandyopadhyay, Sanjay Bhattacharyya, S. Mathur, Nilkant Mondal, S. Divakaran, M. Asif (Hari Shankar Singhania Elastomer and Tyre Research Institute [HASETRI], Kankroli, India), Nibedita Kasyapi, Shib Shankar Banerjee, Moumita Kotal, Vidhi Choudhury, Nabarun Roy, M. Bhattacharya, Dinesh Kumar, Suman Mitra, Pradip Maji, Ganesh Basak, Anusuya Choudhury (Indian Institute of Technology [IIT] Patna and IIT Kharagpur, India), B. Dutta (Indian Rubber Institute [IRI], Kolkata, India), Atanu Banerjee, Sandip Bhattacharya, Niloptal Dey (Tata Steel, Jamshedpur, India), Bharat Singh Mandoloi, A.K. Biswas (Hindustan Aeronautics Limited [HAL], Nasik and Bengaluru, India), and A.T. Khan, Sanka Babu Narendra and K. Neeraj (IIT Patna). Special thanks go to Titash Mondal, senior research scholar (IIT Patna) for his help in the final processing of this book. We appreciate Donald Kierstead for microscopy work and Maryann Smith for preparing the report presented in Section 2.6 (both of Akron Rubber Development Laboratory, Akron, Ohio). Finally, the book would never have been possible without the support of Raghupati Singhania, president, HASETRI, and the management of HASETRI. Anil K. Bhowmick is grateful to the Ministry of Human Resource Development (MHRD), New Delhi, for encouraging industry–institute interaction, which has resulted in this book.

Saikat Das Gupta

Rabindra Mukhopadhyay

Krishna C. Baranwal

Anil K. Bhowmick

The Authors

Saikat Das Gupta is chief scientist in research and development at HASETRI. He earned a BTech in Polymer Science and Rubber Technology from Calcutta University, India, in 1987 and completed the MTech in Rubber Technology from the Indian Institute of Technology (IIT), Kharagpur, India, in 1992. Das Gupta then joined the research and development section of JK Tyre and Industries Limited. He pursued his PhD degree while working, and earned that degree in 2009. He completed a postgraduate diploma in business management in 2001.

Rabindra Mukhopadhyay is director and chief executive, Hari Shankar Singhania Elastomer and Tyre Research Institute (HASETRI); director (research and development), JK Tyre and Industries Limited; and chairman of Indian Rubber Institute. He is associated with several institutes and universities in India as a visiting faculty in the Department of Polymer Science and Rubber Technology. His fields of interest are reverse engineering, environmentally friendly technology, sustainable development, and nanotechnology with special reference to the Rubber and Allied industry including various business excellence models on quality.

Krishna C. Baranwal earned his BS and MS degrees from the University of Allahabad, and his MTech degree from the Indian Institute of Technology, Kharagpur, India. After receiving his PhD in polymer science from the University of Akron, Ohio, he joined BF Goodrich Research Center, where he later became director of tire research and tire testing. After about 30 years with Goodrich, he joined the Akron Rubber Development Laboratory, Ohio, as executive vice president–technical. He has a distinguished professional career of 43 years in the rubber industry, which includes 38 publications, four patents, and two books.

Baranwal was previously active in the Rubber Division, American Chemical Society (ACS). He has been chairman of the Program Planning Committee and also chaired the Science and Technology Awards Committee. He has been the editor of *Rubber Chemistry and Technology*, an internationally recognized rubber journal, and has edited textbooks used in the correspondence courses by the Rubber Division, ACS. In addition to his activities in technical groups, he is involved in community volunteer work for several nonprofit organizations.

Anil K. Bhowmick is a Professor of Eminence and former head of Rubber Technology Centre and Dean of Postgraduate Studies and Dean (Sponsored Research and Industrial Consultancy), IIT Kharagpur. He was previously associated with the University of Akron, Ohio; London School of Polymer Technology, London; and Tokyo Institute of Technology, Japan. His main research interests are thermoplastic elastomers and polymer blends, nanocomposites, polymer modification, rubber technology, failure and degradation of polymers, adhesion and adhesives. He has more than 500 publications in peer-reviewed journals, 35 book chapters, and 7 co-edited books. He was also coeditor of the special issue of Polymer and Composite Characterization of the *Journal of Macromolecular Science.* He was the 2002 winner of the Chemistry of Thermoplastic Elastomers Award and 1997 winner of the George Stafford Whitby Award of the Rubber Division, American Chemical Society, for innovative research, and the 2001 K.M. Philip Award of the All India Rubber Industries Association for outstanding contribution to the growth and development of rubber industries in India. He was also awarded the NOCIL Award (1991), the JSPS Award (1990), the Commonwealth Award (1990), the MRF Award (1989), and the Stanton-Redcroft ITAS Award (1989). He is on the editorial board of the *Journal of Adhesion Science & Technology, Journal of Applied Polymer Science, Journal of Materials Science, Polymers and Polymer Composites, Rubber Chemistry and Technology, Polymers for Advanced Technology,* and *Natural Rubber Research.* He holds 12 patents including one U.S. and one German patent. He guided 40 PhD students. He is a Fellow of the National Academy of Engineering, Indian National Science Academy, and W.B. Academy of Science and Technology. He is currently the director of the Indian Institute of Technology, Patna.

1

Compounding Ingredients and Formulation Construction

1.1 Introduction

Rubbers are described as materials that show elastic properties. These materials are long-chain molecules also known as polymers, and the amalgamation of elasticity and polymers has led to the term elastomers. Rubber is generally used in three different forms:

Raw or base polymers—These are responsible for major characteristics of the final product.

Semi-manufactured product—Different chemicals are added to raw rubbers to achieve desirable properties, also termed compounding. This semi-finished material achieves its performance properties after vulcanization.

Final product—After shaping and curing, rubber compounds impart their useful properties.

A typical rubber article consists of around 50% of elastomers, either synthetic or natural, depending on the requirements of the end use. The others are fillers, vulcanization agents, plasticizers, accelerators, antidegradants, and other chemicals. These additions help to achieve the desired properties of the final product.

The majority of polymers are hydrocarbons. The hydrogen atom is often replaced by other atoms or functional groups/molecules (like $-CH_3$, -CN, -Cl, or -Br, etc.), thereby generating varieties of elastomer. The rubber chains are chemically bonded together by sulfur, peroxides, bisphenol, etc. An exception is silicone rubber, which contains flexible siloxane backbones (Si-O-Si) and can be cured with peroxide or platinum catalyst.

Elastomers offer a wide range of properties that can be balanced to meet specific needs. Their use depends on properties of elasticity, flexibility, toughness, and relative impermeability.

An important property of rubber is elasticity. A rubber band can be stretched to nine or ten times its original length. As soon as the load is released, it retracts immediately to almost its original length. Similarly, a block of rubber can be compressed and as soon as the load is released, the block will assume its original shape and dimensions. The extent to which a rubber can be distorted, the speed of its recovery, and the degree to which it reverts to its original shape and dimensions are unique. Hence these are the prime requirements of "rubberiness."

Many rubbers, however, are not perfectly elastic. The initial rapid recovery is not complete. Part of the distortion is recovered more slowly, and part is permanently retained. The extent of this permanent distortion, or "permanent set," depends on the rate at which the distorting force is applied and the length of time it is maintained. The quicker the force and shorter the time, the less is the permanent set. Even after long periods of time, however, the permanent set is not complete, and there is some pressure to resume the original shape, a quality that is important in gaskets and other sealing members.

Allied with elasticity is flexibility. A thin sheet of rubber is almost as flexible as a handkerchief, but the rubber is elastic, while the handkerchief is not.

Toughness is equally important. The high strength and great toughness of rubber permit the use of its elastic qualities under conditions in which most other elastic materials would fail. Along with these properties, rubber shows excellent resistance to cutting, tearing, and abrasion.

Rubber is relatively impermeable to both water and air. Hence, it can be used to hold water, as in hose and water bottles, or to keep water out, as in raincoats. It can also be used to either contain air and other gases or to provide protection from them.

This combination of useful physical properties is well maintained over a wide range of temperatures from –25°C to 120°C, which covers the usual range of climatic conditions and beyond.

Rubber is relatively inert to the deteriorating effects of the atmosphere and of many chemicals. Hence, it has a relatively long, useful life under a wide variety of conditions. The life span is adversely affected by exposure to coal tar and petroleum solvents, high temperatures, direct sunlight (especially when stretched), and certain chemicals such as ozone, chlorine, some strong acids, and strong oxidizing agents. Special-purpose synthetic rubbers have gained their positions primarily because they are resistant to one or more of these deteriorating influences.

1.2 Elastomers: Properties, Uses, and Vulcanization

Natural rubber was originally the only elastomer available, and compounding was the only method for achieving desired properties. Later, rubber-like elastomers were discovered.

By introducing the blending of elastomers, a common practice to develop products having a wide range of properties, compositions available to tailor-make compounds to engineering needs increased greatly.

Along with natural rubber (NR) as general-purpose elastomers are synthetic polyisoprene (IR), styrene-butadiene rubber (SBR), and polybutadiene (BR). Other commonly used elastomers are high performance or "specialty elastomers" including Butyl (IIR), Nitrile (NBR), Chloroprene (Neoprene, CR), and Ethylene Propylene Methylane (EPM) or Ethylene Propylene Diene Methylane (EPDM).

1.2.1 Elastomer Nomenclature (ASTM D-1418)

The rubber industry, along with other scientific fields, employs many symbols and abbreviations for chemical identification. The American Society for Testing and Materials (ASTM) recommended practice establishes letter symbols as a system of general classification for rubbers based on the chemical composition of the polymer chain. It provides a means for standardization of terms for use in industry, commerce, and government. Symbols should appear first with the polymer name for use in later references in any technical paper or presentation.

"R" is the letter symbol for most of the common rubbers. This is the assigned class for polymers having an unsaturated carbon chain (NR, Natural Rubber; IR, Polyisoprene Rubber) and is readily recognized by even the newest compounder. However, many readers do not often encounter the exotic polymers of classes Q, T, and U and are not familiar with the other designations.

The polymer classes represented by the symbols are as follows:

R—Rubber having unsaturated carbon chains—natural rubber (NR), styrene butadiene rubber (SBR), polybutadiene rubber (BR), acrylonitrile butadiene rubber (NBR), isoprene isobutylene rubber (IIR), and polychloroprene rubber (CR).

M—Rubbers having a saturated chain of the polymethylene type—ethylene propylene copolymer (EPM); terpolymer of ethylene, propylene, and a diene (EPDM); polyacrylate (ACM); chloropolyethylene (CM), chlorosulfonated polyethylene (CSM); fluorocarbon rubber (FKM), etc.

N—Rubbers having nitrogen, but not oxygen or phosphorus, in the polymer chain—acrylonitrile butadiene rubber.

O—Rubbers having oxygen in the polymer chain—epichlorohydrin rubber (CO/ECO).

Q—Rubbers having silicon and oxygen in the polymer chain—silicone rubber (MQ) and fluorosilicone rubber (FVMQ).

T—Rubbers having sulfur in the polymer chain—thiokol rubber (OT, EOT).

U—Rubbers having carbon, oxygen, and nitrogen in the polymer chain—polyurethane polyether (EU) and polyurethane polyester (AU).

The structures and properties of a few polymers are shown later.

1.2.2 Elastomer Properties

1.2.2.1 Natural Rubber (NR)

Natural rubber has very high elasticity, good tensile strength, and good abrasion resistance. The rubber is obtained by coagulation of latex from a rubber tree. Natural rubber is susceptible to degradations due to aging and oil exposure. This is one of the reasons natural rubber is rarely used for seal applications. However, NR can be compounded with other elastomeric compounds like EPDM to improve different physico-chemical properties, like aging resistance, etc. It is extensively used in tires. Improved properties of natural rubber are due to its high molecular weight and molecular weight distribution. Increased use of natural rubber in commercial vehicle tires is mainly due to a reduction in rolling resistance, lower heat generation, better adhesion properties, and higher hot tear strength.

There are different chemical modifications of NR that have been developed by various research scientists. Some important chemical modifications include deproteinized NR (DPNR), oil extended NR (OENR), and epoxidized NR (ENR). Recently ENR has gained application in the tire industry. ENR, because of its polar nature, is showing better oil resistance. It also shows improved air retention properties as compared to NR. ENR, due to the presence of epoxide linkage, gives better interaction with polar silica filler, which is absent in NR. Due to its better interaction with silica filler, ENR is used in passenger car tire tread recipes to manufacture green tires.

1.2.2.2 Styrene-Butadiene Rubber (SBR)

SBR is a synthetic rubber. It is a copolymer of styrene and butadiene. It has good abrasion resistance and good aging resistance compared to NR. This is due to the lower unsaturation level as compared to NR. After suitable protection by additives, SBR is widely used in car tires, where it is blended with natural rubber. Emulsion polymerized styrene-butadiene rubber (E-SBR) is widely used in polymers. E-SBR is available in a wide rage of viscosity grades. Lower-viscosity grades of SBR are easy to process. They allow the forming of a band on the mill and incorporate fillers and oil more readily, showing less heat generation during mixing. Low-viscosity SBR also gives higher extrusion rates and better extrudate appearance than the higher-viscosity grades. However, high-viscosity SBR has better green strength, less porosity in the vulcanizate, and accepts higher filler and oil loadings. Another important variation in SBR is the styrene content. It varies from a very low percentage to a very high percentage up to 50%. The commonly available styrene grades contain 23.5% styrene. In vulcanizates of SBR, as styrene content increases, dynamic properties and abrasion resistance decrease while traction and hardness increase.

Recently, anionic polymerized SBR, termed solution SBR (SSBR), is being widely used in the passenger tire industry. Anionic polymerization offers highly versatile elastomers. The benefit of using SSBR in tire tread is a reduction in hysteresis. This is due to microstructural and macrostructural differences of SSBR as compared to E-SBR. Another difference between S-SBR and E-SBR is that S-SBR gives higher modulus, whereas E-SBR gives higher tensile strength.

1.2.2.3 Polybutadiene Rubber (BR)

Polybutadiene (BR) is the second largest used synthetic rubber, after SBR. The major use of BR is in tire treads and sidewalls. BR shows excellent abrasion resistance because of its low glass transition temperature (T_g). It also gives low rolling resistance properties in tire tread. The low T_g is a result of the high cis content of BR. However, low T_g also leads to poor wet traction properties. BR is also used as an impact modifier for polystyrene and acrylonitrile-butadiene-styrene resin (ABS). Polybutadiene rubber is generally produced by Ziegler-Natta catalyst. There are different Ziegler-Natta catalysts based on titanium, cobalt, nickel, and neodynium being used. Commonly used BR is based on nickel catalyst. It gives high cis-polybutadiene. Neodymium-based BR is a high cis linear polymer. There is another development in BR known as syndiotactic polybutadiene. It behaves as a plastic as well as rubber. Syndiotactic polybutadiene can be used in adhesives, films, golf balls, etc.

1.2.2.4 Ethylene Propylene Rubber (EPDM/EPM)

EPM is a copolymer of ethylene and propylene. EPM is crosslinked only with peroxides. During the copolymerization of ethylene and propylene, a third monomer, a diene, can be added. This addition of a third monomer generated a terpolymer with one of the monomers with unsaturation. The resulting polymer is known as EPDM. This polymer can be crosslinked with sulfur as the polymer chain has unsaturation.

EPDM is famous for its better heat, ozone, and weather resistance properties which are due to a saturated polymer backbone. Its steam aging property is also commendable. It has excellent electrical insulating properties. EPDM copolymer can be extended with very high nonreinforcing filler, which can be as high as 200% of its own weight. The highly extended EPDM compound is generally used in cost-effective rubber products. The above performance properties allow EPDM to be used as a suitable polymer for window seals, gaskets, cable insulation, etc. As EPDM contains ethylene and propylene, it also shows good dimensional stability. Due to this, EPDM-based rubber compounds are widely used for profile calendered product.

1.2.2.5 Nitrile Rubber (NBR)

NBR is a copolymer of acrylonitrile (ACN) and butadiene monomers. It is an unsaturated polymer due to the presence of the butadiene unit. Physical and chemical properties of nitrile rubber depend on the total acrylonitrile content in the rubber. The higher the nitrile content, the higher is its plastic nature. Better oil resistance properties result from increased polarity. The polymer with a low level of acrylonitrile content behaves like a rubber and is cured by a normal sulfur-accelerator system. A polymer with a higher acrylonitrile level shows higher glass transition temperature, T_g, and is not suitable for low temperature application. Rubber grade NBR (medium acrylonitrile polymer) is used in the automotive industry to make fuel and oil handling hoses, seals, and grommets.

Depending on the acrylonitrile content, the temperature range of applications of NBR is also wide. This property helps NBR be an ideal material for automotive use. Nitrile rubber is more resistant than natural rubber to oils and acids but has less strength and flexibility. Nitrile rubber is generally resistant to aliphatic hydrocarbons. However, it is susceptible to attack by ozone, aromatic hydrocarbons, ketones, esters, and aldehydes, due to the presence of unsaturation in its backbone.

On hydrogenation, a new class of elastomer, hydrogenated nitrile butadiene rubber (HNBR), is obtained. The properties of this elastomer are superior to those of nitrile rubbers, but this elastomer is more expensive. Similarly, carboxylated nitrile rubber (XNBR) is popular in many applications.

1.2.2.6 Hydrogenated Nitrile Butadiene Rubber (HNBR)

The properties of HNBR depend on the degree of hydrogenation. Due to hydrogenation, the unsaturation percentage of the polymer is reduced, and as a result, HNBR shows improved resistance to ozone attack and high heat resistance properties.

Due to the low unsaturation level in HNBR, it shows good hot water and steam resistance properties. HNBR, depending upon the hydrogenation level, improves temperature, oil, and chemical resistance. HNBR is suitable for use in methanol and methanol/hydrocarbon mixtures, if the correct acrylonitride (ACN) level is selected. For the best properties, peroxide curing is used, unless low hysteresis is required.

1.2.2.7 Chloroprene Rubber (CR)

Chloroprene rubber is commonly known under the trade name DuPont™ Neoprene (DuPont, Wilmington, Delaware). Among the specialty elastomers, polychloroprene (CR) [poly (2-chloro-1,3-butadiene)] is an important polymer. CR has applications in adhesives, some dipped articles, molded foams, etc. It also shows good mechanical strength, better ozone and weather resistance, improved aging resistance, very low flammability, good resistance toward chemicals, moderate oil and fuel resistance, and good adhesion to many substrates. Due to the presence of the allylic chlorine atom, the unsaturation nature of CR is almost deactivated. This causes very high weather resistance properties. The presence of chlorine improves the adhesion properties of CR rubber. Like NR, it shows strain-induced crystallization, thus resulting in very high green strength, which leads to wide application in different adhesives. The very low flammability of CR is also due to the presence of a Cl atom in the backbone. During decomposition at high temperature, CR liberates HCl gas which reduces the oxygen at the burning area. This helps to improve resistance to flammability properties of CR. As the unsaturation of CR is almost deactivated, the effective crosslinking agent is metal oxide. ZnO is used as the crosslinking agent of CR.

1.2.2.8 Ethylene Vinyl Acetate Rubber (EVA)

Ethylene vinyl acetate (EVA) is a copolymer of ethylene and vinyl acetate. The backbone of EVA is fully saturated. Due to high saturation in the backbone, the polymer shows high temperature stability. Polymer properties of EVA depend on the ratio of ethylene and vinyl acetate content. EVA with higher ethylene content will behave as a plastic, whereas EVA with higher vinyl acetate content will be a rubber. Different grades of EVA are available. The main differences between the grades are in the vinyl acetate content and the polymer viscosity. A higher proportion of vinyl acetate in the copolymer reduces crystallization. Hence, copolymers with high vinyl acetate content are amorphous. EVA vulcanizates display excellent aging resistance and continue to function over extended periods of stress at elevated temperatures. The heat resistance of EVA vulcanizates is considerably better compared to most other common elastomers. The heat resistance of EVA is surpassed only by silicone rubber and fluoro rubber and is equivalent to that of acrylate rubber. The low temperature performance of EVA depends significantly on the VA content. Decreasing VA content results in a lower T_g.

EVA also shows a better flame resistance behavior. Wherever the halogen free flame resistance polymer is required, EVA is chosen. The properties of halogen-free EVA, like low and high temperature resistance, oil resistance, weather and ozone stability, and processability, offer the best performance-to-cost ratio among the specialty elastomers.

1.2.2.9 Butyl Rubber (IIR)

Butyl rubber (IIR) is the copolymer of isobutylene and a small amount of isoprene. IIR shows excellent air retention properties as well as good flex fatigue resistance. It finds wide use as vibration dampers. A major application of IIR is in automotive tubes. The bulky isobutylene group is responsible for the high air barrier performance of IIR. Butyl rubber, due to its very low unsaturation in the backbone, ~1 to 2 mole percent, exhibits superior aging resistance in the elastomer family. This is one of the key performance criteria for curing bladder application. Vibrational movement at ordinary and high temperatures is utilized in the case of IIR compound mixing as well as curing. Due to low unsaturation, IIR requires a booster-type accelerator in the case of sulfur curing.

Another important development in the IIR rubber category is halogenated butyl rubber. Two types of halogenated butyl rubbers are available in the market: chlorobutyl rubber (CIIR) and bromobutyl rubber (BIIR). The development of halobutyl rubbers was due to increased compatibility with other unsaturated polymers. Due to its saturated backbone, butyl rubber is not cure compatible with general-purpose unsaturated rubber. With the introduction of CIIR and BIIR, these polymers show improved compatibility with NR, SBR, etc. Halobutyl rubber finds its application in tubeless tire inner liners. Due to the presence of -Cl and -Br in halobutyl rubber, the air retention properties of halobutyl are further improved along with the heat resistance property. A new polymer, brominated isobutylene p-methyl styrene, has also been developed and marketed by ExxonMobil (Irving, Texas). It provides improved grip properties to winter tire tread and has a very stable bromine group in the structure.

1.2.2.10 Polyurethane (PU)

Commercially, polyurethanes are produced by the exothermic reaction of molecules containing two or more isocyanate groups with polyol molecules containing two or more hydroxyl groups. A unique advantage of polyurethanes is in the very wide variety of high-performance materials. High material performance coupled with processing versatility has resulted in the spectacular growth and wide applicability of the polyurethane family of materials. Polyurethanes can be manufactured in an extremely wide range of grades, in densities from 6 to 1220 kg/m^3 and polymer stiffness from flexible elastomers to rigid, hard plastics.

1.2.2.11 Silicone Rubber (VMQ/MVQ/HTV)

Silicone rubber differs from other polymers, as its backbone consists of Si-O-Si units. Due to its basic structure, it shows excellent high as well as low temperature properties. The general operating temperature of silicone rubber is from –55°C to 230°C. Due to the absence of unsaturation in the polymer backbone, silicone rubber is not attacked by ozone. Its weather resistance property is also very good. Compared to organic rubbers, however, the tensile strength of standard silicone rubber is lower. For this reason, care is needed in designing products to withstand low imposed loads. Silicone rubber compounds with improved tensile strength are available.

1.2.2.12 Acrylic Rubber (ACM)

Acrylic rubber, known by the chemical name alkyl acrylate copolymer (ACM), is a type of rubber that has outstanding resistance to hot oil and oxidation. It has a continuous working temperature of 150°C and an intermittent limit of 180°C. Disadvantages include its low resistance to moisture, acids, and bases. It should not be used in temperatures below –10°C. It is commonly used in automotive transmissions and hoses.

1.2.2.13 Fluorocarbon Rubber (FKM)

Fluoroelastomers, a class of synthetic rubber, provide extraordinary levels of resistance to chemicals, oil, and heat, while providing useful service life up to 200°C. The very high heat stability and excellent oil resistance of these materials are due to the presence of chemically inert fluorine atoms in the backbone of the polymer. The bond strength of C-F is very high, which requires very high energy for dissociation. This is another reason for the high temperature stability of fluoroelastomers. The original fluoroelastomer was a copolymer of hexafluoropropylene (HFP) and vinylidene fluoride (VF2). It was developed by DuPont in 1957 in response to high-performance sealing needs in the aerospace industry. To provide even greater thermal stability and solvent resistance, tetrafluoroethylene (TFE) containing fluoroelastomer terpolymers was introduced in 1959, and lower-viscosity versions of FKMs were introduced in the mid to late 1960s. A breakthrough in crosslinking occurred with the introduction of the bisphenol cure system in the 1970s. This bisphenol cure system offered much improved heat and compression set resistance with better scorch safety and faster cure speed. In the late 1970s and early 1980s, fluoroelastomers with improved low temperature flexibility were introduced by using perfluoromethylvinyl ether (PMVE) in place of HFP.

Fluoroelastomers are a family of fluoropolymer rubbers, and not a single entity. Fluoroelastomers can be classified by their fluorine content, viz., 66%, 68%, and 70%, respectively. Fluoroelastomers having higher fluorine content have increasing fluid resistance derived from increasing fluorine levels. Peroxide cured fluoroelastomers have inherently better water, steam, and acid resistance.

Fluoroelastomers are used in a wide variety of high-performance applications. FKM provides premium, long-term reliability even in harsh environments. A partial listing of current end-use applications (industries like aerospace and automotive) includes O-ring

TABLE 1.1

Comparison of NR, SBR, and BR

Processability	NR	SBR	BR
Internal mixing	Two-step premastication masterbatch or one-step masterbatch	One-step masterbatch hot processing	Mixing possible but general processing difficulties observed, hence always used in blends
Extrusion	Fast, smooth extrusion	Fast, smooth extrusion only if adequately loaded with high carbon black and clay	NR/BR SBR/BR
		Higher die swell	
Calendering	Goes to hot slow roll Higher thermoplasticity	Goes on to cold fast roll	
Building up	Excellent green tack	Green tack poor	Green tack poor
	Excellent green strength	Green strength lower than NR	
Vulcanization	Higher tendency to scorch	Slow curing, marching modulus	Slow curing
	Fast cure rate Shows reversion	Higher curing temperature	Slightly marching modulus

seals in fuels, lubricants and hydraulic systems, shaft seals, valve stem seals, fuel injector O-rings, diaphragms, lathe cut gaskets, and cut gaskets.

The important properties of NR, SBR, and BR are listed in Table 1.1.

1.3 Fillers

Fillers are generally used to modify physical properties of a rubber compound in addition to lowering cost. It is important to a rubber compounder to understand the actual requirements of the rubber product. Cost reduction of the compound with the introduction of non-reinforcing filler results in a compromise of the required performance properties. Optimization of the cost-properties trade-off is very important. Filler is the second most important material in the rubber industry after rubber. Use of reinforcing filler generally improves physical properties such as modulus and other failure properties. There are several fillers available which will be discussed in the next few paragraphs. However, the most commonly used filler in the rubber industry is carbon black.

Filler dose in a rubber compound varies depending upon the application and the corresponding functional requirement. A typical bias tire tread compound generally contains reinforcing carbon black of 45 to 50 phr level. For tread application, one of the important requirements is durability of the tire. However, in the case of engine mounts, the product should carry the load of the engine without deformation. The type of filler used in this application should give better compression set of the rubber compound at high temperature. Accordingly, a suitable grade of filler is needed in this application. In a similar way, several performance requirements are fulfilled with a suitable combination of polymer and filler. Most of the rubber fillers used today offer some functional benefit that contributes to the processability or utility of the rubber product. Styrene-butadiene rubber, for example, has virtually no commercial use as an unfilled compound.

The important properties of a typical reinforcing carbon black are particle size, particle surface area, particle surface activity, and particle shape. Surface activity relates to the compatibility of the filler with a specific elastomer and the ability of an elastomer to adhere to the filler.

Functional fillers help transfer applied stress from a rubber matrix. It seems reasonable that this stress transfer will be better if the particles are smaller. Greater surface area is exposed for a given filler concentration. If these particles are needle-like, fibrous, or platy in shape, they will better intercept the stress propagation through a matrix.

A compound's physical properties are heavily dependent upon the surface activity of the filler and the effective surface area of the filler.

Particle size of the filler plays an important role in filler reinforcement of a rubber matrix. The lower the particle size, the higher is the filler reinforcement. If the particle size of the filler is high as compared to the polymer interchain distance, it causes an area of localized stress. This will result in elastomer chain rupture while flexing or stretching. Filler particles with greater than 10,000 nm (10 μm) diameter are generally not suitable, as they reduce performance rather than reinforce. Fillers with particles between 1000 and 10,000 nm (1 to 10 μm) are used as diluents and have no significant effect on rubber properties. Semi-reinforcing fillers range from 100 to 1000 nm (0.1 to 1 μm). Truly reinforcing fillers, which range from 10 nm to 100 nm (0.01 to 0.1 μm), can significantly improve rubber properties.

1.3.1 Function of Filler

One or more of the functions below are satisfied by filler:

- To improve wear and failure characteristics of vulcanizates (reinforcing fillers)
- To provide better cost/economics of a compound
- To take care of processing characteristics of a compound
- To impart special properties like flame resistance, electrical properties, etc.
- To impart color to a compound as required

The differences in the effects of the fillers are given, in principle, by the size of their particles, their shape, and surface activity.

There are several types of fillers used in rubber compounds. Major fillers used in the rubber industry are carbon black and silica. Carbon black is mainly an elemental carbon with very fine particles having an amorphous molecular structure. Carbon black is produced by the incomplete combustion process of the carbon black feedstock (CBFS). In the process of carbon black manufacturing, first an intensely hot combustion zone is produced with a convenient fuel. After combustion is complete, the CBFS in excess of stoichiometric quantities is injected into that intensely hot zone. At this very high temperature, carbon black will be produced. Following the feedstock injection, the reaction is stopped by either injection of water or by allowing the temperature to decay with time. All furnace grades are made in a continuous process, while thermal grades are made in a cyclical process. The carbon particles produced by this process are separated from the process gas stream by conventional means and pelletized to increase the bulk density.

TABLE 1.2
Approximate Compositional Summary

Source	Element	Channel (gas)	Thermal (gas)	Modern Oil Furnace
	Carbon	96	99	98
Oxidation	Oxygen	3	Nil	<1
Feedstock	Sulfur	<0.1	<0.5	0.5–1.7
Quench water	Ash	<0.1	<0.2	<1
Feedstock	Hydrogen	0.5	0.4	<0.4

Carbon atoms are arranged in layer planes in carbon black. The layer structure is similar to graphite; however, the arrangement of the layer in carbon black is highly irregular. These irregular crystallographic arrangements are responsible for reinforcement in rubber. Composition of the carbon black mainly depends on the process condition of the carbon black manufacturing process. The typical composition and its probable sources are listed in Table 1.2.

1.3.1.1 ASTM Designation of Rubber-Grade Carbon Blacks

- The carbon black designation is done by a letter and three digits. The letter classifies the cure rate, the first digit indicates typical particle size, and the last two digits are assigned arbitrarily to distinguish blacks within a given particle group.
- The letters used to designate cure rate are N and S. The letter N stands for normal cure rate, and S stands for slower cure rate.
- The first number recognizes particle size as a fundamental characteristic of carbon black. The particle size pattern normal to rubber grade carbon blacks is arbitrarily divided into ten groups that are shown in Table 1.3.
- The particle size is expressed in nanometers. The groups are not equal in numerical size but are chosen so that blacks now known as SRF, ISAF, and HAF do not fall into the same classification.

TABLE 1.3
ASTM Classification of Black

Group No.	Typical Particle Size, nm	Present Identification	ASTM Nomenclature
0	1–10	—	—
1	11–19	SAF	N110
2	20–25	ISAF	N220
3	26–30	HAF	N330
4	31–39	FF (EPC, MPC)	N440
5	40–48	FEF	N550
6	49–60	GPF	N660
7	61–100	SRF	N770
8	101–200	FT	—
9	201–500	MT	—

- The last two numbers indicate differences within a group. This difference can be in structure level, modulus level, or any of the physical/chemical properties associated with a particular black.

Thus, N330 and N347 will indicate two blacks having a normal-type cure rate with particle size of 26 to 30 nm.

1.3.1.1.1 *Characterization of Carbon Black*

Carbon black is characterized by the following properties:

- Particle size/surface area
- Structure
- Aggregate complexity
- Surface activity
- Chemical nature of the filler

1.3.1.1.1.1 Surface Area/Particle Size Particle size is inversely related to surface area. An electron microscope is used to measure particle size. Surface area is a macro property and is measured by various methods; iodine adsorption, nitrogen surface area, and cetyltriammonium bromide (CTAB) adsorption are common. Structure is given by the dibutyl phthalate (DBP) and crushed DBP numbers. Typical values of various carbon blacks are given in Table 1.4.

1.3.2 Compounding with Carbon Black

1.3.2.1 Effect of Increased Surface Area

At optimum loading the effects of carbon black particle size and structure on rubber compound processing and properties are shown in Tables 1.5 and 1.6.

The effect of increased loading is as follows:

On processing properties:

- Increase in viscosity of a compound
- Decrease in die swell, better surface finish
- Extrusion rate goes through an optimum
- Decrease in calender shrinkage and better surface finish
- Increase in green strength and decrease in green tack
- Decrease in scorch safety unless the filler is a retarding type

On vulcanizate properties:

- Tensile strength, tear strength, and abrasion resistance go through an optimum
- Elongation at break, resilience, and volume swell in fluids decrease
- Hardness, modulus, and compression set increase

Carbon black changes the hardness of compounds, which is measured by Shore A durometer. Often compounds are designed by following the increase or decrease of hardness.

TABLE 1.4

Summary of Typical Properties of Carbon Black

Industry Name	ASTM No.	Iodine No.(mg/g)	Nitrogen (mg/g)	CTAB (mg/g)	DBP (cc/100 g)	CDBP (cc/g)
SAF	N110	145	143	126	113	98
ISAF LS	N210	118	120	113	78	75
ISAF	N220	121	119	111	114	100
ISAF LM	N231	121	117	108	92	86
ISAF HS	N234	120	126	119	125	100
HAF LS	N326	82	84	83	72	69
HAF	N330	82	83	83	102	88
HAF HS	N339	90	96	95	120	101
HAF HS	N347	90	90	88	124	100
FEF	N550	43	42	42	121	88
GPF HS	N650	36	38	38	122	87
GPF	N660	36	35	35	90	75
SRF LS	N762	27	28	27	65	57
SRF HS	N765	31	31	33	115	86
SRF HM	N774	29	29	29	72	62

Tables 1.7 and 1.8 give the base durometer hardnesses for various rubbers and changes of hardness by the addition of fillers or softeners.

1.3.2.2 Equal Hardness Approach

Table 1.9 can be used to handle any change in carbon black while maintaining the hardness.

To determine the approximate loading of N660 for use as a replacement for MT at equal hardness, multiply the MT loading by 0.60.

To determine the approximate loading of N762 as a replacement for FEF at equal hardness, multiply the FEF loading by 1.22.

TABLE 1.5

Effect of Carbon Black Particle Size and Structure on Compound Processing

Processing Properties	Decreasing Particle Size			Increasing Structure		
Loading capacity	Decreases			Decreases		
Incorporation time			Increases			Increases
Oil extension potential		Little				Increases
Dispersibility	Decreases					Increases
Mill bagging			Increases			Increases
Viscosity			Increases			Increases
Scorch time	Decreases			Decreases		
Extrusion shrinkage	Decreases			Decreases		
Extrusion rate	Decreases				Little	
Extrusion smoothness			Increases			Increases

TABLE 1.6
Effect of Carbon Black Particle Size and Structure on Vulcanizate Properties

Vulcanizate Properties	Decreasing Particle Size	Increasing Structure
Rate of cure	Decreases	Little
Tensile strength	Increases	Decreases
Modulus	Increases to maximum then decreases	Increases
Hardness	Increases	Increases
Elongation	Decreases to minimum then increases	Decreases
Abrasion resistance	Increases	Increases
Tear resistance	Increases	Little
Cut-growth resistance	Increases	Decreases
Flex resistance	Increases	Decreases
Resilience	Decreases	Little
Heat buildup	Increases	Increases slightly
Compression set	Little	Little
Electrical conductivity	Increases	Little

TABLE 1.7
Base Hardness (Shore A) of Rubbers

For 100 Parts of Polymer	Base Durometer
Polychloroprene and nitrile rubber	44
Natural rubber and cold polymerization SBR	40
Hot polymerization SBR	37
Butyl rubber	35
25 parts oil extended cold SBR	31
37.5 parts oil extended cold SBR	26

TABLE 1.8
Effect of Fillers and Softeners on Hardness

Fillers and Softeners	Durometer Change
FEF, HAF, channel blacks	+1/2 part of loading
ISAF black	+1/2 part of loading + 2
SAF black	+1/2 part of loading + 4
SRF black	+1/3 part of loading
Thermal blacks and hard clay	+1/4 part of loading
Whiting (in natural rubber)	+1/7 part of loading
Factice and mineral rubber	–1/5 part of loading
Most liquid softener	–1/2 part of loading

TABLE 1.9

Hardness Conversion Factors for Replacing Carbon Blacks

Replacement Carbon Black		Hardness Conversion Factors for Replacing the Following Carbon Blacks				
Industry Type	**ASTM No.**	**MT**	**SRF**	**GPF**	**GPF-HS/ FEF**	**HAF**
SRF	N762	0.665	1.00	1.10	1.22	1.50
SRF	N774	0.665	1.00	1.10	1.22	1.50
GPF	N660	0.60	0.90	1.00	1.11	1.36
GPFHS	N650	0.545	0.82	0.90	1.00	1.23
FEF	N550	0.545	0.82	0.90	1.00	1.23
HAF	N330	0.445	0.67	0.74	0.82	1.00

1.3.2.2.1 Unit Replacement Factor

This type of change involves changing the loading of a given grade of black and maintaining the hardness by adjusting oil. It is good to remember that for N330, 1 phr of oil is needed for each phr of black (see Table 1.10). For example, the unit replacement factor for N660 is 1.74; this means 0.74 parts of oil must be added for each additional part of N660 in order to maintain compound hardness.

1.3.2.2.2 Varying Hardness Approach

Table 1.11 can be used as a guide for designing new compounds, switching blacks, and/or varying hardness.

1.3.2.2.3 Modulus Compounding

Modulus can be varied by changing the carbon black structure and carbon black/oil loading:

- As a guideline, the addition of 1 phr of carbon black will raise the modulus by 0.14 to 0.28 MPa, depending on the structure of the black and the polymer system.
- As a guideline, the addition of 1 phr of oil will lower the modulus by 0.21 to 0.28 MPa, depending on the structure of the black and the polymer system.

TABLE 1.10

Unit Replacement Factor

Carbon Black (or Other Filler)			
Black Type	**ASTM No.**	**Oil Requirement**	**Unit Replacement Factor**
SRF	N762	0.667	1.667
SRF	N774	0.667	1.667
GPF	N660	0.74	1.74
GPFHS	N650	0.80	1.80
FEF	N550	0.80	1.80
HAF	N330	1.00	2.00
MT	N990	0.435	1.435
Whiting	—	0.435	1.435

TABLE 1.11

Varying Hardness Approach

		Parts Carbon Black Required for 10-Point Increase in Compound Hardness						
Industry Type	**ASTM No.**	**NR**	**SBR**	**IIR**	**CR**	**BR**	**NBR**	**EPDM**
SAF	N110	15	18	13	12	22	17	24
ISAF	N220	17	20	15	13	25	19	27
HAFLS	N326	21	26	19	17	32	24	34
HAF	N330	19	23	17	15	28	21	30
HAFHS	N347	17	21	16	14	26	20	28
FEF	N550	23	28	21	18	34	26	37
GPF	N660	25	31	23	20	38	29	41
SRF	N774	28	25	25	22	42	32	45

1.3.2.2.4 Hysteresis Compounding

A novel approach to compounding for equal hysteresis can be taken by using the following equation:

$$(\text{phr black/phr total})^2 \times N_2SA = \Phi\gamma' \text{ factor}$$

The $\Phi\gamma'$ factor of the existing compound and the compound containing the new carbon black of different surface area is solved and then compared.

Table 1.12 will give an idea about the requirements of different carbon blacks in the different components of tires.

Control of quality during the production of carbon black is mandatory to get the actual performance in rubber compound. Various quality control tests of carbon black are shown in Table 1.13.

In the last few years, research into changing the basic properties of carbon black has taken place. Modification of the carbon black surface is one such research activity. This modification is generally being done through process modification. Some of the post-process modifications are oxidation, plasma treatment, and polymer grafting. There are also a few in-process modifications that were studied by several workers who had developed blacks called inversion black, carbon-silica dual phase fillers.

1.3.3 Non-Black Fillers

A number of fillers other than carbon black are also used in the rubber industry:

- Silica
- Kaolin clay or china clay (hydrous aluminum silicate)
- Mica (potassium aluminum silicate)
- Talc or french chalk (magnesium silicate)
- Ground chalk or whiting
- Limestone (calcium carbonate)
- Titanium dioxide
- Fibrous fillers like asbestos, cellulose fiber, wood flour, etc.
- Barytes ($BaSO_4$)
- Aluminum hydroxides

TABLE 1.12

Use of Different Carbon Blacks in Tire Components

Tire Component	Component Property Requirement	Required Carbon Black
Passenger tread compound	Treadwear, dry and wet traction, low hysteresis, resistance to cutting and chipping	N220, N234, N299, N339, N351
Light truck tread compound	Same as passenger tire	Same as passenger tire
Truck/OTR tread compound	Less heat generation, resistance to chipping and chunking	N110, N121, N134, N220, N231, N330, N339
Carcass compound	Good adhesion, green strength, resistance to tear and flex fatigue, good calendering properties	N330, N326, N351, N660
Belt-skim compound	Good adhesion to the textile or steel cord, resistance to tear and flex fatigue, low hysteresis, relatively high modulus and retention of properties after prolonged aging	N330, N326, N351, N550
Sidewall compound	Resistance to flexing, cutting, and weathering and good extrusion properties	N330, N326, N351, N550, N650
Inner-liner compound for tubeless tire	Air retention, resistance flex and aging, compound green strength, good calendering characteristics	N550, N650, N660
Bead insulation/bead filler compound	High modulus and hardness, low die swell, good tack, and adequate adhesion to the wire both in green and cured conditions	N326, N550, N660

TABLE 1.13

Quality Control Tests of Carbon Black and Its Criticality

Test Parameters	Critical Use	Significance
Iodine adsorption number	Required for specification	Surface area
Nitrogen surface area	Required for critical application	Total surface area
Oil absorption number	Required for specification	Structure
Compressed oil number	Required for critical application	Compressed structure
STSA*	Required for critical application	External surface area
CTAB surface area	Required for critical application	External surface area
Pellet hardness	Required for specification	Pellet strength
Fines content	Required for specification	Dust level, handling issue
Heat loss	Required for specification	Moisture level
Sieve residue	Required for specification	Presence of contaminants
Ash content	Required for critical application	Presence of inorganic materials
Tinting strength	Required for critical application	Fineness of black

*Statistical Thickness Surface Area

Among the various non-black fillers, silica plays a dominant role. The nature, synthesis, properties, and applications of various fillers are briefly mentioned here.

1.3.3.1 Silica Fillers

- Silicas are highly active light-colored fillers.
- Chemically these are made from silicic acids.
- Silica can be manufactured by two processes:
  - Solution or precipitated process giving rise to precipitated grades
  - Pyrogenic process giving rise to fumed or pyrogenic grade
- For the rubber industry use, mostly precipitated types are used since pyrogenic silica is too active and expensive.

1.3.3.1.1 *Precipitated Silicas*

- Alkalisilicate solutions are acidified under controlled conditions.
- The precipitated silicic acid (silica) is washed out and dried.
- The activity of the silica fillers depends on the condition of preparation; the products with highest activity are pure silicic acids with large surface area.
- These silicas are silicon dioxide containing 10 to 14% water, with particle size in the range of 10 to 40 nm.

1.3.3.1.2 *Pyrogenic or Fumed Silicas*

- Silicon tetrachloride is reacted at high temperature with water.

$$SiCl_4 + 2H_2O \rightarrow SiO_2 + 4HCl$$

- The reaction products are quenched immediately after coming out of the burner.
- Fumed or pyrogenic silica thus produced contains less than 2% combined water and particle size smaller than precipitated grades.

1.3.3.1.3 *Characteristics of Silica Fillers*

- Silica is an amorphous material having a tetrahedral structure of silicon and oxygen.
- Particle size ranges from 1 to 40 nm, and surface area ranges from 20 to 300 m^2/g.
- Silica is hygroscopic and hence requires dry storage conditions.
- Surface silanol concentration (-Si-O-H) influences the degree of surface hydration.
- Surface activity is controlled by hydroxyl groups on the surface of the silica. This surface activity has an effect on peroxide curing but has no significant effect on sulfur curing.
- All precipitated silicas contain a certain amount of moisture since the time of manufacture. This moisture content appreciably influences the processing and vulcanizing properties of rubber compounds.

All of the above parameters are also true for silicate fillers like calcium silicate, aluminum silicate, magnesium silicate, and sodium aluminum silicates.

1.3.3.1.4 Effect of Silica Fillers on Rubber Compound

1.3.3.1.4.1 Processing Properties

- With increasing surface area, the incorporation of fillers in a rubber compound becomes difficult, and compound viscosity increases considerably.
- If substances that are adsorbed by the silicas are added in silica compounds, compound viscosity reduces and processing becomes easier. Additive of this kind includes accelerators like Diphenyl guanidine (DBG), Diorthotolyl guanidine (DOTG), Hexamethylene tetramine (HMT); glycols like diethylene and triethylene glycols; amines like triethanolamines, dibutylamines, cyclohexyl, and dicyclohexylethyleneamines; etc. These additives not only facilitate processing but also reduce acceleration adsorptions.
- As amounts of additives are normally calculated in terms of rubber, the filler activators are also calculated in terms of filler amount. For example, when Mercaptobenzothiazyl disulfide (MBTS) is used as the main accelerator, hexamethylene tetramine, DPG, or DOTG is recommended as additional accelerator with the amount being 4 to 6 phr on 100 phr of silica.
- Silanes are also suitable silica activators. The difference is that while the earlier activator used to block the adsorptive centers on surfaces of the filler, silanes take part in a chemical reaction with the silanol groups of the silica. This influences the filler-rubber interaction markedly, due to which the compound viscosity is reduced and a fairly large amount of silica fillers can be incorporated in the rubber matrix.

1.3.3.1.4.2 Vulcanizate Properties

- There is comparatively high hardness along with low modulus values.
- Tear property improves. The situation is exploited in off-the-road (OTR) tires, where 10 to 15 parts of silica are added along with reinforcing blacks.
- Precipitated silica fillers play an important role in bonding systems, catalyzing the resin formation or resin-rubber interaction.
- Precipitated silicas often give better hot air resistance than carbon blacks.
- Under dynamic conditions, compounds containing precipitated silicas have lower loss factors than carbon blacks, and this result is exploited in tires with a carbon black/silica blend along with a silane activation system, where improved reversion resistance, better retention of tear, and tear propagation resistance are obtained.
- The benefits of reinforcement of silicas/silanes in certain blends of solution SBR and BR are translated into a new generation of tires having considerable lower rolling resistance. These tires are called "green tires."

1.3.3.2 Other Filler Systems

1.3.3.2.1 China Clay

China clay can be classified into four varieties:

Soft clay (particle size greater than 2 millimicron)

- Used mainly in mechanical goods as semi-reinforcing fillers
- Hardness and tensile strength are greater but resilience is less as compared to calcium carbonate
- Not used in bright-colored articles

Hard clay (particle size less than 2 millimicron)

- Give better hardness, tensile strength, tear and abrasion properties than soft clays
- Compounds have high electrical resistivity
- Used in mechanical goods, hose, flooring, and shoe soling

Calcined clay

- Hard clays calcined to remove combined water
- Hardness, tensile strength, and electrical resistivity higher than hard clays
- Used where color or electrical properties are important

Treated clay

Three types are available:

- Amine coated
- Silane coated
- Polybutadiene coated

This variety of the clay gives greater reinforcement than untreated grades and finds application in high-grade mechanical goods.

1.3.3.2.2 Ground Chalk or Whiting

- A white powder of particle size below 30 nm
- Used in low-cost compounds
- Gives moderate hardness and fairly high resilience at high loadings with poor tensile strength and tear properties

1.3.3.2.3 Limestone (Calcium Carbonate)

- Particles below 100 mesh
- Used where low cost is the prime consideration
- High loadings can give moderate hardness

1.3.3.2.4 Barytes ($BaSO_4$)

- Used as inert filler to reduce cost
- Purer grades produced by precipitation of barium salts are used in pharmaceutical products
- Used in chemical-resistant compound for tank lining as an inert filler

1.3.3.2.5 Titanium Dioxide

- Finds extensive use in white and other colored products where appearance is important
- Used in white sidewall for passenger tires, hospital accessories, floor tiles, etc.

1.3.3.2.6 Mica Powder

- Natural mica, washed and ground to pass through 200 to 300 mesh
- A filler imparting good resistance to heat and lower permeability to gases
- Finds use in inner tube for tires

1.3.3.2.7 *Talc or French Chalk (Magnesium Silicate)*

- Used as an inert filler in heat-resistant compounds like gaskets
- Talc also widely used as a lubricant to prevent uncured stocks from sticking to themselves or other surfaces

1.3.3.2.8 *Aluminum Hydroxides*

This material is used less for its filler effect but more for its ability to split off water at high temperatures as a flame retardant and finds application in cable industries.

1.3.3.2.9 *Fibrous Filler*

1.3.3.2.9.1 *Asbestos*

- Asbestos is a naturally occurring siliceous fiber
- Used as short fibers or ground material in flame-resistant or heat-resistant compounds, in gaskets, brake shoes, etc.

1.3.3.2.9.2 *Cellulose Fiber*

- Hardwood fibers chemically surface treated to make them compatible with rubber
- Capable of producing very high modulus vulcanizates at 60 phr loading

1.3.3.2.9.3 *Flocks*

- Short fibers of cotton, rayon, or nylon
- Increase tensile strength, tear resistance, and abrasion resistance of vulcanizates
- Used in shoe soles and similar products

1.4 Protective Agents

The unsaturated group in diene rubbers allows these rubbers to be cured with sulfur, but the presence of the free double bonds even after vulcanization makes them vulnerable to attack by oxygen, ozone, and other reactive substances. The changes that occur in the rubber/rubber product may be in the following forms:

- Hardening and embrittlement
- Softening and becoming sticky
- Loss of resilience and strength
- Cracking (with or without orientation)
- Different surface changes like chalking, frosting, crazing, change of color and gloss, etc.
- Other changes like rotting and development of odor

High temperature makes these changes more noticeable. The presence of oxidation catalysts like copper and manganese (rubber poison) accelerates the process of degradation. When overvulcanization (reversion) occurs, these effects become more apparent.

Rubbers containing high amounts of unsaturation—general-purpose rubbers like NR, SBR, and BR—are more prone to attack by degradative forces than the rubbers containing low amounts of unsaturation like IIR and EPDM.

Basically the aging in a rubber article may be divided into the following processes:

- Oxidation at lower or higher temperatures (aging in the real sense)
- Oxidation accelerated by metal/heavy metal compounds (rubber poison)
- Effect of heat in the presence of moisture (steam aging, hydrolysis)
- Oriented crack formation by dynamic stress (fatigue)
- Oriented crack formation by ozone under static and dynamic conditions
- Random crack formation by high-energy ultraviolet light and oxygen (crazing, chalking effect)
- Changes in surface luster (frosting)
- Heat aging with simultaneous exposure to swelling in chemicals

1.4.1 Oxygen Aging at Lower or Higher Temperature (Aging in the Real Sense)

The chemical changes that a polymer experiences due to the attack of atmospheric oxygen are of the following types:

1. Chain breakage (chain scission), mainly in NR, leading to
   - Reduction in hardness and modulus
   - Loss in failure properties and wear (tensile strength, tear strength, abrasion resistance)
   - Loss in resilience
   - Increase in elongation at break
2. Chain crosslinking mainly in synthetic rubbers containing vinyl groups leading to
   - Hardening (increase in hardness and modulus)
   - Loss in tear strength and elongation at break
   - Loss in fatigue resistance
3. Chain modification of the polymer chain: Oxygen combines to the chain giving rise to sidegroups without breaking or reforming the polymer chain. There are no losses in performance properties.

The above three reactions may occur simultaneously.

1.4.2 Accelerated Oxidation in the Presence of Heavy Metal Compounds (Rubber Poison)

Many heavy metal compounds like those of copper and manganese have a catalytic action on the oxidation of rubber vulcanizates. In particular, even a trace (0.001 wt%) of Cu and Mn compounds in NR are able to accelerate the auto-oxidation of the rubber and the vulcanizate, and hence they are called *rubber poison*.

Besides these rubber poisons, other heavy metal compounds like Fe^{++} are especially poisonous in SBR. Co and Ni compounds also accelerate aging phenomena. Standard antioxidants are not used as protecting agents against degradation by heavy metal ions. In the case of protection against degradation by metal ions, chelating agents are commonly used which react with the metal ion to give a stable coordination complex.

1.4.3 Fatigue

When a rubber article is subjected to prolonged mechanical stress such that the article periodically bends back and forth, cracks will slowly develop in the surface which will grow and lead to the breakage of the article. Fatigue failure is initiated at minute flaws where stresses are high and mechanical rupture at such points can lead to development of cracks. The cracks develop perpendicular to the stress direction. NR vulcanizates form these cracks relatively quickly, but they grow slowly. SBR shows slower start of crack, but once the cracks are formed they grow faster. This is related to the low tear resistance of the SBR vulcanizates. Higher temperature and higher frequency of stress change accelerate crack formation. If higher ozone concentrations are also present, dynamic crack formation is accelerated. The fatigue crack resistance depends not only on the type of rubber but also on the crosslink density and type of crosslink (higher crosslink density and sulfur-rich crosslink structures are preferred).

1.4.4 Crazing

When an unsaturated vulcanizate is exposed to prolonged sunlight, small unorientated cracks can develop on the surface. The surface appears like an elephant's skin. The surface on prolonged irradiation becomes brittle, and the filler can chalk. The vulcanizate in this case is not destroyed. This phenomenon is found only in light-colored articles. The carbon black filled and dyed articles do not exhibit this phenomenon. The crazing is more severe when the stock is thin.

1.4.5 Frosting

The light-colored articles on exposure to warm, moist, and ozone-containing atmosphere appear to be dull colored as compared to the previous gloss. Whitish surface discoloration is known as frosting and generally occurs during the summertime as ozone concentration in the atmosphere is highest. Frosting can be differentiated from bloom by heating the rubber article. The chemical bloom will redissolve in the warm rubber, while frosting is permanent.

1.4.6 Heat Aging in the Absence of Oxygen

In the presence of heat and in the absence of oxygen, various reactions can take place (e.g., in steam or in immersed oil), where the changes are like

- Thermal decomposition of crosslinks and also hydrolysis of water-sensitive structures (softening)
- Continuation of inter- or intramolecular network formation (hardening)
- Shifting in crosslinks without change in the total number (no change)

In hydrolyzable rubbers (i.e., polyurethane, EVA, silicone, and others), steam aging proceeds faster because of additional breaking of C-N, C-O, and Si-C bonds.

1.4.7 Ozone Crack Formation

Like in oxidation, the double bonds in the rubber backbone provide the sites for attack to ozone. Most rubbers that can be vulcanized with sulfur are therefore susceptible to attack, with the exception of EPDM where the double bond is present as a pendant group in the side chain and not in the main chain. The physical damage due to ozone attack is manifested in the formation of cracks in the direction perpendicular to the applied stress. The reaction of ozone with rubber is very rapid and occurs at the surface of the rubber/rubber article. In the case of unstretched rubbers, the reaction ceases when all surface unsaturations are consumed. However, when the rubber is stretched again, new unsaturation sites are exposed at the surface and ozone reacts, resulting in the formation and growth of cracks. Under ordinary circumstances, ozone has no effect on saturated rubbers. The speed of ozone crack formation depends strongly on the temperature and humidity of the air and strain level. A comparison of oxygen and ozone attack is shown in Table 1.14.

1.4.8 Auto-Oxidation

Auto-oxidation depends on the following parameters:

1. Nature of the polymer
   - Presence of unsaturation in the polymer backbone—The saturated polymers are less affected by oxygen than are unsaturated polymers.
   - Reactivity of the double bond—The presence of electropositive groups (e.g., -CH_3) activates the double bond leading to fast oxidation in NR; the presence of electronegative groups like -Cl, -CN, etc., deactivates the double bond (e.g., in CR and NBR) causing slow oxidative degradation.
   - In the presence of vinyl side chains (e.g., in CR, SBR, NBR), crosslinks are formed during aging.

2. Type of crosslinkage
   - Nonsulfur crosslinkages are comparatively more stable.
   - Among the sulfur crosslinkages, mono and to some extent disulfidic crosslinkages are more stable compared to polysulfidic crosslinkages. Polysulfidic crosslinkages degrade under heat and produce free radicals that start an oxidative chain reaction.

TABLE 1.14

Comparison of Oxygen and Ozone Attack

Effect	Ozone, O_3	Oxygen, O_2
Observed as	Discoloration, cracking	Hardening or softening
Degradation occurs at	Surface only	Throughout
Effect of unsaturation	Diene only	Diene < saturated
Effect of cure system	No effect	Sulfur < sulfur donor < peroxide
Temperature	Rate enhances	Rate doubles per 10°C (rate enhances by 2.8 times for NR and 2.2 for SBR)
Strain level	Increases crack growth	No effect

3. Degree of cure
   - In NR and IR, overcure gives higher degradation.
   - In synthetic rubbers like SBR and CR, overcure improves resistance to oxidative degradation.

1.4.9 Mechanism of Oxygen Attack

Oxidation is a free radical chain reaction. It is now well established that the attack by oxygen takes place at the α methylene carbon atom in the chain. A hydrogen atom is abstracted, and in the presence of oxygen an oxidative chain reaction is initiated which, if unchecked, propagates autocatalytically.

1. $RH \xrightarrow{h\nu} R^{\cdot}$

2. $R^{\cdot} + O_2 \longrightarrow ROO^{\cdot}$

3. $ROO^{\cdot} + RH \longrightarrow ROOH + R^{\cdot}$

4. $ROOH \xrightarrow{h\nu} RO\cdot + ROO\cdot + H_2O$

5. $RO\cdot + RH \longrightarrow ROH + R^{\cdot}$

6. $ROO\cdot + RH \longrightarrow ROOH + R^{\cdot}$

Even if a small amount of combined oxygen is present, considerable loss in performance properties occurs (e.g., if 1% combined oxygen is present in a NR gum vulcanizate, 50% loss in tensile strength occurs).

1.4.10 How an Antioxidant Performs Its Function

In order to prevent the autocatalytic reaction, an inhibitor must break the chain, either

1. By capturing the free radical formed
2. By ensuring that the peroxides and hydroperoxides produced decompose into harmless fragments without decomposing a polymer and also ensuring that no new free radicals (capable of propagating the reaction chain) are formed.

The reaction scheme is as follows:

$$R^{\cdot} + AH \longrightarrow RH + A^{\cdot}$$

$$ROO^{\cdot} + AH \longrightarrow ROOH + A^{\cdot}$$

$$ROOH + AH \longrightarrow \text{Harmless fragments}$$

1.4.11 How an Anti-Ozonant Performs Its Action

The mechanism of ozone attack on a double bond is shown below. It is believed that the anti-ozonant reacts either with the (1) Zwitter ion or the (2) ozonide formed to make an inert protective film that every time when broken, is repaired by formation of a fresh film produced by the reaction of rubber, ozone, and anti-ozonant (protective film theory). The other concept (scavenger theory) is that the anti-ozonant reacts with ozone at a much faster rate than ozone reacts with the double bond of rubber.

$$>C{=}C< + O_3 \longrightarrow >C\overset{O_3}{—}C< \longrightarrow >C^{+}OO^{-} + >C{=}O$$

(1) $\rightarrow$ (2): ozonide $>C—C<$ bridged by O--O above and O below

(2)

1.4.12 Antidegradants

The chemical names of various antidegradants are given in Table 1.15.

1.4.12.1 Classification of Antidegradants

Classifications of various antidegradants are given in Table 1.16.

1.4.12.2 Application of Different Antidegradants

The following points must be taken into consideration before using an antidegradant:

- What are antioxidants for rubber?
  - Rubbers having fairly good amount of unsaturations like NR, IR, SBR, BR, and NBR: Mainly IPPD, 6PPD, 77PD, DPPD, and TMQ

TABLE 1.15
Chemical Names of Different Antidegradants

Common Abbreviation	Chemical Name
IPPD	*N* Isopropyl *N′* phenyl para phenylene diamine
6 PPD	*N* 1,3 Dimethyl butyl *N′* phenyl para phenylene diamine
77 PPD	*N,N′* Di (1,4 dimethyl pentyl) para phenyl diamine
DPPD	*N,N′* Diphenyl para phenylene diamine
TMQ	Polymerized 1,2 dihydro 2,2,4 trimethyl quinoline
SP	Styrenated phenol
ODPA	Octylated diphenyl amine
BPH	2,2 Methylene bis (4-methyl-6-tert butylphenol)
BHT	2,6-Di-tert butyl-4-methyl phenol
MBI	2-Mercaptobenzimidazole
MMBI	4- and 5-methyl mercaptobenzimidazole

TABLE 1.16

Classification of Different Antidegradants

Physical Protectants	Chemical Protectants
Paraffin wax	Staining antioxidants with anti-flex cracking and antiozonant effects: IPPD, 6PPD, 77PD, DTPD
Microcrystalline wax	Staining antioxidants with anti-flex cracking effect but without antiozonant effect: PBN, ODPA, SDPA (styrenated), ADPA (acetonated)
	Staining antioxidants without anti-flex cracking or antiozonant effect: TMQ
	Nonstaining antioxidants with anti-flex cracking effect but without antiozonant effect: styrenated phenol (SP)
	Nonstaining antioxidants without anti-flex cracking and antiozonant effect: BHT, BPH, MBI
	Nonstaining antioxidants without anti-flex cracking but with antiozonant effect: cyclic acetals, enol ethers

- Rubber with low unsaturation (EPDM, IIR, partially unsaturated HNBR): TMQ, ODPA, and MBI
- CR: ODPA, MBI acts as an accelerator for CR. Tables 1.17 and 1.18 give the reference antioxidant doses and their effectiveness as protecting agents in rubber compounds

- Is freedom from staining or contact staining demanded?
  - Products that stain heavily are the paraphenylene diamine types, and the non-staining types are BHT and SP types
- What must be protected?
  - Surface effects (cracking, chalking, embrittlement) or bulk effects (hardening, softening, loss of strength)
- What part does anaerobic (non-oxidative) aging play?
  - Antidegradants do not protect from heat as such. Special vulcanization systems are required here, especially low sulfur systems, use of thiurams, peroxides, etc.
- Is resistance to hydrolysis possibly required in addition to resistance to oxidative degradation?
  - MBI and MMBI have given good results in the steam aging of diene rubbers. Rubbers with hydrolysable groups like EVA and AU are preferably protected with carbodimides. All the ester grades are prestabilized.
- Are there food and drug applications?
- IPPD, DPPD, ODPA, BHT, SP, and BPH are all approved by the U.S. Food and Drug Administration (FDA).

TABLE 1.17

Application and Recommended Doses for Various Antidegradants

Application	IPPD	6PPD	77PD	DPPD	ODPA	TMQ	SP	BPH	MBI	BHT
As anti-flex cracking agent	1.0–3.5 IPPD/TMQ = 2:1	1.3–4.0 6PPD/TMQ = 2:1	1.0–3.0 77PD/TMQ = 1:1	1.0–3.0 DPPD/6PPD = 1:2	1.3–4.0 ODPA/6PPD = 1:1 to 1:2	1.0–4.0 TMQ/6PPD = 1:1 to 1:2	1.5–3.0 SP/MMBI = 1:1	—	—	—
As antiozonant	1.0–3.5	1.3–4.0	0.8–3.5	1.5–3.5 DPPD/6PPD = 1:2	0.7–1.7	0.5–1.5	—	—	—	—
As antioxidant	1.0–2.6 IPPD/MMBI = 1:1 0.5–1.5 IPPD/TMQ = 1:1	1.0–2.6 6PPD/MMBI = 1:1 0.5–1.5 6PPD/TMQ = 1:1	0.7–1.5 6PPD/MMBI = 1:1	0.5–1.5	0.7–1.7	0.5–1.5	0.6–2.0	0.7–1.7	0.6–1.5	0.5–2.0
As protective agent against rubber poison	0.5–1.5	0.5–1.5	—	0.5–1.5	1.0–2.0	1.5–3.0 TMQ /MMBI 1:1	—	0.5–1.5 BKF/ MMB = 1:1	0.5–1.0 6PPD/ MBI = 1:1	—
As antiozonant for CR	0.4–1.5	0.4–1.5	—	0.4–1.5	1.0–2.0	—	1.0–2.0	—	—	—
As antioxidant for EPDM	—	—	—	—	—	1.0–2.0	—	—	1.5–2.5	—
As antioxidant for EVA	—	—	—	—	0.5–1.5	—	—	—	—	—
For heat-resistant NR, SBR, NBR	—	—	—	—	—	—	—	—	2.0–3.0 MBI/TMQ = 1:1	—
As antioxidant for HNBR	—	—	—	—	0.5–1.5	—	—	—	—	—

TABLE 1.18

Protective Effects of Antidegradants

Antidegradant Type	I Autoxidation	II Heat	III Flex Cracking	IV Ozone Cracking	V Metal Poisoning	VI Crazing	VII Cyclization	VIII Staining[a]	IX Contact Staining
IPPD	2	2–3	1	1–2	2	6	—	5–6	5
6PPD	2	2–3	1–2	2[f]	2	6	—	5–6	5
77PD	3–4	3–4	2	1	—	6	—	5	5
DPPD	2	2–3	2	3[f]	2	6	—	5	4
ODPA	2–3	2[d]	4	6	3	6	—	2–3	1–2
TMQ	2	1–2[c]	4–5	6	3–4	6	—	3	2
SP	4	4	4	6	—	3	—	0	0
BPH	2–3	3	6	6	3[c]	1	1–2	1	0
BHT	3–4	4–5	6	6	4–5	1	2–3	0	0
MBI	4[b]	3[e]	6	6	6[h]	6	6	0[i]	0
MMBI	4[b]	3[e]	6	6	6[h]	6	6	0[i]	0

Notes: Rating (columns I–VII): 1 = best and 6 = worst; rating (columns VIII–IX): 0 = no staining and 6 = very heavy staining.

[a] NR and IR.
[b] Compounds accelerated with dithiocarbamates.
[c] In conjunction with MBI or MMBI: 1.
[d] In CR: 1.
[e] In conjunction with 6PPD, TMQ: 1.
[f] In CR: 1.
[g] In conjunction with MBI or MMBI: 2.
[h] In conjunction with BPH: 1.
[i] Has a brightening effect.

1.5 Vulcanizing Chemicals

The main drawback of rubber goods manufactured in earlier days (prior to 1830) is that they were susceptible to variation of temperature (e.g., sticky in hot weather and brittle in cold weather). In 1839 Charles Goodyear by chance found that heating a mixture of rubber and sulfur produced an elastic mass that is neither sticky nor brittle with change of temperature. In the same period, Thomas Hankcock, working independently, observed and reported the same phenomenon. As a result there was a dispute as to the degree of credit; however, there is no doubt that the invention of these two men contributed to the development of the rubber industry as we know it today. Vulcanization is named after Vulcan, the Roman god of fire. Vulcanization is a process where a physically soft viscous compounded rubber material is converted into a hard elastic engineering product.

Elastomers are macromolecules having double bonds or some other active sites located at intervals along the molecular chains. Vulcanization or cure refers to a chemical process by which the long-chain rubber macromolecules are bound together to form a three-dimensional elastic network by the insertion of sulfur or other crosslinking agents at their active sites.

$$\text{Raw rubber} + \text{Crosslinking agent} \xrightarrow{\Delta H} \text{Vulcanized rubber}; \quad \Delta H = \text{Heat input}$$

1.5.1 Changes Achieved in a Product Due to Vulcanization

- A rubber product becomes non-sensitive to changes in temperature.
- Long chains of rubber molecules (molecular weight usually between 10^5 and 10^6) crosslinked by reaction with a crosslinking agent form a three-dimensional network structure, which in turn transforms the soft weak plastic-like material to a strong elastic product. Table 1.19 shows the effects of crosslink structure and crosslink length on vulcanized rubber properties.
- A rubber loses its tackiness and becomes insoluble in solvents.
- It becomes more resistant to deterioration normally caused by heat, light, oxygen, ozone, and other aging processes.
- There is an improvement in wear and tear properties.
- Permanent deformation due to stress is reduced.

1.5.2 Vulcanizate Structure and Property Relation

A rough correlation between vulcanizate structure, namely crosslink number and crosslink length, and properties is given in Table 1.19. Five salient features are observed as follows:

- Modulus or stiffness will increase as the crosslink densities increase. The three-dimensional structure, formed during vulcanization, becomes more resistant to deformation under load. The higher the crosslink density, the lower will be the elongation.
- When the vulcanizate is overcured, the crosslink structure degrades (in the case of NR) and modulus drops. This is prominent in the cases of NR and IIR. Another case may occur, where the modulus continues increasing due to overcure. This

TABLE 1.19
Effect of Crosslink Structure and Crosslink Length on Vulcanizate Properties

Property	Crosslink Number	Crosslink Length (S_x)
Tensile strength	>... then <...	>
Elongation	<	—
Flex life	<	>
Modulus	>	—
Resilience	>	<
Set	<	>
Abrasion	>	>
Rolling resistance	<	—
Hysteresis	<	>
Thermal stability	—	<
Oxidative stability	—	<
Tear strength	>... then <...	>

Notes: >, increase; <, decrease.

phenomenon is known as the marching modulus. This is common with SBR and BR types of rubbers.

- Tensile strength, tear strength, and fatigue life frequently pass through a maximum as cure progresses and then drops. Beside crosslink densities, these failure properties are highly dependent on the presence of flaws in the material. As the test proceeds, any applied load acting on a flaw (due to poor dispersion, small nicks or tears, localized overcure, etc.) will concentrate the stresses at the flaw and failure will take place consequently, reducing the above properties.
- Crosslink density has a prominent effect on permanent set and hysteresis properties of the vulcanizate. These properties are related to the viscous component of the rubber vulcanizate. A lower viscous component will improve the permanent set as well as the hysteresis properties. With increasing crosslink density, the viscous component of the vulcanizate reduces and results in improving the set as well as hysteresis properties.
- During the crosslinking process, different crosslinks are formed depending on the type of curatives. These are monosulfidic, disulfidic, polysulfidic, etc. All of these crosslinks are of different bond energies. (See Table 1.20.) The lower the polysulfidic linkage, the higher will be the bond strength. This will result in better heat aging properties of the vulcanizate.

TABLE 1.20
Dissociation Energy of Bonds

Type of Crosslink	Dissociation Energy (kcal/mol)
-C-C-	84
-C-S-C-	68
-C-S-S-C-	64
-C-S_X-C-	<64

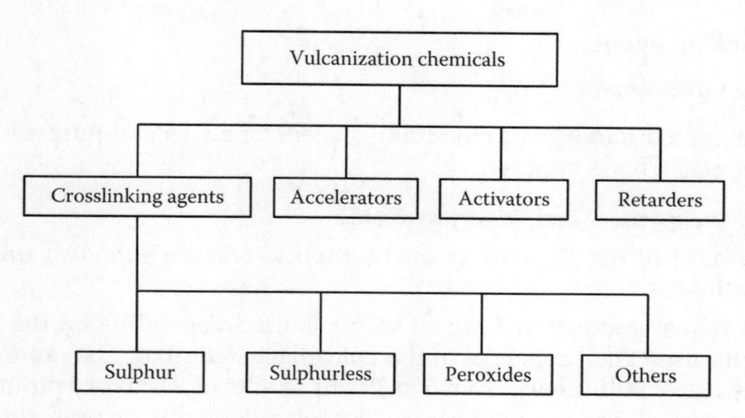

FIGURE 1.1
Vulcanization systems.

1.5.3 Vulcanization Chemicals

Vulcanizing agents are the combinations of crosslinking agents, accelerators, activators, and retarders, as displayed in Figure 1.1.

The following points are interesting to note:

- Sulfur was the first agent used to vulcanize the first commercial elastomer, NR.
- Vulcanization was achieved by mixing 8 phr of sulfur at 140°C; the time required was 5 hours.
- The addition of 5 phr of ZnO reduced the time to 3 hours.
- The use of certain rate-enhancing ingredients known as accelerators in concentrations as low as 0.5 phr reduced the time to 2 to 5 minutes.
- In the above example sulfur is considered as the crosslinking agent. The chemical used in low concentration to enhance the vulcanization is known as the accelerator. ZnO activates the accelerating effect of the accelerator and is known as the activator.
- Very often one has to retard the onset of vulcanization to facilitate sufficient processing safety. This is achieved by adding certain chemicals known as retarders.

1.5.4 Parameters of Vulcanization

- Cure time
- Scorch time
- Vulcanization
- Optimum cure
- Plateau effect
- Reversion
- Rate of cure

1.5.5 Crosslinking Agents

1.5.5.1 Sulfur Vulcanization

Sulfur is the most commonly used crosslinking agent for all general-purpose rubbers (NR, BR, SBR, NBR, etc.). This is because:

- Sulfur is inexpensive and widely available.
- NR and most of the low-cost synthetic rubbers contain sufficient unsaturation for crosslinks.
- For soft rubber goods 0 to 5 phr of sulfur is used depending on the amount of accelerator used and properties of the vulcanizate required. The product derived from the use of sulfur level from 5 to 15 phr results in a leathery product (practically not useful). For hard rubber articles (ebonite) sulfur is used up to 25 phr. During crosslinking of rubber with sulfur, the sulfur gets chemically bound with the rubber network in different ways.

1.5.5.1.1 *Soluble and Insoluble Sulfur*

- *Soluble sulfur* is the rhombic form of elemental sulfur and exists as a cyclic (ring) structure of eight atoms of sulfur as S_8. This form of sulfur causes blooming in the vulcanizate, which reduces building tack of compounds. This form is more in use as compared to its insoluble counterpart as it is much cheaper.
- The *insoluble sulfur* is the amorphous form of elemental sulfur, which is insoluble in CS_2 and is polymeric in nature with a molecular weight of 100,000 to 300,000. This form of the sulfur is insoluble in most of the solvents and rubbers, and hence the name is insoluble. Because of its insolubility, it resists migration to the surface prior to cure, and hence blooming of sulfur does not take place. It is used in compounds where blooming will not occur to deteriorate building tack. Precautions must be taken while mixing insoluble sulfur so that the mixing temperature (dump temperature) does not exceed 100°C, otherwise this costlier insoluble sulfur will be converted to soluble sulfur.

1.5.5.1.2 *Effect of Vulcanization of Rubber with Sulfur Alone*

- The rate of reaction is very slow, taking a long time.
- The cure rate varies leading to undercure or overcure—poor quality of a product.
- At optimum cure conditions the original (initial/unaged) physicals and aged physicals are far from satisfactory.
- Maximum crosslinks are polysulfidics (number of sulfur atoms is more than 2) and therefore aging characteristics are very poor.
- The vulcanizates are dark in color and show sulfur blooming.

1.5.5.1.3 *The Chemistry of Accelerated-Sulfur-Vulcanization*

- The accelerator reacts with sulfur to give monomeric polysulfides of the type: AC-S_X-AC.
- The polysulfides can interact with the rubber to give polysulfides of the type: Rubber-S_X-AC.
- The rubber polysulfides then react, either directly or through a reactive intermediate to give crosslinks of rubber polysulfides of the type: Rubber-S_X-Rubber.

1.5.5.2 Sulfur Donor Vulcanization

In this case, vulcanization is done without elemental sulfur, by the use of sulfur donor systems. When the sulfur level is below 0.5 phr, it is not possible to match the modulus simply by increasing the accelerator level, unless the accelerator is a sulfur donor type. Sulfur donors liberate sulfur at vulcanization temperatures and can be directly substituted for sulfur. They can be subdivided into two types:

1. One type can act as a sulfur donor as well as an accelerator. An example is TMTD (tetramethyl thiuram disulfide).
2. The other type does not act as an accelerator and must always be used in conjunction with an accelerator. An example is DTDM (dithiodimorpholine). Because of toxicity DTDM is not used. A suitable alternative with no toxicity is available in the market. CLD 80 is one such alternative, manufactured by Rhein Chemie (Mannheim, Germany).

Generally, sulfur donors are used when the free sulfur level is reduced with the objective that the vulcanizates produced will primarily have mono- and disulfidic crosslinks and hence higher thermal and oxidative aging.

1.5.5.3 Peroxide Vulcanization

Saturated rubbers (Ethylene Propylene Rubber (EPR), silicone rubbers, etc.) cannot be crosslinked by sulfur and accelerator. Organic peroxides are useful for vulcanizing these rubbers. Peroxides are the most common crosslinking agents after the sulfur because of their ability to cure a number of diene and non-diene containing elastomers. When peroxides decompose, free radicals are formed that, in turn, react with polymer chains to produce free radicals in the polymer matrix that then combine to form crosslinks. Crosslinks of this type only involve carbon-to-carbon bonds and are quite thermally stable (very good aging resistance). The main disadvantage of the system is the higher cost involvement and products having low abrasion and tear properties.

1.5.5.4 Vulcanization with Resins

Certain di-functional compounds form crosslinks with elastomers by reacting with two polymer molecules to form a bridge. Epoxy resins are used with NBR, quinone di-oximes and phenolic resins are used with butyl rubber, and dithiols and diamines are used with fluorocarbons. One of the most important curing agents for butyl rubber is phenolic resin. This cure system is widely used for bladders and the curing bags used in tire curing and the retread industry. The low levels of unsaturation of butyl rubber require cure activation by halogen-containing materials like $SnCl_2$, CR, etc.

1.5.5.5 Metal Oxide Vulcanization

Polychloroprene rubber (CR or Neoprene) and chlorosulfonated polyethylene (CSM) are vulcanized with metal oxides. The reaction involves active chlorine atoms. Usually a combination of zinc oxide and magnesium oxide is used for the purpose of controlling the vulcanization rate and absorbing the chloride formed.

Table 1.21 provides a summary of the above discussion.

TABLE 1.21
Crosslinking Agents for Different Elastomers

Class	Elastomers	Advantage	Disadvantage
Accelerated sulfur and sulfur donor	Diene; i.e., NR, SBR, BR, EPDM	Versatile	Heat resistance and set properties
Peroxides	Especially saturated types like silicone, EPR	Excellent heat resistance and set properties	Control of cure rate and poor fatigue resistance
Resins	Primarily butyl rubber	Heat resistance, stable modulus	Slow cure
Metal oxides	Halogenated elastomers		Water resistance

1.5.5.6 Radiation Curing

Radiation curing is a process where the final curing is being carried out at ambient temperature under closely controlled conditions, such as radiation dose, irradiation dose rate, penetration depth (in case of electron beam curing), etc. Radiation can produce a degree of crosslink like normal sulfur curing. The type of crosslink formed by the radiation technique is mainly carbon-carbon (-C-C-), as compared to -C-S_x-C- link in the case of sulfur cure. The -C-C- crosslink gives better thermal aging properties at high temperature. However, the flexing properties are reduced with -C-C- crosslink. Other properties like compression set, abrasion resistance, etc., are also improved with radiation cure. Radiation cure is widely used in the tire industry.

1.5.5.7 Conventional, Semi-Efficient, and Efficient Vulcanization Systems

- For the most frequently used conventional vulcanization (CV) systems for a NR-based compound, roughly 2.0 to 3.5 phr of sulfur and 0.4 to 1.2 phr of accelerator are used. These typically provide high initial physical properties—tensile, tear, and good fatigue properties—but with a greater tendency to lose these properties after heat aging.
- When accelerator dosage is increased to 1.2 to 2.5 phr, less sulfur (1.0 to 1.7) is required to achieve the same crosslink density. This results in the formation of crosslink with a lower number of sulfur atoms. This system is known as the semi-efficient vulcanization (Semi EV) system and produces heat and reversion resistant vulcanizates.
- A further increase in accelerator dose and reduction in sulfur content (or even in the absence of elemental sulfur and in the presence of a sulfur donor) results in a system called the efficient vulcanization (EV) system. This leads to mono- and disulfidic crosslink structures and the result is very good heat and reversion resistance.

Table 1.22 briefly illustrates the above systems in natural rubber:

- For conventional vulcanization of NR, a slightly higher sulfur dose and a somewhat lower accelerator dose are employed than for synthetic rubbers.
- SBR in particular can take advantage of EV curing. If a conventional system is employed, it does not exhibit fatigue loss as observed in natural rubber, and the

TABLE 1.22

Cure Types of Natural Rubber (NR) Based Rubber Compound

Type	Sulfur (S, phr)	Accelerator (A, phr)	(A/S) Ratio	Monosulfide (%)	Di- and Polysulfide (%)
CV	2.0–3.5	0.4–1.2	0.1–0.6	5	95
Semi EV	1.0–1.7	1.2–2.5	0.7–2.5	50	50
EV	0.4–0.8	2.0–5.0	2.5–12	80	20

average number of sulfur atoms per crosslink is significantly lower in SBR than that in NR. The use of SEV and EV systems is advantageous for SBR.

- For butyl and EPDM rubbers having very limited unsaturations, larger amounts of sulfur and accelerators are required than for other diene rubbers.

1.5.6 Accelerator

An accelerator is a chemical that increases the rate of vulcanization of rubber with a crosslinking agent like sulfur.

The features required for an ideal accelerator are:

- Fast cure
- High activity (crosslinking efficiency)
- Soluble in rubber (no bloom, good dispersion)
- Good processing safety (delayed action)
- Good storage stability (as material and as compounded)
- Flat plateau (no reversion, particularly for NR)
- Effective over a wide range of temperatures
- Compatible with other ingredients
- No safety or handling problems
- No side effects on other properties (aging, adhesion, etc.)

The advantages of the use of accelerators for vulcanization include:

- Increased output—the cure time is reduced from hours to minutes
- Better physical properties (tensile, tear, abrasion, etc.)
- Reduced variability of degree of cure with different batches of rubber
- Greater control over vulcanization—correct cure times can be predicted, the occurrence of overcure and undercure is reduced, and products with specific properties can be tailor-made
- Better colored products
- Better aging properties

An ideal accelerator should give rise to a cure curve having sufficient scorch safety, and once the curing starts, it would cure very rapidly and no reversion would occur.

TABLE 1.23
Different Accelerator Types

Accelerator Type	Abbreviations	Speed
Guanidines	DPG, DOTG	Slow
Dithiocarbamates	ZDMC, ZDEC, ZDBC	Ultra
Thiurams	TMTM, TMTD, TETD, DPTT	Ultra
Thiazoles	MBT, MBTS, ZMBT	Moderate
Sulfenamides	CBS, TBBS, MBS, DCBS	Fast (delayed action)
Thioureas	ETU, TMTU, DETU, DBTU	Fast
Thiophosphates	ZDBP, ETPT, D8 ITO	Semi-fast

1.5.6.1 Classification of Accelerators

According to the chemical composition and speed of vulcanization, the accelerators may be classified into seven main groups as shown in Table 1.23.

The accelerators that have the widest application worldwide and are produced in the greatest volumes are the thiazoles and the sulfenamides. The sulfenamides have the greatest range due to a combination of:

- Fast cure
- Good processing safety
- High activity
- Flat plateau (less reversion, particularly for NR)
- Solubility

The thiazoles, however, offer slightly greater resistance to thermal and oxidative aging to the vulcanizates.

1.5.6.2 Accelerated Sulfur Vulcanization for General-Purpose Rubber (GPR)

Table 1.24 shows the effects of various accelerators on scorch safety, cure rates, and crosslink length.

TABLE 1.24
Characteristics of Accelerators

Accelerator Type	Scorch Safety	Cure Rate	Crosslink Length
None	—	Very slow	Very long
Guanidines	Moderate	Moderate	Medium–long
Thiazoles	Moderate	Moderate	Medium
Sulfenamides	Long	Fast	Short–medium
Thiurams	Short	Very fast	Short
Dithiocarbamates	Least	Very fast	Short

1.5.6.3 Primary Accelerators

A single accelerator system is capable of producing sufficient activity to produce satisfactory cures within specified times.

Take for example an accelerator in a lug-type cap compound (bias tire):

Soluble sulfur	2.50 phr
N-Oxydienthylene-2-benzothiazole	0.50 phr
Prevulcanization inhibitors (PVI)	0.10 phr

1.5.6.4 Secondary Accelerator or the Boosters

In a vulcanization system two or more accelerators are present: the primary accelerator is present as the largest amount and the secondary or the booster in 10 to 20% of the total. The secondary accelerator activates and improves efficiency of the primary accelerator. Due to the synergistic effect, the final properties of the vulcanizates are better than those produced by either accelerator separately. Primary accelerators are generally the thiazole and sulfenamide classes, while the secondary or the boosters are thiurams, dithiocarbamates, and guanidines. As a general rule, sulfenamides exhibit faster cure rate than thiazoles, and among the secondary accelerators dithiocarbamate is the scorchiest, followed by thiurams and then guanidines.

1.5.7 Activators

Activators are chemicals used to increase the rate of vulcanization by reacting first with the accelerators to form rubber-soluble complexes. These complexes then activate the sulfur to effect vulcanization. Common activators are combinations of zinc oxide and stearic acid for sulfur-accelerator curing systems for general-purpose rubbers. Natural rubber usually contains sufficient fatty acid to solubilize zinc salt. However, if the fatty acids are first extracted by acetone, then the resultant clean natural rubber exhibits a much lower state of cure. Therefore, fatty acids are usually added for assurance. Sulfenamide accelerators generally require less fatty acid because they release an amine during vulcanization which acts to solubilize the zinc. Guanidines as amines also serve both to activate as well as to accelerate.

1.5.8 Retarders

Retarders are added to the formulation specifically to obtain longer processing safety to prevent premature cure during factory processing. The function of a retarder is to retard the rate of cure, which means it has the opposite function of an accelerator.

The choice of a retarder depends on:

- The accelerator system
- Cost
- Side effects

Examples are:

- Benzoic acid
- Phthalic anhydride
- Salicylic acid

The limitation of the above retarders is that they reduce effectiveness of accelerators and thus prolong vulcanization time. The advent of new generation PVIs increased the processing safety of the compound without sacrificing the rate of cure. The example is cyclohexyl thiophthalimide (CTP) used at 0.10 to 0.30 phr. The mechanism of functioning of a CTP is that it scavenges MBT which is an auto-catalyst for the breakdown of the accelerator to form cyclohexyldithiobenzothiazole and phthalimide.

1.5.9 Effect of Other Compounding Ingredients on Vulcanization

1.5.9.1 Antidegradants

The amine type of antioxidants will affect curing. Due to the alkaline nature of these antioxidants, these reduce scorch safety. This effect is significant in metal oxide curing of polychloroprene.

1.5.9.2 Fillers

- Furnace blacks like HAF, FEF, GPF, etc., are slightly basic and have an activating effect on sulfur curing rates. Also, carbon blacks promote the formation of mono- and disulfidic crosslinks. This helps minimize reversion and enhances aging properties.
- Precipitated silica generally slows the sulfur cure. Silica has a tendency to absorb some sulfenamide accelerator, which results in slowness in the curing process. The use of materials like diethylene glycol or triethanol amine prevents this competition and maintains the desired cure characteristics.
- Neutral fillers like calcium carbonate (whiting) and clays have little or no effect on the cure properties.
- Most rubber process oils contain some unsaturated components, and this unsaturation can compete with a polymer for the curatives. This process leads to slowness in the curing reaction, where a large quantity of processing oils is used. In order to avoid this slowness, fast accelerators are used in the curing process.

1.6 Processing Aids

Processing aids are materials used in rubber compounds to improve the processing properties without affecting the performance characteristics. The processing aids are generally used in low doses. Processing additives are generally characterized on the basis of their chemical structure and on the basis of their application. There are numbers of naturally occurring materials used as processing aids in rubber compounds.

Process oils serve as a primary processing aid in rubber compounds. There are three main categories of oils, *viz.*, paraffinic, naphthenic, and aromatic, which are common rubber process oils. The selection of oils for a particular rubber formulation depends on the compatibility criteria of the oil. Compatibility of the oils depends on the solubility parameters of the oils. If the oil is not compatible with the polymer, it will migrate out of the compound and cause deterioration of physical properties of the compounds. The compatibility of oil also depends on its viscosity, molecular weight, and molecular composition.

Since January 2010, aromatic oils, which are also known as distillate aromatic extract (DAE), were banned in European countries. This was integrated into Annex XVII of REACH (Registration, Evaluation, Authorization and Restriction of Chemicals). This DAE contains a high percentage of polyaromatic hydrocarbons (PAHs), which are carcinogenic in nature. The EU Reach Regulation set out a prohibition on the use of "high-PAH" extender oils in the production of tires. With a growing focus on environmental and safety standards, both at international and European levels, the European Commission has over the past few years introduced an increasing number of stringent health, safety, and environmental requirements that are applicable in the tire sector as in others. As per EU law, the process oils containing more than 3% polycyclic aromatic (PCA) content cannot be used as rubber process oils in tire and other rubber products. Suitable alternatives of DAE, with low PCA content (<3%) are already available in the market in the form of treated distillate aromatic extract (TDAE), mild extracted solvent (MES), residual aromatic extract (RAE), etc.

Besides processing oils, there are several other processing aids being used with different rubber compounds. Some of these processing aids are as follows:

Chemical peptizer: It is used to reduce polymer viscosity through the chain scission. It is mainly used in the case of NR-based compounds. Pentachlorothiophenol, 2.2′ dibenzamidodiphenyl disulfide are common examples of chemical peptizers.

Physical peptizer: It is used to reduce the polymer viscosity through internal lubrication, like zinc salt of higher fatty acids.

Homogenizing agents: These chemicals are used to improve the compatibility of polymer blends, mainly where polymers with varying viscosities are blended. Blended resins are used as homogenizing agents.

Plasticizers: To improve product performance at low as well as high temperatures, plasticizers are being used in rubber compounds. There are varieties of plasticizers available in the market, both polar as well as non-polar in nature. The plasticizers are selected depending on the types of polymers. Aliphatic, aromatic esters are widely used as plasticizers.

Dispersing agent: These materials are used for better dispersion of filler in the rubber matrix. Dispersing agents also help to reduce the incorporation time of filler in rubber compounds. Mineral oils, fatty acids, esters, and metal soaps are commonly used as dispersing agents.

Tackifiers: It helps to improve the tack of the rubber compounds. Hydrocarbon resins are an example of the tackifiers.

Coupling agents: These are materials capable of reacting with both the reinforcement and the resin matrix of a composite material.

There are many other minor compounding ingredients that are not discussed here. Readers are advised to refer to Chapter 1 on compound design by A.D. Thorn and R.A. Robinson in *Rubber Products Manufacturing Technology* (A.K. Bhowmick, M.M. Hall, and Benary, Eds., Marcel Dekker, New York, 1994).

1.7 Formula Construction

In earlier sections, various compounding ingredients have been described. In formula construction, these ingredients are to be judiciously used. There are two types of situations faced by a compounder:

1. To design a formula when there is no specification
2. To design a compound when the specification exists

These two types are described below. The following sequence should be followed for formula construction:

- Design a formulation. This may be the current formulation used or a new design based on the application requirements.
- Test rubber properties of the available products. For an unknown compound, this is not possible. For a known compound, check which properties need to be improved.
- Determine and analyze if the formulation uses correct compounding ingredients which would meet the desired properties.
- If not, make changes and suggest an alternate formulation by using the design of experiments. Calculate cost.
- Mix cure samples and measure properties in laboratory.
- Select one or two of the best compounds with improved properties and match with the specification available.
- Run a plant trial for mixing, processing, and curing the products.
- Test factory-mixed compound in the laboratory for specified properties.
- Test the actual product and check whether it meets the requirements.
- Select the best compound. Calculate cost.
- If any of the steps do not work, go to the appropriate step and start all over again.

A common order for formula development should be:

1. Considering the service environment or the specification, the type of rubber (natural rubber, nitrile rubber, etc.) should be selected.
2. The next choice is for the most appropriate filler and its amount.
3. As soon as the elastomer and the filler are chosen, oil type and its level can be decided with reference to the requirement/specification.

4. Next the vulcanizing agent such as sulfur, sulfur donors, metal oxides (in case of CR), etc., should be chosen.
5. With appropriate selection of the above items, accelerators and activators are selected.
6. Depending on the elastomer type, next the age resistor/antidegradant package is chosen.
7. Special compounding ingredients like fire retardant, color, etc., are selected next.

The following paragraphs explain how a compound is designed with or without specifications.

We explain here how to design a tire tread compound when the specifications are not given (usually in normal practice specifications of this compound are available; for the sake of illustration, this is taken).

First we have to know the general requirements for this product. For a typical tire tread, consideration should be given to abrasion, heat buildup, wet grip, tear strength, ease of processing, maximum service temperature, etc. As in the rubber formulation, when an attempt is made to improve one property, the others may deteriorate. Hence the properties listed must be prioritized. For example, in the above case, abrasion resistance is the most important property. There are various kinds of tires, including truck tires, passenger car tires, racing car tires, and tires used in the mountains (which describes the environmental condition). The formula for each is different, as the conditions/specifications for use are different. Here we describe a general procedure for a truck tread compound.

As mentioned, selection of the polymer is done first. The major requirement is high abrasion resistance. The properties of various elastomers are reviewed first. Natural rubber, styrene-butadiene rubber, and polybutadiene rubber would be the logical choices. Out of them, polybutadiene displays maximum abrasion resistance. However, it suffers from poor processing. Further, the cost is to be considered. Natural rubber also has good abrasion resistance and low heat buildup (two key properties). Hence a blend of natural rubber and polybutadiene rubber is chosen.

	phr
Natural rubber	75.0
Polybutadiene rubber	25.0

Major factors for choosing any elastomer are now described. Some principal properties to consider include maximum service temperature and service life, minimum service temperature, processability conditions, cure system, cure rate, capacity for extension by added filler, and chemical and environmental resistance.

The next choice is the filler or reinforcement. Because of the criticality of abrasion properties, maximum reinforcement is needed. Hence carbon black is the only choice. Abrasion resistance increases as the particle size decreases. Again, when the particle size decreases, processing and dispersion are affected. Hence the choice is HAF or ISAF. SAF is eliminated for poor processing, and other carbon blacks are omitted for poorer abrasion resistance. As the wear resistance displays maximum at 45 to 50 phr loading for these blacks, ISAF at 50 phr loading is chosen. In case there is any doubt, variations of carbon black loading and type may be carried out using a design of experiments.

Such a loading of carbon black would be difficult to process, so the formula needs plasticizer or softener. In this case, a plasticizer that would help processing and ease dispersion may be selected, so naphthenic rubber processing oil is chosen. The practice is to take 1 part of oil for 10 parts of this black; i.e., for 50 parts, 5 parts may be taken. The resulting formulation at this stage is:

	phr
Natural rubber	75.0
Polybutadiene rubber	25.0
ISAF	50.0
Process oil	5.0

The next question is the curative package. Based on earlier discussion, only a sulfur cure system is recommended, as peroxide would not give enough flexibility to tire tread. Sulfur donors are also used in some tread formulation. Using sulfur, three types of systems are possible: conventional, efficient, and semi-efficient. As we require maximum flexibility of the tread, a conventional system is recommended. However, a semi-efficient system is also practiced. The typical package for a conventional vulcanization system is:

	phr
Sulfur	2.0
Accelerator	0.8

The amount of accelerator would also depend on its nature. Generally, sulfenamides are used. There are four types of sulfenamides that are popular. *N*-Cyclohexyl benzothiazole sulfenamide (CBS) or *N-tert* butyl-2-benzothiazole sulfenamide (TBBS), 2-(4-morpholinothio)benzothiazole (MBS), or *N,N′* dicyclohexyl-2-benzothiazole sulfenamide (DCBS) may be used. Of all these, DCBS has the slowest curing and should be avoided here.

The above system needs activation by activators. Using the standard practice, 5 parts of zinc oxide and 2 parts of stearic acid may be used. The formula at this stage is:

	phr
Natural rubber	75.0
Polybutadiene rubber	25.0
ISAF	50.0
Process oil	5.0
Sulfur	2.0
Accelerator	0.8
Zinc oxide	5.0
Stearic acid	2.0

A final step is to choose an antidegradant. As the application requires flexibility, aging resistance, and ozone resistance, two to three types of age resistor (physical as well as chemical) can be used. The following package may be used:

	phr
Antioxidant	2.0
Antiozonant 1 (chemical)	1.5
Antiozonant 2 (physical)	1.0

Antiozonants 1 and 2 may be alkyl-aryl *p*-phenylene diamine type and paraffin wax, respectively. The antioxidant may be 2,2,4 trimethyl-1,2-dihydroquinoline or an acetone aldehyde condensation product. Often a retarder and a few other ingredients are added to meet the desired properties and processing.

When the formulation is written, the ingredients are arranged in order of mixing schedule. The complete formulation is:

	phr
Natural rubber	75.0
Polybutadiene rubber	25.0
Stearic acid	2.0
Zinc oxide	5.0
ISAF carbon black	50.0
Process oil	5.0
Antioxidant	2.0
Antiozonant 1	1.5
Antiozonant 2	2.0
Sulfur	2.0
TBBS	0.8
PVI	0.1
Total	**170.4**

Once the formulation is designed, it is easy to check the density of the stock. The stock can be cured at 150°C for 15 minutes in the laboratory. The following properties are obtained:

Tensile strength	22 MPa
Elongation at break	600%
300% Modulus	9 MPa
Hardness	60 Shore A
Tear strength	95 kN/m
Abrasion loss (DIN)	90 mm^3

In order to derive the right formulation, experimental design methods are often used. Single-ingredient variability, two-ingredient variability, and multi-ingredient variability are nicely described by Bhowmick et al. (1994). Software packages are also available to manage compounding ingredients and specifications.

If the specifications are given, it is easier to design a compound based on the literature data and steps described earlier. The job is to design a particular formulation and match the properties with those in the specification. If there is variation, the objective should be to improve without much compromise in important properties. Formulations of a few compounds are provided here so that readers are acquainted with the types of ingredients and their amounts.

To continue our discussion on tire tread, let us examine how a cycle tire tread formula is to be designed. Cost is the most important factor here. In order to meet the cost, compounds must be cheaper, and the time for vulcanization (cure time) should be short. Following these guidelines, the compound given below is proposed.

Cycle Tire Tread Formula

	phr
Natural rubber	60.0
Peptizer	0.06
Reclaim	80.0
Zinc oxide	5.0
Stearic acid	2.0
Antioxidant	1.0
GPF black	35.0
China clay	40.0
Naphthenic oil	10.0
Wood rosin	2.0
Mercapto benzothiazole accelerator	1.2
Tetramethyl thiuram disulfide accelerator	0.4
Sulfur	2.5

Note the use of black, and china clay as filler, and reclaim providing 40 phr rubber to the formulation. In order to reduce the cure time, a MBT/TMTD combination is used. Typical properties of a cycle tire tread would be:

Tensile strength	10 MPa
Elongation at break	400%
300% Modulus	4.0 MPa
Hardness	60 Shore A
Abrasion loss (DIN):	250 mm^3

The cure time of the above compound is 5 minutes at 150°C. Compare the difference in properties of the earlier compound with those of the present compound.

Let us now look at an entirely different formulation.

Hawai Sheets

	phr
Natural rubber	80.0
SBR-1958	20.0
Hawai crumb	70.0
Zinc oxide	5.0
Stearic acid	2.0
Antioxidant styrenated phenol	1.0
Aluminum silicate	10.0
Naphthenic oil	5.0
Blowing agent (DNPT)	5.0
Diethylene glycol	2.0
MBT	0.40
ZDC	0.60
Sulfur	2.5

The typical hardness of such a compound is 35 Shore A and specific gravity is 0.60. Note the use of a blowing agent.

The formula for a heat-resistant cover of a conveyor belt may be written as follows:

Heat-Resistant Conveyor Belt Cover Compound

	phr
SBR 1502	100.0
Stearic acid	2.0
Zinc oxide	5.0
Processing aid	5.0
HAF carbon black	45.0
IPPD antidegradant	1.5
TDQ antioxidant	2.0
Aromatic oil	5.0
MC wax	2.0
CBS	2.0
PVI	0.3
Sulfur	0.8

Note that SBR is used in place of NR for heat resistance purposes, and more accelerator is used to make an efficient vulcanization system with lower polysulfidic linkages in the crosslinks.

Similarly, the formulation of representative oil seal, oil and fuel hose, and sealing rings for water mains is given below:

Oil Seal

	phr
Nitrile rubber	100.0
Stearic acid	2.0
Zinc oxide	5.0
Antioxidant	2.0
FEF black (N550)	35.0
Thermal black (N990)	75.0
Plasticizer-ester type	3.0
TMTD	1.5
TETD	1.5
MBTS	1.5
Sulfur	0.2

Oil and Fuel Hoses

	Tube (phr)	Cover (phr)
Medium NBR	100.0	—
Polychloroprene	—	100.0
Light Calcined MgO	—	4.0
Zinc oxide	4.0	5.0
Stearic acid	1.0	0.5
Antioxidant (dihydrotrimethyl-quinoline)	1.5	—

FEF	30.0	40.0
SRF	35.0	—
Silica	—	15.0
Ppt. Calcium carbonate	20.0	35.0
DBP	20.0	15.0
Ethylene thiourea (ETU)	—	0.75
CBS	1.0	—
TMTD	0.5	0.5
Insoluble sulfur	1.5	—

Sealing Rings for Water Mains

	phr
Natural rubber	100.0
Stearic acid	1.0
Zinc oxide	5.0
SRF black	60.0
Process oil	3.0
Antioxidant	1.0
MBTS	0.2
TMTD	0.5
Sulfur	1.0

In earlier formulation, properties and specifications were not given. Readers may refer to texts listed in the Bibliography. In the following formulation, a set of desired properties are also listed.

Bridge Bearing Pad

	phr
CR (Neoprene WRT)	100.0
Magnesia	4.0
Octylated diphenylamine	2.0
Para phenylene diamine	1.0
Stearic acid	0.5
ISAF carbon black	25.0
Precipitated silica	10.0
Dioctyl sebacate	10.0
Zinc oxide	5.0
ETU	0.5
CBS	1.0

Cure time: 15 minutes at 150°C.

Properties of the Compound in the Laboratory

Tensile strength	21 MPa
Elongation at break	550%
Hardness	60 Shore A

Now correlate with the specifications as shown here:

Property	Standard	Equivalent Indian Standard	Specific Value
Hardness (Shore A durometer)	ASTM D 2240	IS:3400 (Part II)	60 + 5 Shore A
Tensile strength	ASTM D 412	IS:3400 (Part I)	Min. 17 MPa
Ultimate elongation	ASTM D 412	IS:3400 (Part I)	Min. 350%
Heat resistance (70 h at 100°C)	ASTM D 573	IS:3400 (Part IV)	
Change in durometer hardness			Max. +15 Shore A
Change in tensile strength			Max. –15%
Change in ultimate elongation			Max. –40%
Compression set	ASTM D 395	IS:3400 (Part X)	Max. 35%
22 h at 100°C, Maximum %			
Ozone	ASTM D 1149	IS:3400 (Part XX)	No cracks
100 ppm ozone in air by volume, 20% strain 38°C ± 1°C 100 h mounting procedure D518, Procedure A			
Elastomer content as polychloroprene	ASTM D 297		Min. 60%

Belt Formulation for Recorder Tape Disk Mechanism

Formulation and Properties of Belt Compound	
Formulation	*phr*
CR	100.0
Stearic acid	0.5
MgO	4.0
OCD	2.0
N770	45.0
DOP	10.0
NA/22	1.0
ZnO	5.0
Properties	
Tensile strength	23 MPa
Elongation at break	390%
Hardness	63 Shore A
Properties of the Belts	
Load at 10% stretch before aging	3.2 N
Load at 10% stretch after aging at 70°C for 72 h	2.6 N
Belt cutoff strength, original	55 N
Belt cutoff strength, after aging at 70°C for 72 h	58 N
Set property after 10% stretching	1.7%
Tensile strength	22 MPa

An extrusion compound is designed as follows:

Extrusion Compound

	phr
EDPM	100.0
GPF	100.0
SRF	25.0
Oil	120.0
Zinc oxide	5.0
Stearic acid	2.0
TMTD	1.6
MBT	0.5
Sulfur	1.5
Total	**355.6**

The above compound would be expected to have these properties:

Density (cured)	1.10
Tensile strength	12 MPa
Elongation at break	450%
Hardness	60 Shore A
Compression set[a]	20%

[a] ASTM method B 22 h at 70°C

A few things may be noted from the above formulation. There are a number of ways to derive a formulation or to achieve a specification. The larger the item in the specification, the more difficult it is to match formulation by doing reverse engineering. Also, in the above formulation, the properties are dependent on grades of rubber, carbon black/silica, antidegradant, or the way the compounds are mixed in the factory/laboratory.

1.7.1 Cost of Compounds

While formulating any compound, cost is one of the most important parameters to introduce the formulation in the market. Even if the performance of a compound is excellent, higher-cost compounds do not have any commercial value. The magic of a compounder is to strike a balance between cost and performance. Here we describe how cost is typically calculated for a compound and optimized if necessary.

PROBLEM 1

Calculate the specific gravity and cost per kilogram of the following compound.

	Part by Weight	Specific Gravity	Cost/kg
SBR	100	0.94	142.83
ZnO	5	5.56	105.24
Stearic acid	1	0.85	64.25
FEF	40	1.80	76.00
Whiting	50	2.36	19.48
CBS	1	1.27	286.29
Antioxidant	1	1.06	194.27
Sulfur	3	2.05	21.52
Oil	5	0.98	41.93
Total weight	**206 kg**		

$$\text{Volume} = \frac{\text{Weight}}{\text{Specific gravity}}$$

	Part by Weight	Specific Gravity	Volume, l	Cost per Weight	
SBR	100	0.94	106.38	14283.00	
ZnO	5	5.56	0.90	526.20	
Stearic acid	1	0.85	1.18	64.25	
FEF	40	1.80	22.22	3040.00	
Whiting	50	2.36	21.19	974.00	
CBS	1	1.27	0.79	286.29	
Antioxidant	1	1.06	0.94	194.27	
Sulfur	3	2.05	1.46	64.56	
Oil	5	0.98	5.10	209.65	
		Total volume	**160.16**	**19642.22**	**Total cost**

$$\text{Specific Gravity} = \frac{\text{Mass}}{\text{Volume}} = \frac{206}{160.16} = 1.286$$

$$\text{Cost per kg} = \frac{19642.22}{206} = 95.35$$

The cost per kg of above said compound is Rs. 95.35 (≡ USD 1.766).

$$\text{Cost per liter} = 95.35 \times 1.286 = 122.62/1.286$$

The cost per liter of the above compound is Rs. 122.62 (≡ USD 2.271).

Rs. 1.00 = 0.018519 USD

PROBLEM 2

How much furnace black must be replaced by whiting of specific gravity 2.6 at Rs. 20 per kg to reduce the cost of the following compound by Rs. 10 per kg?

Compound Parts by Weight	phr	Specific Gravity	Cost in Rs./kg
Smoked rubber sheet	100	0.92	150
Zinc oxide	5	5.6	105
Furnace black	100	1.8	90
Stearic acid	1	0.85	65
Accelerator	0.5	1.3	300
Softener	3	0.9	40
Sulfur	2.5	2.1	20
Calculation of Total Cost			**Cost in Rs.**
Smoked rubber sheet			15000
Zinc oxide			525
Furnace black			9000
Stearic acid			65
Accelerator			150
Softener			120
Sulfur			50

Total cost = Rs. 24,910

Total weight = 212 kg

Cost/kg = 24,910/212 = Rs. 117.5

Weight of the carbon black to be replaced by whiting = x

New weight = 212

Old cost = a + 9000, where a is the cost of rubber, zinc oxide, stearic acid, accelerator, softener, and sulfur

New cost:

$$\text{New cost} = \text{a} + (100 - x)9000 + x.20$$

$$\frac{(a + 9000) - a - 9000 + 90x - 20x}{212} = 10$$

$$\frac{70x}{212} = 10$$

$$x = 212 \times 10/70 = 30.28$$

Amount of carbon black to be replaced = 30.28 phr

PROBLEM 3

What is the effect on cost and performance for replacing 80 phr of N234 by 85 phr of N339 in a typical passenger tire tread compound?

The objective of this problem is to optimize the cost of the compound without sacrificing the properties. The compound was modified with respect to the polymer type. The original compound contained N234 black. N234 black was replaced by N339 black. While replacing N234 black by N339, the dose of N339 was also increased to maintain the physical properties suitable for application.

Ingredients	Specific Gravity	Cost/kg (in Rs.)	Compound A	Compound B	Compound A: Material Cost (in Rs.)	Compound B: Material Cost (in Rs.)	Compound A: Material Volume (in Liters)	Compound B: Material Volume (in Liters)
PBR	0.91	108.90	30.00	30.00	3267.00	3267.00	32.97	32.97
SBR 1712	0.95	63.70	96.25	96.25	6131.13	6131.13	101.32	101.32
N234	1.80	64.66	80.00	0.00	5172.80	0.00	44.44	0.00
N339	1.80	52.97	0.00	85.00	0.00	4502.45	0.00	47.22
ZnO	5.57	93.96	2.00	2.00	187.92	187.92	0.36	0.36
Stearic acid	0.85	39.54	1.00	1.00	39.54	39.54	1.17	1.17
Antioxidant	1.20	234.47	1.00	1.00	234.47	234.47	0.83	0.83
Processing aid	1.15	102.12	3.00	3.00	306.36	306.36	2.61	2.61
Ar. oil	0.99	28.48	8.00	8.00	227.84	227.84	8.10	8.10
Sulfur	2.05	12.23	1.90	1.90	23.24	23.24	0.93	0.93
Accelerator	1.27	224.76	1.20	1.30	269.71	292.19	0.94	1.02
Retarder	1.30	286.91	0.25	0.25	71.73	71.73	0.19	0.19
Total			**224.60**	**229.70**	**15931.73**	**15283.86**	**193.86**	**196.72**
Calculated Density							**1.16**	**1.17**
Cost/kg					**70.93**	**66.54**		
Cost/lt.							**82.18**	**77.69**

Mooney Viscosity at 100°C		
ML_{1+4} at 100°C	65	70
Mooney Scorch at 135°C		
T_5 (min)	20.53	17.69
Rheometric Properties at 160°C/30′		
Min. TQ. (lb-in)	2.81	2.95
Max.TQ. (lb-in)	14.04	15.48
Final TQ. (lb-in)	13.39	14.83
tS2 (min)	5.68	4.98
tC10 (min)	5.09	4.56
tC40 (min)	6.53	5.83
tC50 (min)	6.92	6.19
tC90 (min)	10.25	9.47
Physical Properties		
100% MOD. (MPa)	1.8	2.3
300% MOD. (MPa)	7.8	10.4
T.S. (MPa)	16.8	17.7

E.B. (%)		539	492
Hardness (Sh-A)		63	65
Tear strength (N/mm)		47.1	47.1
Other Properties			
Rebound resilience at RT		35	34
Rebound resilience at 70°C		45	44
Abrasion loss (mm³)		64	61
HBU (°C)		36	31

Dynamic Properties Comparison of the Compound

		E′ (MPa)	E″ (MPa)	Tan Delta
Compound A	@ –25°C,11 Hz, 5% strain	16.50	8.24	0.498
Compound B		17.90	8.56	0.478
Compound A	@ 0°C,11 Hz, 5% strain	10.30	4.48	0.433
Compound B		11.80	4.90	0.417
Compound A	@ 30°C, 11 Hz, 5% strain	7.11	2.52	0.354
Compound B		7.97	2.86	0.359
Compound A	@ 70°C, 11 Hz, 5% strain	5.50	1.51	0.274
Compound B		6.27	1.73	0.276
Compound A	@ 100°C, 11 Hz, 5% strain	4.88	1.09	0.222
Compound B		5.61	1.27	0.226

2

Principal Chemical and Analytical Methods Used in Reverse Engineering

2.1 Chemical Methods

2.1.1 Introduction

It is important to gain as much information as possible about any material. Analysis of elastomeric materials should be based upon a comparison of the material to a properly chosen control material. The values obtained on controls are often used for quantitative calculations of the value of the unknown.

Material analysis is also an important part of testing, whether for quality assurance or developmental activities. Analyses of raw materials are mainly of two types: chemical methods and instrumental methods. In a quality assurance check of raw materials, results are compared with the standard specification available. A specification of raw material is drawn based on the requirements or particular application. Such standard specifications for raw material used in the rubber industry have been laid down by standards institutions at national and international levels.

During analysis, preparation of the sample is important for getting a representative specimen from the sample. A number of sample preparation techniques are available. All are different in nature, depending upon the analysis to be performed. In this section two major types of testing are discussed: chemical analysis and physical analysis.

2.1.2 Chemical Analysis

Chemical analysis of materials is the first step in any rubber and allied industry. This helps the user to understand the behavior of rubber during processing, storing, and end use. Chemical analysis is generally performed for a representative sample taken from the bulk; hence the sample preparation technique plays an important role. In general, sample preparation for chemical analysis is done by homogenization of the material.

A number of chemical tests have to be performed for different rubbers to check quality. Tables 2.1 and 2.2 present a brief overview of the variety of chemical tests required to characterize different raw materials, their definitions, significance, and methods used, various critical controlling parameters, basic necessary tools for analysis, and widely used instrumental techniques. Chemical analysis can be controlled by some critical parameters, which are listed in Table 2.3.

In addition to the chemical tests mentioned above, a number of instrumental analyses are gaining significance in characterizing raw materials. The analytical instrumental methods of analysis are found to be the most accurate and less tedious. Among the

TABLE 2.1

Chemical Analysis of Different Raw Materials at a Glance

Sl Number	Raw Materials	Chemical Analysis	ASTM Standards
1	Rubber	Specific gravity	D1298
		Ash (%)	D297
		Dirt content of NR (%)	D1278
		Volatile matter (%)	D1416, D1278
		Nitrogen content (%)	D3533
		Stabilizer and bound styrene (%)	D1416
		Organic acid and soap (%)	D1416
		Oil content (%)	D1416
		Microstructure	D3677
		Copper and manganese content of NR (ppm)	D4075
2	Carbon black	Specific gravity	D1817
		Ash (%)	D1506
		Heat loss (%)	D1509
		DBP absorption (cc/100 gm)	D2414
		Pour density (lb/cft)	D1513
		Iodine adsorption number	D1510
		Residue on sieve (%)	D1514
		Pellet hardness	D5230
		Tint strength	D3265
		Bulk attrition (%)	D4324
		Toluene extract (%)	D1618
		pH	D1512
		Sulfur content (%)	D1619
		Nitrogen surface area (sqm)	D4818
3	Processing oils	Specific gravity	D1298[b]
		Saybolt viscosity (sus)	D88[b]
		Flash-and-fire point (°C)	D92[b]
		Pour point (°C)	D97[b]
		Aniline point (°C)	D611[b]
		API gravity	D1298[b]
		Clay gel analysis (%)	D2007[b]
		Viscosity gravity constant	D2501[b]
		Refractive index	D1218[b]
4	Antioxidant	Specific gravity	D1817
		Ash (%)	D4574
		Heat loss (%)	D4571
		Softening point (°C)	E28[a]
		Melting point (°C)	D1519
5	Fabric	Spin finish (%)	D885[c]
		Dip pick-up (%)	D885[c]
6	Zinc oxide	Zinc oxide content (%)	D3280
		Heat loss (%)	D280
		Acidity (%)	D4569
		Sieve residue (%)	D4315
		Total sulfur (%)	D3280
		Lead and cadmium content (%)	D4075
		Particle size	D3037
		Nitrogen surface area (sqm)	D4620

TABLE 2.1

Chemical Analysis of Different Raw Materials at a Glance *(Continued)*

Sl Number	Raw Materials	Chemical Analysis	ASTM Standards
7	Stearic acid	Specific gravity	D1817
		Final melting point (°C)	D1519
		Acid number	D1980[a]
		Iodine number	D1959[a]
		Ash (%)	D4574
		Saponification number	D1962[a]
		Fe, Ni, Cu, Mn content (ppm)	D4075
		Titer (°C)	D1982
8	Accelerator	Specific gravity	D1817
		Final melt point (°C)	D1519
		%Assay of sulfenamide	D4936
		Accelerators	D5044
		Free MBT in MBTS (%)	D4574
		Ash (%)	D4571
		Heat loss (%)	D4934
		Solubility/insolubility (%)	D4315
		Sieve residue (%)	D1514
9	Retarder	Specific gravity	D1817
		Ash (%)	D4574
		Heat loss (%)	D4571
		Solubility (%)	D4934
		Melting point (°C)	D1519
10	Soluble/ insoluble sulfur	Specific gravity	D1817
		Ash (%)	D4574
		Heat loss (%)	D4571
		Oil content (%)	D4573
		Sieve residue (%)	D4572
		Acidity (%)	D4569
		Total sulfur (%)	D4578
		Insoluble sulfur (%)	D4578
11	Wax	Specific gravity	D1817
		Ash (%)	D4574
		Melting point (°C)	D1519
12	Resin	Ash (%)	D4574
		Heat loss (%)	D4571
		Softening point (°C)	E28[a]
		Acid number	D1980[a]
		Solubility/insolubility (%)	D4934
13	Silica	Specific gravity	D1817
		Heat loss (%)	D4571
		pH	D1512
		Nitrogen surface area (sqm)	D1993
		Residue (%)	D5461
14	Peptizer	Specific gravity	D1817
		Ash (%)	D4574
		Sieve residue (%)	D5461
		Heat loss (%)	D4571

(Continued)

TABLE 2.1

Chemical Analysis of Different Raw Materials at a Glance (*Continued*)

Sl Number	Raw Materials	Chemical Analysis	ASTM Standards
15	Adhesion promoter	Specific gravity	D1817
		Final melting point (°C)	D1519
		Heat loss (%)	D4571
		Insolubility/solubility (%)	D4934
16	Clay	Specific gravity	D1817
		Heat loss (%)	D4571
		pH of aqueous extract	D1512
		Sieve residue (%)	D5461
		Manganese content (ppm)	D4075
		Copper content (ppm)	D4075
17	Latex	pH	D1417
		Specific gravity	D1417
		Total solid (%)	D1417
		Dry rubber content (%)	D1417
		Chemical stability	D1417
		Total coagulam (%)	D1417
		Brookfield viscosity (cPs)	D1417
		Surface tension (dynes/cm)	D1417
		Mechanical stability	D1417

[a] These methods are from ASTM Volume No. 06.03.
[b] These methods are from ASTM Volume No. 05.01.
[c] These methods are from ASTM Volume No. 07.01.
Note: All methods are from ASTM Volume No. 09.01.

instrumental methods, the spectroscopic, chromatographic, and thermal techniques are gaining wide acceptance as analytical tools for qualitative and quantitative analysis. An overview of some principles and applications of different analytical instruments are tabulated in Table 2.4.

2.2 Infrared Spectroscopy

2.2.1 Introduction

When infrared (IR) light is passed through a sample, some of the frequencies are absorbed while other frequencies are transmitted through the sample. If absorbance or transmittance is plotted against frequency or wave number, the result is an infrared spectrum. For a nonlinear molecule (with n number of atoms), three degrees of freedom describe rotation and three describe translation. The remaining $3n$–6 degrees of freedom are vibrational degrees of freedom, or fundamental vibrations. In addition to fundamental vibrations, other frequencies can be generated by modulations. Two frequencies may interact to give beats which are combination or difference frequencies. Absorptions at x cm^{-1} and y cm^{-1} interact to produce two weaker beat frequencies at $(x \pm y)$ cm^{-1}. So all peaks of the spectrum are not of analytical importance. IR spectroscopy is basically vibrational spectroscopy. There are three regions in IR spectra: near infrared (12,500 to 4000 cm^{-1}), middle infrared

TABLE 2.2

Chemical Analysis: Definition, Significance, and Procedure of Measurement

Sl Number	Chemical Analysis	Definition, Significance, and Measurement Procedure
1	Specific gravity	• The ratio of the weight of unit volume of a material and the weight of the same volume of water at a given temperature. • It is an important parameter for checking accuracy in compounding and serves as a guide in comparing relative compound costs. • Three major ways of determining specific gravity are : i) Hydrometer method: used for liquid materials ii) Direct weight method: used for solid materials iii) Liquid displacement method: used for polymeric materials
2	Ash content	• Residue obtained after ignition of materials at high temperature. • Ash is mainly inorganic impurities like copper, manganese, etc. These impurities can have serious implications on the aging properties of final products. • A weighed amount of sample is ignited in a muffle furnace at 900°C. Residual material is calculated as ash percentage.
3	Heat loss	• Weight loss of materials measured at around 60 to 70°C for rubber chemicals and at 125ºC for carbon black. • Heat loss of rubber chemicals indicates low boiling organic materials. Carbon black heating loss consists primarily of moisture, but other volatile matter may also be lost. • A weighed amount of sample is heated under controlled temperature; weight loss is the measure of heat loss.
4	Melting point	• Temperature at which the material changes its phase from solid to liquid, under specified conditions. • Melting point is an important tool to identify the materials and to check the purity of the materials. This also helps in predicting the processing characteristics of materials. • Two methods—capillary method and differential scanning calorimetry—are mainly used for the determination of melting point of the raw materials.
5	Sieve residue	• The sieve residue test is used to perform the qualitative evaluation of aggregate size of particulate material, namely, different fillers, and distribution of aggregate size of powdery chemicals. • This test generally analyzes suitability of the use of the powdered rubber chemicals in rubber compounds, where fine distribution of the small particle clusters is needed to achieve a uniform crosslinked network. This test is also used to ascertain that there will be no excessively large particles present in the powdery rubber chemicals which would result in "physical flaws" in the rubber matrix. • Powdered rubber chemicals are wetted with a dilute aqueous solution containing defoamer. The sample is passed through stacked sieves arranged in order of decreasing mesh size with the help of water flow. The residue retained on each sieve is dried in an oven. The dried mass of residue is obtained for each sieve. The percent residue is calculated on the basis of the original sample mass.
6	Acidity	• Determines the presence of acid material, which dissociates in distilled water. • Acidity indicates the effect of raw materials on the vulcanization system and on the rate of vulcanization. • Determined by titration.

(*Continued*)

TABLE 2.2

Chemical Analysis: Definition, Significance, and Procedure of Measurement *(Continued)*

Sl Number	Chemical Analysis	Definition, Significance, and Measurement Procedure
7	Insolubility	• Generally determines the insoluble impurities present in rubber chemicals with the use of suitable organic solvents. • Solubility is an important parameter to characterize sulfenamide-type accelerators. Since MBTS is a primary degradation product of sulfenamides, the determination of MBTS is a means of assessing possible degradation of sulfenamides. Insolubles are a means of MBTS content of sulfenamide. This also indicates the purity of the materials. • A specimen is dissolved in a prescribed solvent, stirred, and filtered through a gooch crucible. The insoluble content is calculated from the amount of residue.
8	Oil content	• The oil content test determines the amount of hydrocarbon oils present in any oil extended materials. • This parameter helps to control fly loss and proper dispersibility of materials in the rubber matrix. • Measured by extracting the oil by suitable solvents.
9	Bulk attrition	• Determines the pellet attrition of pelleted carbon black. • This attrition of carbon black gives some indication as to the amount of fines that may be expected to be created by pellet breakdown in conveying and handling or in a bulk shipment while in transit. • A test sample of carbon black is placed on a 120 mesh sieve and shaken in a mechanical sieve shaker for 5 min to remove the fines. The same test sample is shaken for an additional 15 min to determine the amount of pellet breakdown created during this additional shake interval. The attrition is expressed in percent.
10	Softening point	• Softening point is the temperature at which a solid sample gets softened, when the sample is heated under constant atmospheric pressure and at a particular heating rate, generally at 5°C/min in a suitable heating media. The commonly used heating media is a glycerin bath. • Generally measured for polymeric materials. It indicates the average molecular weight as well as processing temperature. It also indicates the dispersibility of the materials at processing temperature. • Measured by ring and ball method.
11	Acid number	• The number of milligrams of KOH required to neutralize the fatty acids in 1 gm of sample. • This analysis is required for determining the acidity of certain rubber chemicals which may affect the vulcanizing reaction. • This analysis is done by titrimetric method.
12	Saponification number	• The number of milligrams of KOH to react with 1 gram of sample. • The measure of the alkali reactive group. A higher saponification number from the normal indicates the fatty acid has been oxidized. • Measured by titrating with KOH.
13	Iodine number	• The number of milligrams of iodine required for 1 gram of sample. • Iodine number is used to measure the unsaturation in organic material which indicates stability of materials. • The determination of iodine value is based on the absorption of iodine under suitable conditions selected to promote stoichiometrical relations.

TABLE 2.2

Chemical Analysis: Definition, Significance, and Procedure of Measurement *(Continued)*

Sl Number	Chemical Analysis	Definition, Significance, and Measurement Procedure
14	Toluene discoloration	• This test method measures the amount of toluene discoloration when the carbon black is extracted using toluene. • Toluene discoloration value obtained by UV-Visible spectrophotometer gives an estimate of toluene-soluble discoloring materials present in carbon black. These are generally low boiling organic materials. • Measured by the UV-Visible spectrophotometer at 425 nm wavelength. The percent transmittance of the toluene extract of carbon black is determined.
15	Pour density	• The pour density of carbon black is defined as the mass per unit volume of pelleted carbon black. • The pour density of carbon black is used to determine the weight-volume relationship for certain applications like automatic batch loading systems, and for estimating total weights permissible for bulk shipments. • Measured by estimating the amount of carbon black (in gm) in a fixed-capacity (624 cm^3) cylindrical container.
16	Iodine adsorbtion	• The number of grams of iodine adsorbed per kilogram of carbon black under specified conditions. • Measures the surface area and is generally in agreement with multipoint nitrogen surface area. The presence of impurities like volatiles, surface porosity, or extractables affects the iodine adsorption number. • A weighed amount of carbon black is treated with a portion of standard iodine solution and the mixture is shaken and centrifuged. During this process, a certain amount of iodine is adsorbed by the surface of carbon black and the rest is evaporated off. The excess iodine was treated with standard sodium thiosulfate solution, and the adsorbed iodine is expressed as a fraction of the total mass of carbon black.
17	DBP absorption	• This test method determines the *n*-dibutyl phthalate absorption number of carbon black. • The DBP absorption number is related to the processing and cured properties of rubber compounds containing carbon black. It also indicates the structure of carbon black. • In this test method, DBP is added in a controlled way through a burette to a sample of carbon black in the mixer chamber of an absorptometer. As the sample absorbs the DBP, the mixture changes its physical form, from powder to semi-plastic agglomeration. As a result the viscosity of the mixture increases. The increased viscosity is sensed through the torque-sensing system of the absorptometer. When the viscosity reaches a torque level, the absorptometer and burette will stop immediately. The volume of DBP per unit mass of carbon black is the DBP absorption number. • The DBP has been listed as SVHC (substance of very high concern) under REACH (Registration, Evaluation, Authorisation, and Restriction of Chemical substances). As a result the DBP absorption test is replaced by the OAN (oil absorption number) test.

(Continued)

TABLE 2.2

Chemical Analysis: Definition, Significance, and Procedure of Measurement *(Continued)*

Sl Number	Chemical Analysis	Definition, Significance, and Measurement Procedure
18	Aniline point	• This is the minimum temperature at which equal volumes of aniline and the desired oil should be completely miscible. • This is an important characterizing tool for rubber processing oils. A high aniline point oil indicates lower aromaticity and vice versa. The aniline point of aromatic oil is 25 to 50°C, naphthenic oil is 55 to 75°C, and paraffinic oil is 100 to 150°C. Aniline point also indicates the compatibility of oils with different types of polymer. • Sample oil and aniline are taken in a U-tube where two separate layers are observed. Both the materials are then heated, and at a particular temperature, both the materials are miscible to each other. That particular temperature is expressed as the aniline point.
19	Pellet hardness	• The force required to fracture or crush a carbon black pellet. • Pellet hardness is related to the mass strength and attrition characteristics of the carbon black. Pellet hardness also plays an important role in obtaining proper dispersion of carbon black in a rubber matrix. High pellet hardness black needs higher shearing energy for better dispersion. Also, high pellet hardness black acts as physical flaws in the rubber network. • Measured by a pellet hardness tester.
20	Organic acids	• This method is intended to determine the organic acid remaining in a synthetic rubber. • Organic acids in the polymer may affect the cure rate of compounded stock. • Thin, narrow strips of dried rubber are extracted twice in hot extraction solvent. The solvent extracts are titrated against sodium hydroxide solution using a chosen indicator. The titration and sample mass are used to calculate the organic acids.
21	Nitrogen content	• This test method determines the total nitrogen in natural and synthetic rubbers and other nitrogen containing raw materials. • The determination of nitrogen in natural rubber is usually carried out in order to estimate the protein content. This is required to assess the quality of the material and its processing characteristics. • The common technique for determination of total nitrogen content is the Kjeldahl method. In this method rubber is digested with catalytic mixture followed by distillation from a strong alkaline solution. The distillate is absorbed in a boric acid solution, and the excess base is titrated against the acid.
22	Assay of accelerators	• Percent active content is known as assay. • This test method is designed to assess the purity of accelerators. These products are used in combination with sulfur for the vulcanization of rubber. • For most commonly used sulfenamide accelerators, a weighed specimen is dissolved in the appropriate solvent, the "free amine" blank is titrated with standard acid, and the sulfenamide is reduced with H_2S. That is, $BtSNR_2 + H_2S \rightarrow BtSH + HNR_2 + S$ where Bt is benzothiazole radical; BtSH is 2-mercaptobenzothiazole; HNR_2 is free amine; and $BtSNR_2$ is sulfenamide accelerator.

TABLE 2.2

Chemical Analysis: Definition, Significance, and Procedure of Measurement (*Continued*)

Sl Number	Chemical Analysis	Definition, Significance, and Measurement Procedure
23	Pour point	• The temperature at which the oil starts flowing is known as pour point. • Pour point identifies the processing characteristics of the rubber processing oils. • Measured by pour point apparatus under specified conditions.
24	Solvent extractable	• The amount of material extracted from raw materials by organic solvents. • Organic solvent extractables of different raw materials used in the rubber industry are measured by this method. The actual chemical composition of the extractables can vary significantly with the nature of raw materials to be tested. This is an important parameter to use as a quality check of the raw materials. These extractables may affect the processing, storing, as well as curing reaction. • A specimen of carbon black is extracted with a suitable solvent in a soxhlet extractor to obtain the equilibrium extraction. The solvent is removed by controlled temperature evaporation, and the extracted residue is determined gravimetrically.
25	Tint strength	• The ratio expressed as tint units, of the reflectance of a standard paste to a sample paste, both prepared and tested under specified conditions. • Tint strength of the carbon black depends on the particle size of the black. Tint strength can be used as an indication of particle size. However, tint strength is also dependent on structure and aggregate size distribution of carbon black. • A carbon black sample is mixed with a white powder (ZnO) and soyabean oil epoxide to produce a black gray paste. This paste is spread to produce a surface suitable for measuring the reflectance of the mixture by means of a photo-electric reflectance meter. The reflectance of the tested sample is compared to the reflectance of the Industry Reference Black (IRB) prepared in the same manner. The tint strength of the tested sample is expressed as units of the reflectance of the IRB divided by the reflectance of the sample multiplied by 100.
26	Flash and fire point	• The flash point of a volatile material is the lowest temperature at which it can vaporize to form an ignitable mixer in air. Measuring a flash point requires an ignition source. At the flash point, the vapor may cease to burn when the source of ignition is removed. • Fire point is the temperature at which the vapor continues to burn even after the source of ignition is removed. • Flash and fire point indicates the process safety of the materials. It also indicates the nature of the processing oils. • Measured by Cleveland open cup apparatus.
27	pH	• The concentration of hydrogen ions in a medium. In short it indicates the acidity and basicity of the materials. • Measurement of pH value of a material determines whether it is acidic or basic in nature. It is better to be aware of the pH since basic material may accelerate vulcanization while acidic material may delay the vulcanization time of a rubber compound. • Solid sample is taken in a beaker and stirred with distilled water. The whole mixture is then allowed to settle, and the filtered pH of the filtrate is determined by using glass electrode.

(*Continued*)

TABLE 2.2

Chemical Analysis: Definition, Significance, and Procedure of Measurement *(Continued)*

Sl Number	Chemical Analysis	Definition, Significance, and Measurement Procedure
28	Dip pick-up	• This method is used to determine the amount of adhesive present on the fabric. • Textile yarns and cords are treated with an adhesive dip to improve the adhesion of elastomer to textile materials. The amount of dip solids pick-up on the yarns or cords is used for process control. • The fiber is dissolved in an appropriate solvent and the residue of dip solids is recovered by filtration after which the residue is dried and weighed. The amount of adhesive solid on a sample is reported as percentage dip pick-up based on the oven dry weight of the dip free sample.
29	Titer value	• The temperature at which the stearic acid starts solidifying. It is also identified as solidification temperature. • Titer value indicates the purity of the materials and also the processing characteristics of stearic acid. • Measured by monitoring the temperature during cooling of the melted stearic acid.
30	Spin finish	• Spin finish determines the extractable materials on most fibers, yarns, and fabrics. • This test method is used for the determination of oily or waxy impurities that were applied to yarns or fabric during manufacturing. These are generally the impurities that generate dip fumes and dip dipositions. • The specimen is extracted with appropriate solvent. The solvent is evaporated, and residue and the specimen are dried and weighed separately. The amount of extracted material is reported as percent extractables or percent spin finish.
31	Soaps	• This method is intended to measure the soap content in emulsion polymer (SBR). • Since soap is a by-product of the emulsion process, it may affect the cure rate of compounded stock, tack, and adhesion properties. • One portion of the solvent extract is titrated with hydrochloric acid, using the chosen indicator. The percentage of soap is calculated from the weight of the original sample and the titrations.
32	Viscosity	• The internal friction between layers of fluid as they pass over each other while moving with different velocities. Viscosity is defined as the internal force per unit area required to maintain a unit velocity gradient. • Viscosity signifies the processing characteristics of different liquid materials used along with rubber. It also helps to identify the flowability of the materials. • A number of methods are available to determine the viscosity of a liquid. Of these, the Saybolt viscometer and Brookfield viscometer are widely used for viscosity measurement.
33	American Petroleum Institute (API) gravity	• This is a special way of representing relative density, represented by deg = (141.5/5p.gr.60/60°F) – 131.5. • API gravity governs the quality of the crude petroleum. It is also necessary for conversion of measured volumes to volumes at standard temperatures. • API gravity is calculated after measuring the specific gravity of oils by the hydrometer method.

TABLE 2.2

Chemical Analysis: Definition, Significance, and Procedure of Measurement *(Continued)*

Sl Number	Chemical Analysis	Definition, Significance, and Measurement Procedure
34	Viscosity gravity constant (VGC)	• VGC is a function which is calculated • From kinematic viscosity at 40°C and density at 15°C: $$VGC = \frac{G - 0.0664 - 0.1154 Log(V-5.5)}{0.94 - 0.109 Log(V-5.5)}$$ where G is the density of oil at 15°C, g/mL and V is the kinematic viscosity of oil at 40°C, cSt. • From kinematic viscosity at 100°C and density at 15°C: $$VGC = \frac{G - 0.108 - 0.1255 Log(V' - 0.8)}{0.90 - 0.097 Log(V' - 0.8)}$$ where G is the density at 15.5°C, g/mL and V′ is the kinematic viscosity at 100°C, cSt. • VGC is an indication of the nature of oils. If VGC is around 0.8, oils are classified as paraffinic; those with VGC above 1 are classified as extremely aromatic. • VGC is calculated after measuring the specific gravity by hydrometer and viscosity by Saybolt viscometer.
35	Surface tension	• Expressed as the force in dynes per unit length acting at right angles to the line along the surface of the liquid. • It signifies the stability of the liquid or liquid mixtures. In the rubber industry, surface tension is generally an important characteristic for latex. It indicates the storage and handling characteristics of the latex. • Measured by Torsion ring method using du Nouy principle.
36	Mechanical stability	• The amount of material creamed off during high-speed stirring of the latex. • Latices which are mainly used in dip units are kept under constant stirring condition. This test helps the user to determine whether the latex will bear the high speed without generating excess cream. It is an important processing parameter. • Measured by a mechanical stability tester.
37	Surface area	• The square meter area per gram. • Surface area indicates the effective area available for interaction with rubber. This is generally measured for fine particles, *viz.*, carbon black, ZnO, precipitated silica, etc. The higher the surface area, the higher will be the efficiency of the materials. • Measured by BET method.
38	Clay-gel analysis	• This analysis determines the % *n*-pentane insolubles, % polar compounds, % aromatics, and % saturates. • The concentrations of the characteristics of hydrocarbon groups as determined by this method are used to classify petroleum oil types as used for extending and processing rubbers. Compatibility and certain finished product properties can often be correlated with the composition as determined by this method. • The sample is diluted with *n*-pentane and purged to a glass percolation column containing clay in the upper section and silica gel in the lower section. The *n*-pentane is then passed to the double column until a definite quantity of effluent has been collected. The upper section is removed from the lower section and washed with *n*-pentane which is discarded. A benzene-acetone (50:50) mixture is then charged to the clay section, and a specified volume of effluent is collected. From this collection, saturates and polars are measured. Aromatics are calculated by differences.

TABLE 2.3

Chemical Analysis: Critical Parameters

Sl Number	Chemical Analysis	Critical Parameters
1	Specific gravity	Weight of the sample, calibration of the hydrometer, analytical balance, and specific gravity bottle
2	Ash content	Sample weight, temperature of muffle furnace, and time
3	Heat loss	Sample weight, analysis temperature, time for analysis
4	Sieve residue	Sample weight, type of sieve
5	Melting point	Sample preparation, rate of heating
6	Acidity	Reagent (for analysis) purity and preparation of sample
7	Solubility	Sample weight, temperature, and solvent type
8	Oil content	Reagent (for analysis) purity and preparation of sample
9	Bulk attrition	Preparation of sample, speed of mechanical shaker, and time
10	Softening point	Sample preparation, rate of heating
11	Acid number	Preparation of sample, purity of reagents
12	Iodine number	Preparation of sample, purity of reagents
13	Saponification number	Preparation of sample, purity of reagents
14	Toluene discoloration	Purity of toluene, preparation of sample, time for extraction
15	Pour density	Calibration of the fixed volume container
16	Dibutyl phthalate (DBP) absorption	Calibration of absorptometer, preparation of sample, purity of DBP
17	Aniline point	Purity of aniline, calibration of thermometer
18	Iodine adsorption	Purity of iodine, calibration of glassware, preparation of sample
19	Pellet hardness	Preparation of sample, analysis method
20	Organic acid	Purity of reagents, time for extraction
21	Nitrogen content	Time for digestion, time for distillation, rate of distillation, purity of reagents
22	Assay of sulfenamide accelerators	Reduction time, purity of reagents, preparation of sample
23	pH	Sample preparation, glass electrode stabilization, calibration of pH meter
24	Dip pick-up	Preparation of sample, time for stirring, purity of reagents
25	Spin finish	Purity of solvent, time for extraction
26	Solvent extractables	Purity of solvents, extraction time, rate of heating, preparation of sample
27	Tint strength	Sample preparation, calibration of the equipment
28	Soaps	Preparation of sample, time of extraction, purity of reagents
29	Flash and fire points	Calibration of thermometer, preparation of sample
30	Titer value	Calibration of thermometer, preparation of sample, rate of heating
31	Pour point	Heating rate, purity of sample, calibration of thermometer
32	API gravity	Determination of specific gravity
33	Viscosity gravity constant (VGC)	Determination of Saybolt viscosity, determination of specific gravity
34	Surface tension	Sample preparation, calibration of equipment, temperature for analysis
35	Mechanical stability	Rate of stirring, type of sieve, sample weight
36	Surface area	Preparation of sample, test method
37	Clay-gel analysis	Purity of reagents, purity of column materials, azobenzene value of clay materials

TABLE 2.4

Some Critical Instrumental Analysis : Principles and Applications

Sl Number	Name of Technique	Brief about Techniques	Applications
1	Thermal analysis	A series of techniques that measures changes in physical or reactive properties of materials as a function of temperature and time. Provides information that characterizes polymers, organic and inorganic chemicals, etc. Thermal analysis systems are sophisticated yet easy to use, producing thoroughly analyzed data quickly and with minimal operator involvement. Broad utility, high reliability, and high productivity have made thermal analysis an essential analytical tool for material research and selection, product design, process optimization, and quality control. Two widely used thermal analysis techniques are thermogravimetric analyzer and differential scanning calorimeter.	For differential scanning calorimetry (DSC): Phase diagrams, reaction kinetics, dehydration reactions, heats of reaction, heats of polymerization, purity determination, thermal stability, oxidation stability, glass transition temperature, thermal conductivity, specific heats, etc. For thermogravimetric analysis (TGA): Thermal decomposition, roasting and calcining of minerals, proximate analysis of coal, dehydration, oxidative degradation of polymeric materials, reaction kinetics, etc.
2	Spectroscopy	Spectroscopy is one of the most important characteristic properties of a compound. It provides a fingerprint identification and is a powerful tool for the study of molecular structure. A variety of spectroscopic techniques are available, of which the following are widely used in rubber and allied industries. 1. Infrared spectroscopy: Used to characterize different motions of atoms in a molecule like twisting, bending, rotating, and vibrational. 2. UV-Visible spectroscopy: In this technique a given molecular species absorbs radiation only in specific regions of the spectrum where the radiation has the energy required to raise the molecules to some excited state. A display of absorption versus wavelength is called an absorption spectrum of that molecular species and serves as a fingerprint for identification.	Infrared spectroscopy: Compound identification, quantitative estimation of materials, purity of the materials UV-Visible spectroscopy: Qualitative and quantitative estimation of materials

(*Continued*)

TABLE 2.4

Some Critical Instrumental Analysis : Principles and Applications *(Continued)*

Sl Number	Name of Technique	Brief about Techniques	Applications
		3. Atomic absorption spectroscopy: In this technique, combustion flame provides a means of converting analytes in solution to atoms in the vapor phase freed of their chemical sorroundings. The free atoms absorb radiation from a source external to the flame. The incident radiation absorbed by the free atoms in moving from the ground state to an excited state provides the analytical data.	Atomic absorption spectroscopy: Quantitative estimation of metal content in ppm level
3	X-ray diffraction	X-ray diffraction is gaining significance, as in this technique a material can be analyzed in solid state. X-ray powder diffraction techniques for identification of crystalline substances have been applied to a wide variety of analysis with useful results. This technique is also used to characterize crystallite size, orientation, etc. The main advantages of x-ray diffraction methods are that they are non-destructive and results can often be obtained in lesser time than by routine chemical analysis.	X-ray diffraction: Determination of crystal structure, polymer characterization, particle size determination, determination of cis-trans isomerism, determination of linkage isomerism
4	Chromatography	In chromatography, two mutually immiscible phases are brought into contact, out of which one phase is stationary and the other is mobile. A sample introduced into a mobile phase is carried along through a column containing a stationary phase. Components in the sample undergo repeated interactions between the mobile phase and the stationary phase.	Chromatography: Material purity, material identification, and quantitative estimation

(vibration-rotational region, 4000 to 400 cm^{-1}), and far infrared (rotational region, 400 to 10 cm^{-1}). Generally, middle IR is commonly used for application in the rubber industry.

Modern Fourier transform infrared (FTIR) instruments are normally based on the Michelson's interferometer, where the radiation from an infrared source is split into two beams by a half silvered 45° mirror in such a way that the resulting beams are at right angles to each other. Michelson's interferometer is used to measure length or change in length, with great accuracy by means of interference fringes.

FTIR is used in the polymer industry for qualitative as well as some quantitative applications. The following sections deal with the quantitative use of FTIR in the rubber industry.

2.2.2 Estimation of Vinyl Acetate Content in Polyethylene Vinyl Acetate (EVA)

EVA is widely used in the rubber industry as a wrapping material for different rubbers like styrene-butadiene rubber (SBR), isoprene isobutylene rubber (IIR), etc. EVA is also used as a weighing bag for different rubber chemicals. Another important area of application of EVA is ladies shoe soles. The property of EVA is dependent on vinyl acetate content. FTIR is used to quantify vinyl acetate content in EVA. A representative spectrum of the scanned EVA film is shown in Figure 2.1.

The methyl ($-CH_3$) group of vinyl acetate absorbs at wavenumber 1370 cm^{-1}, and the paraffin group of the ethylene block ($-CH_2-$) absorbs at wavenumber 720 cm^{-1}. The absorbance ratio (A_{1370}/A_{720}) of the standard materials of known composition is plotted against the ratio of vinyl acetate content (%) and ethylene content (%). From this graph, the vinyl acetate content of the unknown sample can be calculated on the basis of the absorbance ratio (A_{1370}/A_{720}).

From the calibration graph, the slope (*m*) and intercept (*c*) are calculated and represented as

$$y = mx + c \tag{2.1}$$

where y is the absorbance ratio, and x is the ratio of vinyl acetate to ethylene content (A/B).

With the value of $x(A/B)$ and the relation of $A + B = 100$, the percentage (%) of vinyl acetate content can be calculated. The vinyl acetate content of EVA polymer can also be determined by the same method.

2.2.3 Determination of Microstructure of Polybutadiene (BR) Rubber

The Indian standard IS 10016 describes the method for the microstructure determination of polybutadiene rubber. Dry rubber (2.5 g) was prepared in small pieces and placed into

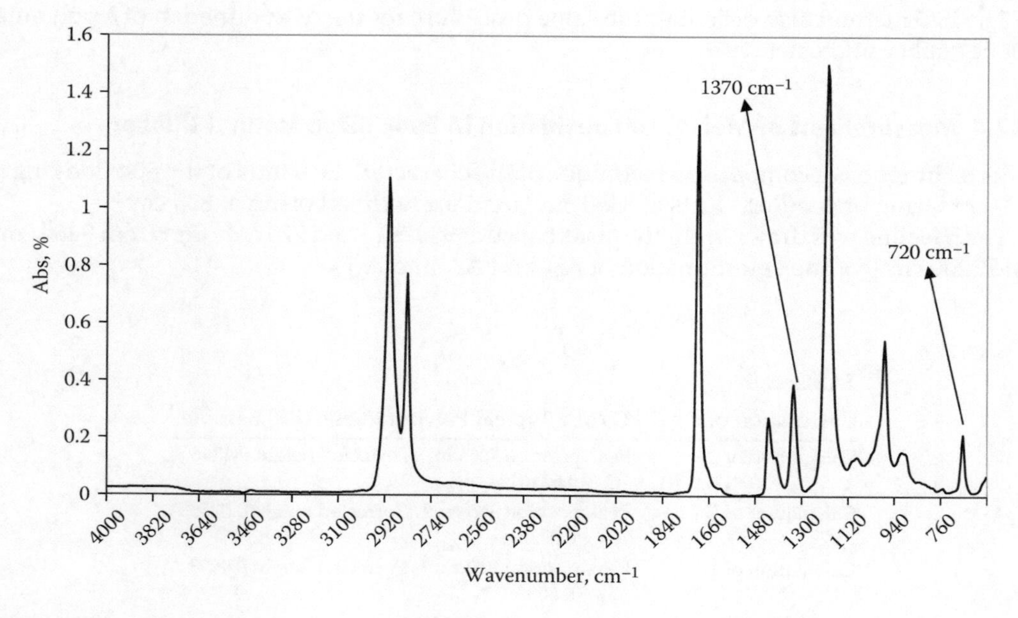

FIGURE 2.1
Spectrum of EVA film.

a conical flask to which a measured quantity of dichloromethane was added. The whole mixture was shaken for around 6 to 8 h.

After complete dissolution of the rubber, a few drops of polybutadiene rubber (BR) solution were taken in potassium bromide (KBr) pellet and a uniform film was prepared. The film was then dried and the spectrum was taken. Absorbance at 965 cm^{-1}, 910 cm^{-1}, and 735 cm^{-1} represents the trans, vinyl, and cis peaks, respectively.

This spectrum was corrected using baseline correction. The corrected absorbance at the specific wavenumber was measured. The relative concentration of the three components is given in the following equations:

$$C_C = 1 \times \text{cis peak} \tag{2.2}$$

$$C_T = 0.118 \times \text{trans peak} \tag{2.3}$$

$$C_V = 0.164 \times \text{vinyl peak} \tag{2.4}$$

where C_C, C_T, and C_V represent the concentrations of cis, trans, and vinyl groups, respectively.

Factors 0.118 and 0.164 were derived from the certified standard of high cis-polybutadiene rubber with the help of nuclear magnetic resonance (NMR spectroscopy).

$$C_X \text{ percent by mass} = \frac{(C_C / C_T / C_V) \times 100\%}{C_C + C_T + C_V} \tag{2.5}$$

where C_X is the concentration of cis, trans, and vinyl in percent. An example of the calculations of cis, trans, and vinyl structures of polybutadiene rubber (BR) is given in Table 2.5, and the polybutadiene rubber spectrum is represented in Figure 2.2.

The ISO method also describes the same procedure for the determination of a polybutadiene rubber microstructure.

2.2.4 Measurement of Mole% of Epoxidation in Epoxidized Natural Rubber

This technique is a comparison technique of the characteristic bands of the epoxide ring at 870 cm^{-1}, ring opened products at 3460 cm^{-1}, and unmodified olefin at 835 cm^{-1}.

The baseline was drawn from the absorbance at 950 cm^{-1} and 770 cm^{-1} and from 3460 cm^{-1} and 3200 cm^{-1} for the determination of A_{870} and A_{835} and A_{3460}.

TABLE 2.5

Calculation of C_C, C_T, C_V of a Typical Polybutadiene (BR) Sample

Calculation of C_C	Peak position 735 cm^{-1}, Corrected height 0.6246 A, $C_C = 0.6246$
Calculation of C_T	Peak position 965 cm^{-1}, Corrected height 0.0269 A, $C_T = 0.0032$
Calculation of C_V	Peak position 910 cm^{-1}, Corrected height 0.0498 A, $C_V = 0.0082$
Cis% by mass	Cis% by mass = 98.22%

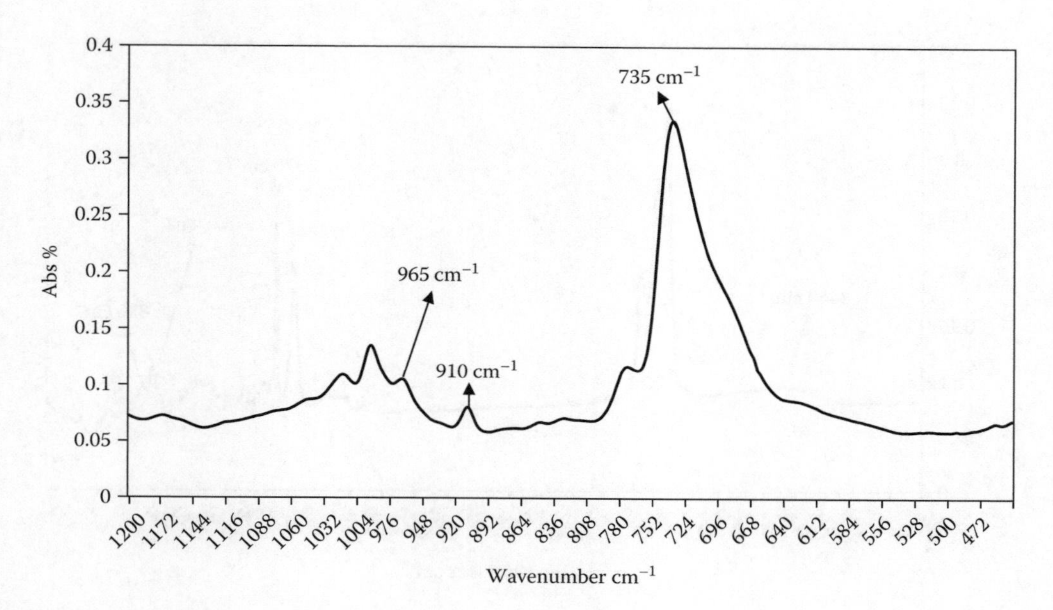

FIGURE 2.2
Spectrum of BR rubber (98.2% cis content). (Plot of absorbance vs. wavenumber.)

The corrected absorbances are:

$$A_{870\,corr} = A_{870} - 0.14 \times A_{835} \tag{2.6}$$

$$A_{3460\,corr} = A_{3460} - 0.019 \times A_{1375} \tag{2.7}$$

The correction factors were estimated as 14% and 1.9%, respectively, due to the interference of the corresponding bonds. These were determined from NMR.

From knowledge of the absorption value, the percentages of the three products, viz., epoxide (*E*), olefin (*O*), and ring opened (*R*), can be calculated as

$$K_1 = \frac{A_{835} \times E}{A_{870\,corr} \times O} \tag{2.8}$$

$$K_2 = \frac{A_{835} \times R}{A_{3460\,corr} \times O} \tag{2.9}$$

K_1 and K_2 are the constants and their values were determined as 0.77 and 0.34, respectively, by NMR.

Now, $E + O + R = 100$:

$$E = \frac{100 \times K_1 \times A_{870\,corr}}{A_{835} + A_{870\,corr} + K_2 \times A_{3460\,corr}}\ \text{mol\%} \tag{2.10}$$

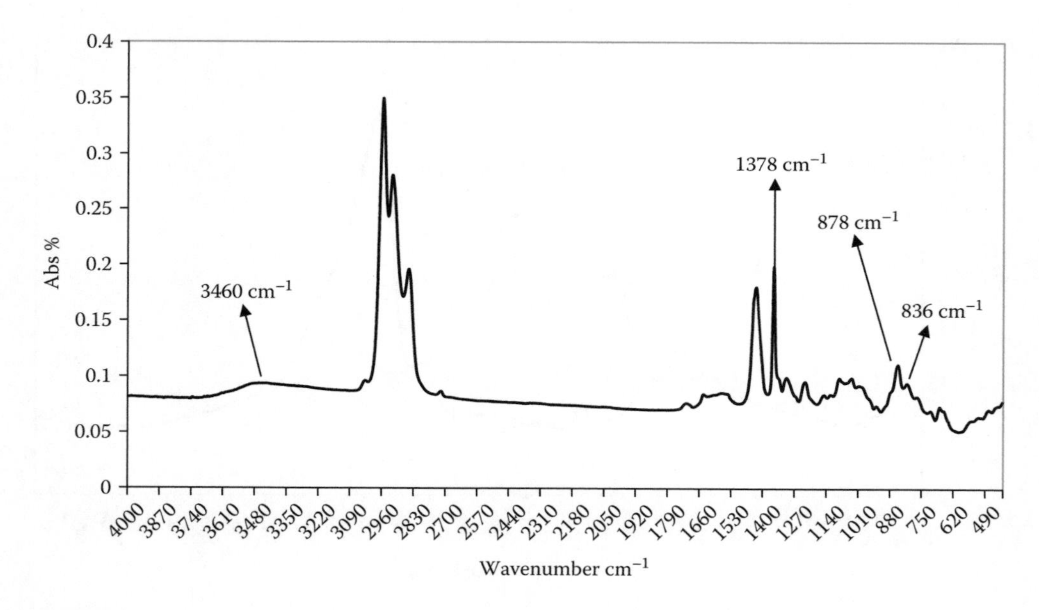

FIGURE 2.3
Spectrum of ENR (50 mol% epoxidized).

It should be noted that FTIR spectrum and baseline drawing are two important factors for ensuring accuracy in this method. The above equations have been correlated and corrected by NMR. One can use these equations directly to measure the mol% epoxidation.

A typical spectrum of 50 mol% epoxidized natural rubber (ENR) is presented in Figure 2.3.

2.2.5 Determination of Ethylene Content in Ethylene Propylene Co-polymers (EPM) and in Ethylene Propylene Diene Monomer (EPDM)

ASTM D3900 describes the determination method of ethylene units in ethylene propylene copolymers (EPM) and in ethylene propylene diene monomer (EPDM).

This method covers the determination of proportion of ethylene and propylene units in EPM and EPDM over the range from 35 to 85% of ethylene. However, the method is not applicable to oil extended EPDM unless oil is removed. It is always better to extract the additives by acetone as per ASTM D297.

2.2.5.1 Test Method A (for Mass% Ethylene in EPM and EPDM between 35 and 65)

A small piece of sample was placed in between two pieces of aluminum foil and pressed for 30 to 60 seconds at 70 MPa (10,000 psi) at 150°C. Pressure was released and the polymer film was carefully removed. The film was taken in a sample holder and the FTIR spectrum was determined. The correct sample thickness is an important variable to give good results. The ratio of the absorbance of the methyl groups from propylene units at 1156 cm^{-1} versus the absorbance of the methylene sequence from an ethylene unit at 722 cm^{-1} was used for analysis.

A calibration curve was drawn with A_{1156}/A_{722} and mass% of ethylene of the EPDM sample with known ethylene content. From this calibration curve, the unknown sample can be analyzed.

2.2.5.2 Test Method B (for Mass% of Ethylene in EPM and EPDM between 60% and 85% Except for Ethylene/Propylene/1,4-Hexadiene Terpolymers)

The sample was prepared in the same way as was method A. In this process, the calibration curve was drawn with an IR absorbance ratio at 1379/722 cm^{-1} and the mass% of ethylene of known standard sample.

2.2.5.3 Test Method C (for All EPM and EPDM Polymers between 35 and 85 Mass% Ethylene Using Near Infrared)

In this method the calibration curve was drawn with infrared absorbance ratio A_{1156}/A_{4255} and mass% ethylene of the known standard EPDM sample.

2.2.5.4 Test Method D (for All EPM and EPDM Polymers between 35 and 85 Mass% Ethylene, Except for Ethylene/Propylene/1,4 Hexadiene Terpolymer)

This method is basically a solution casting method. The EPM or EPDM is first extracted to remove oil, specifically for oil extended polymer. After extraction, the dried sample was taken and dissolved in toluene. Polymer solution was then taken on the salt plate and dried to evaporate the solvent. The absorbance ratio (1379/1460 cm^{-1}) and mass% ethylene were plotted in a standarad calibration curve for different known ethylene contents of EPM or EPDM. The ethylene content of the unknown sample was measured on the basis of the calibration curve.

The following best-fit regression equations from calibration curves have been found using most IR instruments:

Test method A, B, C

$$\text{Mass\% Ethylene} = a - b \ln (\text{absorbance ratio}) \tag{2.11}$$

Test method D

$$\text{Mass\% Ethylene} = a - b (\text{absorbance ratio}) \tag{2.12}$$

EPDM standards are available from Exxon Mobil Chemical Co. (Baytown, Texas).

A typical EPDM spectrum is represented in Figure 2.4.

2.2.6 Determination of Acrylonitrile Content (ACN) of Acrylonitrile Butadiene Rubber (NBR)

The nitrogen content of nitrile rubber is measured by the Kjeldhal method according to ASTM D3533.

The calculation of acrylonitrile content was done on the basis of the following formula:

$$\text{Acrylonitrile content (\%)} = 3.79 \times N \tag{2.13}$$

N is the percent nitrogen determined.

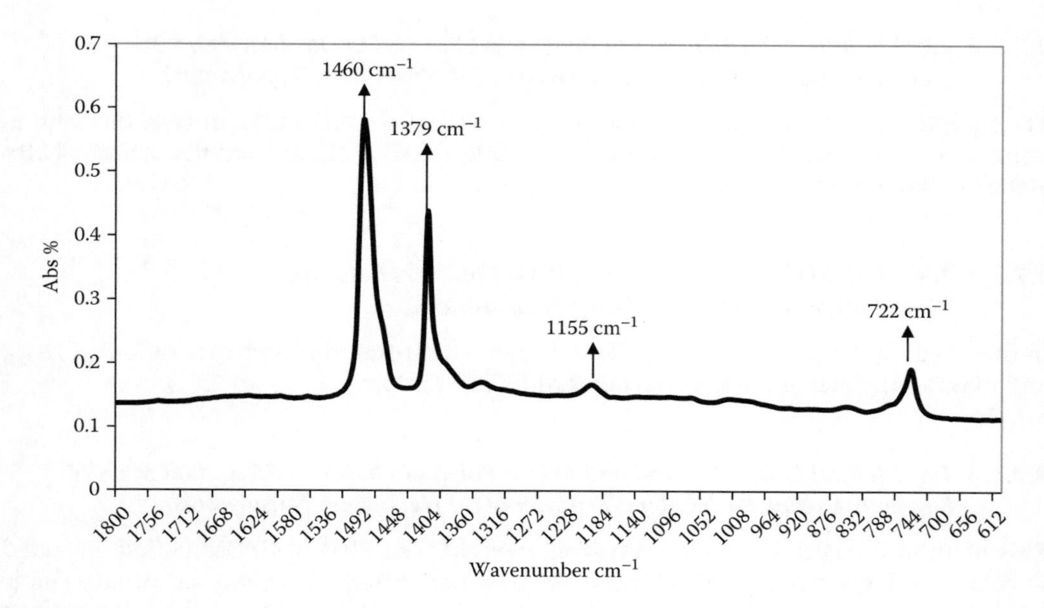

FIGURE 2.4
Spectrum of EPDM.

FTIR can also be used to determine the acrylonitrile content of NBR. In this process the sample was dissolved in di-methyl formamide (DMF). The rubber solution was taken in the NaCl plate. It is better to use the spacer between the salt plates to control the sample thickness. The FTIR spectrum is shown in Figure 2.5.

The ratio of -C equals *N* stretching at 2238 cm^{-1} to the out of plane stretching vibration of the C-H bond of the trans butadiene at 969 cm^{-1}. This ratio was plotted against the

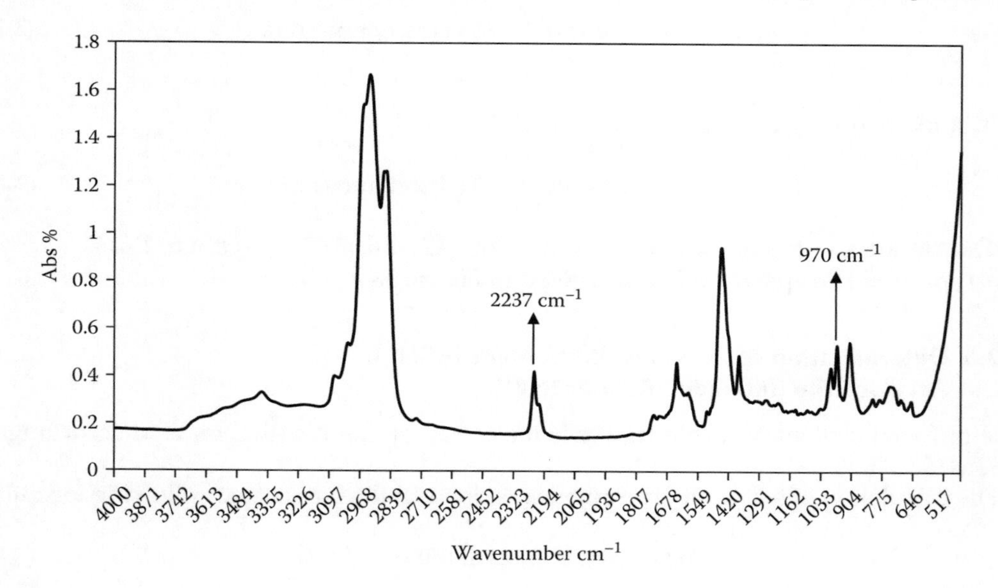

FIGURE 2.5
Spectrum of nitrile butadiene NBR (33% ACN content).

calculated acrylonitrile content from the Kjeldhal method. The correlation coefficient was obtained as 0.99. The ACN content of the unknown NBR was calculated from the correlation equation. The measured variation of the acrylonitrile content using this method was within ±5%.

One can prepare in-house standard NBR samples by measuring the ACN content of commercially available NBR as discussed.

2.2.7 Quantitative Determination of the Microstructure of the Butadiene Units in Solution-Polymerized SBR (S-SBR)

ISO 21561 specifies procedures for the quantitative determination of the microstructure of the butadiene units in solution-polymerized SBR (S-SBR) by proton nuclear magnetic resonance (^{1}H-NMR) spectroscopy as an absolute method and by IR spectroscopy as a relative method.

Initially, the sample was extracted with suitable solvent to extract any oil present in the rubber. ASTM D297 or ISO 1407 can be used as the standard test method for extraction. The sample is dried in a vacuum oven at 50°C.

A background spectrum of KBr plate is taken from 1200 cm^{-1} to 500 cm^{-1}. Extracted and dried sample was dissolved in cyclohexane, and a thin film was cast on a KBr salt plate for FTIR analysis.

A typical spectrum is shown in Figure 2.6.

The absorbance peak at 967 cm^{-1} corresponding to the 1,4-trans bond (D_{trans}), the peak at 911 cm^{-1} for the 1,2-vinyl bond (D_{vinyl}), and the absorbance peak at 724 cm^{-1} corresponding to the 1,4-cis bond (D_{cis}) are used for analysis. The absorbance peak at 699 cm^{-1} corresponds to the styrene ($D_{styrene}$).

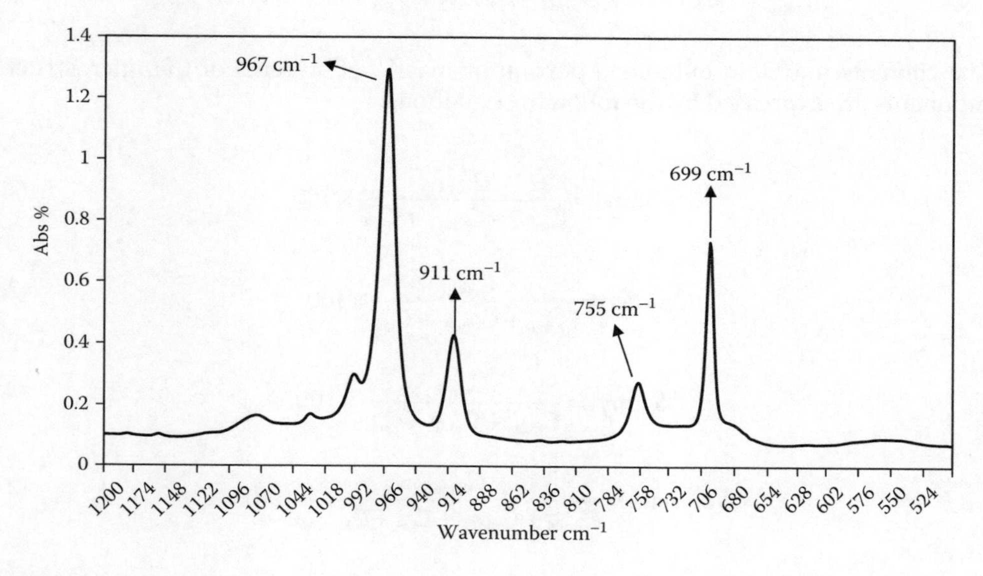

FIGURE 2.6
Spectrum of SBR (23.3% styrene content).

Normalization of the above absorbances is carried out according to the following equations:

$$\% D_{trans} = \frac{D_{trans}}{D_{trans} + D_{vinyl} + D_{cis} + D_{styrene}} \times 100 \quad (2.14)$$

$$\% D_{vinyl} = \frac{D_{vinyl}}{D_{trans} + D_{vinyl} + D_{cis} + D_{styrene}} \times 100 \quad (2.15)$$

$$\% D_{cis} = \frac{D_{cis}}{D_{trans} + D_{vinyl} + D_{cis} + D_{styrene}} \times 100 \quad (2.16)$$

$$\% D_{styrene} = \frac{D_{styrene}}{D_{trans} + D_{vinyl} + D_{cis} + D_{styrene}} \times 100 \quad (2.17)$$

The concentrations of the microstructure components ($C_{styrene}$, C_{cis}, C_{trans}, C_{vinyl}) are expressed by the following equations:

$$C_{trans} = 0.3937 \times \%D_{trans} - 0.0112 \times \%D_{vinyl} - 0.0361 \times \%D_{cis} - 0.0065 \times \%D_{styrene} \quad (2.18)$$

$$C_{vinyl} = -0.0067 \times \%D_{trans} + 0.314 \times \%D_{vinyl} - 0.0154 \times \%D_{cis} - 0.0071 \times \%D_{styrene} \quad (2.19)$$

$$C_{cis} = -0.0044 \times \%D_{trans} - 0.0274 \times \%D_{vinyl} + 1.8347 \times \%D_{cis} - 0.0251 \times \%D_{styrene} \quad (2.20)$$

$$C_{styrene} = -0.0138 \times \%D_{vinyl} - 0.260 \times \%D_{cis} + 0.3739 \times \%D_{styrene} \quad (2.21)$$

The contents (mol% in butadiene portion or mass% of styrene) of the microstructure components are expressed by the following equations:

$$\% \text{ trans} = \frac{C_{trans}}{C_{trans} + C_{vinyl} + C_{cis}} \times 100 \quad (2.22)$$

$$\% \text{ cis} = \frac{C_{cis}}{C_{trans} + C_{vinyl} + C_{cis}} \times 100 \quad (2.23)$$

$$\% \text{ vinyl} = \frac{C_{vinyl}}{C_{trans} + C_{vinyl} + C_{cis}} \times 100 \quad (2.24)$$

$$\% \text{ styrene} = \frac{C_{styrene}}{C_{trans} + C_{vinyl} + C_{cis} + C_{styrene}} \times 100 \quad (2.25)$$

The % trans, % cis, or % vinyl is the mol% of the respective structure in the butadiene portion of SBR. The % styrene is the styrene content in the SBR in mass%.

The absolute value of the microstructure can be obtained from IR measurements by calibration of the IR method using SBR with known styrene and vinyl contents determined by ^{1}H-NMR.

2.2.8 Determination of Residual Unsaturation in Hydrogenated Acrylonitrile Butadiene Rubber (HNBR)

A 1 gm extracted sample was dissolved in methyl ethyl ketone. A smooth film was casted on KBr plate and dried. The spectrum was collected in absorbance mode. Typical spectra are given in Figures 2.7 through 2.9.

For acrylonitrile content, the baseline was drawn from 2280 to 2200 cm^{-1} with the peak at 2236 cm^{-1}. For butadiene content, the baseline was drawn from 1010 to 910 cm^{-1} with the peak at 970 cm^{-1}. For hydrogenated butadiene content the baseline was drawn from 840 to 670 cm^{-1} with the peak at 723 cm^{-1}.

The molar concentration of each unit is calculated as follows:

$$C\ (\text{ACN}) = \frac{1}{\Sigma_i A(i)} \tag{2.26}$$

$$C\ (\text{Butadiene}) = \frac{\tilde{A}\ (970)}{K\ (970)} \times \frac{1}{\Sigma_i A(i)} \tag{2.27}$$

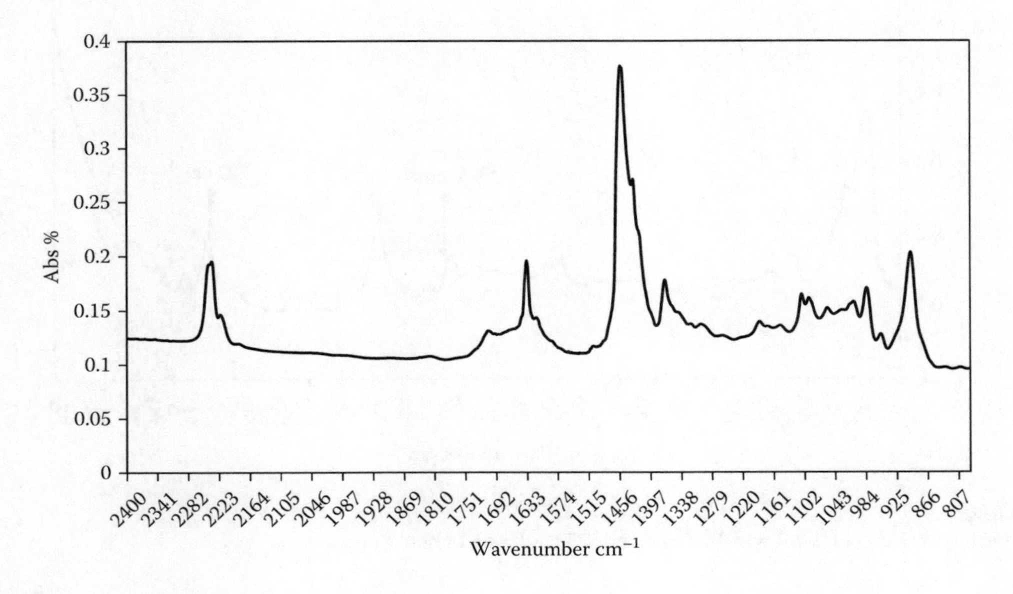

FIGURE 2.7
Typical spectrum of HNBR.

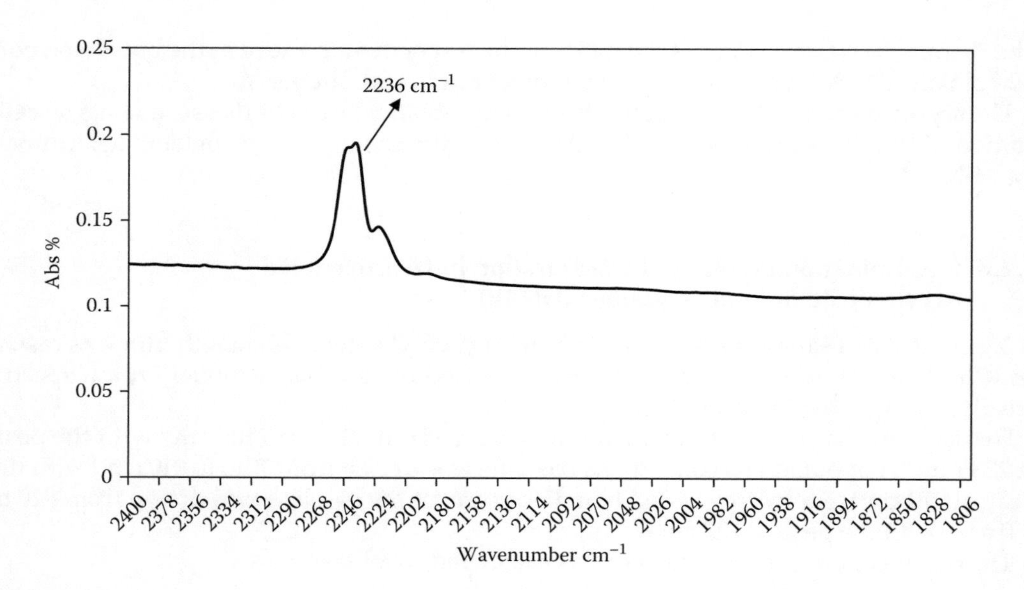

FIGURE 2.8
Typical spectrum of HNBR with 2236 cm^{-1} peak of acrylonitrile.

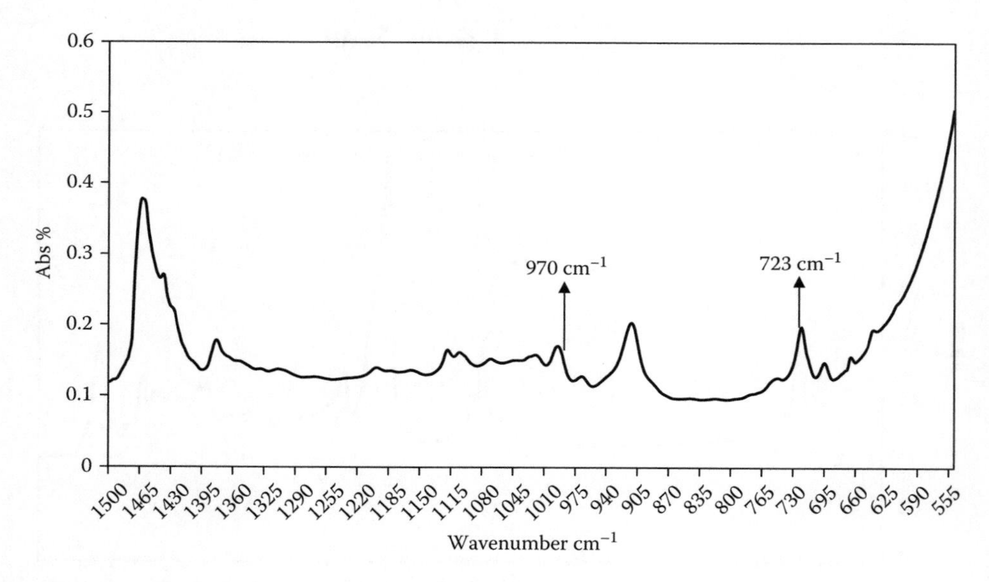

FIGURE 2.9
Typical spectrum of HNBR with 970 cm^{-1} and 723 cm^{-1} peak of butadiene.

$$C\ (\text{Hydrogenated butadiene}) = \frac{\tilde{A}\ (723)}{K\ (723)} \times \frac{1}{\Sigma_i\ A(i)} \tag{2.28}$$

$$\Sigma_i\ A(i) = 1 + \frac{\tilde{A}\ (970)}{K\ (970)} + \frac{\tilde{A}\ (723)}{K\ (723)} \tag{2.29}$$

$$\%\ \text{un-saturation} = \frac{C\ (\text{Butadiene})}{C\ (\text{Butadiene}) + C\ (\text{Hydrogenated butadiene})} \times 100\% \tag{2.30}$$

$\tilde{A}$ (970), normalized absorbance = A (970)/A (2236), and $\tilde{A}$ (723) = A (723)/A (2236). A (970), A (723), and A (2236) are the absorbances of the corresponding bonds. K is the molar absorbance factor, K (2236) = 1, K (970) = 2.3 + (–0.03), K (723) = 0.255 + (–0.002). A typical calculation is given in Table 2.6.

2.2.9 Carbon Type Analysis (C_A, C_P, and C_N) of the Rubber Process Oil by Fourier Transform Infrared (FTIR)

Processing aids are widely used in the rubber industry for easing the processability of the rubber compound in different processing equipment. The commonly used processing aids for rubber are aromatic, paraffinic, and naphthenic oil. These oils are characterized on the basis of the aromatic, paraffinic, and naphthenic carbon content.

The FTIR technique can be used to characterize the C_A, C_P, and C_N contents of the processing oils. The infrared spectrum of the sample was recorded in the two regions 1750 cm^{-1} to 1500 cm^{-1} and 850 cm^{-1} to 600 cm^{-1}. The intensities of the peaks at about 1600 cm^{-1} and 720 cm^{-1} are used to calculate the aromatic and paraffinic carbon contents, respectively. The naphthenic carbon is determined by the difference. The paraffinic carbon type cannot be determined directly if the aromatic content of the oil exceeds 20%. This is because neighboring peaks may mask the peak at 720 cm^{-1}. In such cases a dilution technique is used. A typical spectrum of paraffinic oil is shown in Figure 2.10.

TABLE 2.6

Calculation of Percent (%) Unsaturation of a Medium ACN, Hydrogenated Acrylonitrile Butadiene Rubber (HNBR)

Corrected absorbance, A (2236)	0.278
Corrected absorbance, A (970)	0.033
Corrected absorbance, A (723)	0.117
Normalized absorbance, $\tilde{A}$ (970)	0.119
Normalized absorbance, $\tilde{A}$ (723)	0.421
$\Sigma_i A$ (i)	2.742
C (ACN)	0.370
C (Butadiene)	0.019
C (Hydrogenated butadiene)	0.611
% unsaturation	3%

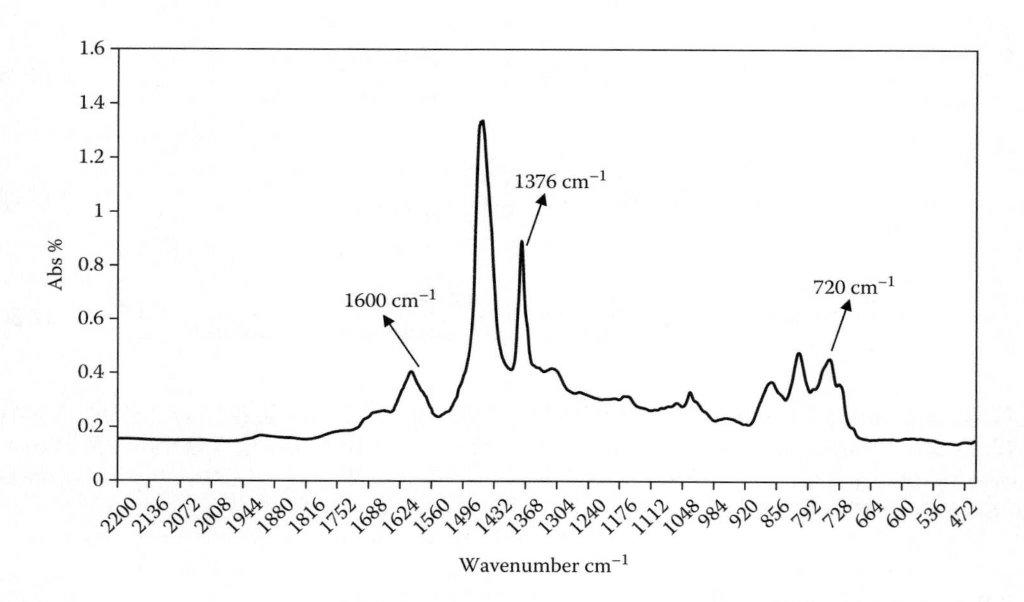

FIGURE 2.10
Spectrum of paraffinic oil.

The path length of the cell is calculated from the formula

$$d = (n \times 10)/2\ (\nu_1 - \nu_2) \tag{2.31}$$

where n is the number of fringes between two wave numbers ν_1 and ν_2.

If no fringes are obtained, either the cell windows are in bad condition or they are not parallel. The absorbance is measured at the corresponding peaks after the baseline correction.

Assuming C_A and C_P are the concentrations of aromatic and paraffinic oils, respectively, the following equations are used:

$$C_A = 1.2 + 9.8\ E_A \tag{2.32}$$

$$C_P = 29.9 + 6.6\ E_A \tag{2.33}$$

$$E_A = A/\text{cd} \tag{2.34}$$

where A is the corrected height in the absorption unit of the corresponding peak; d is the path length of the cell in millimeters; and c is the concentration factor (1 for undiluted oils).

$$C_N = 100 - (C_A + C_P) \tag{2.35}$$

where C_N is the concentration of the naphthenic oil. Results below 10% are reported to one decimal.

In the case of high aromatic oil (aromatic content more than 20%), it should be diluted with paraffinic oil in such a manner that aromatic content becomes about 5%.

TABLE 2.7

Calculation of C_A, C_P, C_N of a Typical Paraffinic Oil

Calculation of cell path length (*d*)	Wavelength range: 2271 cm^{-1} to 491 cm^{-1} Number of fringes: 27 $d = 0.0758$ mm
Calculation of C_A	Peak position: 1604.22 cm^{-1} Corrected height: 0.0865 Absorption unit (*A*): $C_A = 12.4\%$
Calculation of C_P	Peak position: 724.57 cm^{-1} Corrected height: 0.3592 Absorption unit (*A*) $C_P = 61.2\%$
Calculation of C_N	Peak position: 724.57 cm^{-1} Corrected height: 0.3592 Absorption unit (*A*) $C_N = 100\% - 12.4\% - 61.2\% = 26.4\%$

The C_P of the blend is calculated by means of the following formula:

$$C_P = \frac{C_P\ (\text{blend})\ (w_1 + w_2) - C_P\ (\text{diluents})\ w_2}{w_1} \tag{2.36}$$

where w_1 is the weight of the sample in the blend, and w_2 is the weight of the diluents in the blend.

A typical example of the calculation of C_A, C_P, and C_N in paraffinic oil is given in Table 2.7. In this case the authors have taken low aromatic fraction (C_A) containing paraffinic oil.

2.2.10 Miscellaneous Quantitative Applications of FTIR

For quantitative identification of a rubber microstructure, a comparative method based on the FTIR has been established taking ^{1}H-NMR as an absolute method.

FTIR is widely used in the quantification of rubber in blends. Pyrolysis FTIR of the finished rubber products is an important technique used to identify the polymer on the basis of the characteristics peak. In the reverse engineering chapters of this book, this method is widely used. Characteristic IR peaks of different functional groups are described in Appendix B.7.

2.3 Thermal Analysis

2.3.1 Introduction

In the analysis of elastomer vulcanizate, composition is an important quality control requirement for the rubber industry. Practical rubber formulations are complex mixtures of polymers, fillers, oils, curatives, antioxidants, and processing aids. Thermogravimetric analysis not only permits the analysis to be completed in a short time but also requires only a small amount of sample. Thermogravimetry (TG) and differential thermogravimetry (DTG) are powerful analytical techniques for the composition analysis of vulcanizates.

TG-DTG is straightforward; it provides reasonably accurate data for most vulcanizates and it is faster than the classical extraction method. Thus, TG-DTG remains the method of choice for compositional analysis of vulcanizate wherever the equipment is available.

Thermal behavior is a single property by which material can be identified. Knowledge of thermal property helps in selecting proper processing and fabrication conditions of a material. It is also important to characterize the physical properties of a material. This is true for polymers. Thermal methods are widely employed to characterize macromolecules. One of the primary properties for characterizing the polymer is its thermal properties. Thermal properties show good practical relation to other fundamental properties. Differential scanning calorimetry (DSC), thermogravimetic analysis (TGA), and differential thermal analysis (DTA) measurements are widely employed in day-to-day analysis of polymers.

The most commonly used and apparent definition of thermal analysis is: "A group of techniques in which physical properties of a substance and/or its reaction products are measured as a function of temperature while the substance is subjected to a controlled temperature program."

2.3.2 Polymers

It is well known that in the last two decades, polymers have gained more significance than any other class of materials. The exponential growth of polymer application has led to the development of several new techniques for polymer characterization.

Structurally, polymers have characteristic high molecular weight and repeating units. Moreover, except for a few polymers such as silicone rubber, these are mostly based on organic hydrocarbon. These structural characteristics of polymers are greatly affected by heat that leads to changes in several properties that can be measured only by thermal analysis. So, different methods of thermal analysis have already proven its usefulness in characterizing polymers. A typical TGA curve of a rubber compound and the derivative of the thermogravimetric curve (DTGA) are shown in Figures 2.11 and 2.12.

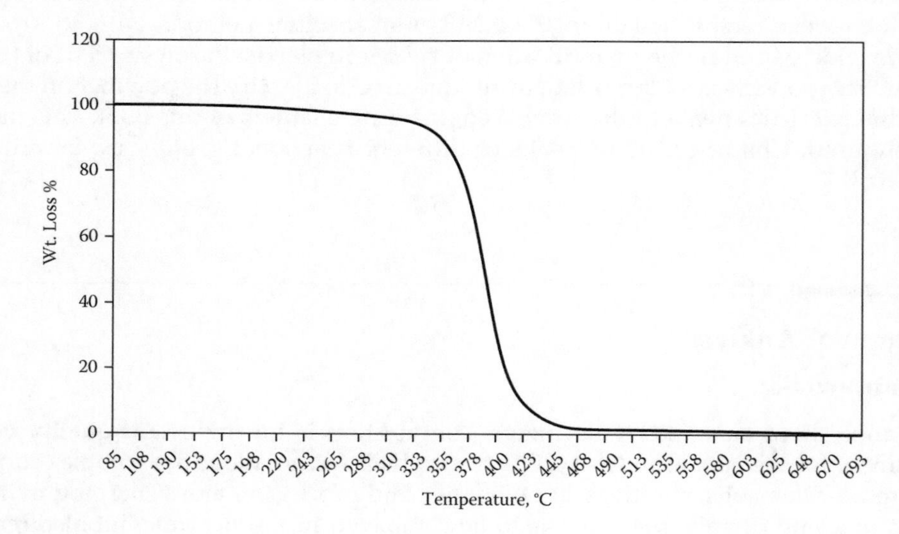

FIGURE 2.11
TGA thermogram of NR.

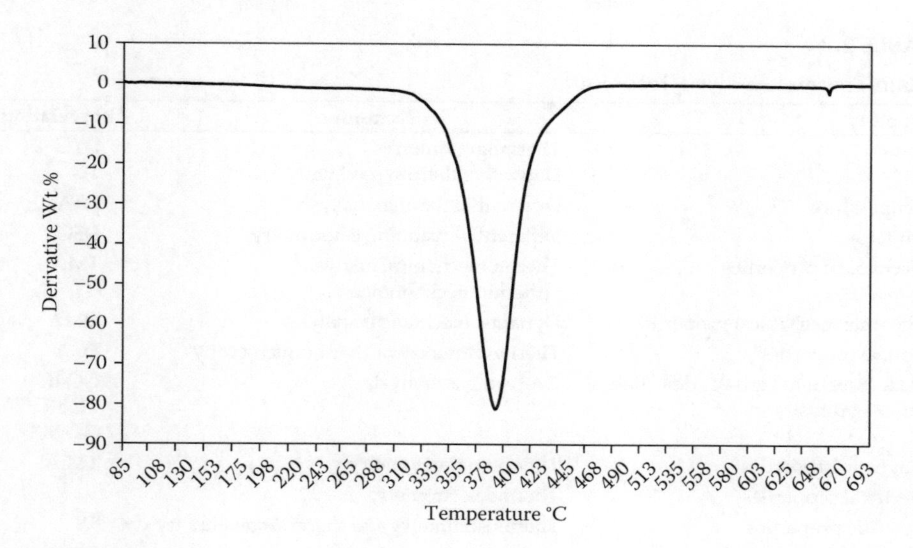

FIGURE 2.12
DTGA thermogram of NR.

For a polymeric material, the following major structural changes occur during the process of heating from a sub-ambient temperature:

- At sufficiently low temperature, polymer chain flexibility ceases and the polymer behaves like a hard plastic.
- Just above the glass transition temperature (T_g), chain flexibility increases.
- For a crystalline polymer, this chain flexibility helps polymer chains to align and respond to any interchain interaction (H-bonding, etc.), resulting in polycrystallites in the polymer phase.
- After achieving a sufficient temperature, the crystallites (ordered structure) melt because of the high internal energy of the chains, resulting in the melting point (T_m) for crystalline polymers.
- For amorphous/semi-crystalline polymers, viscous flow occurs because of increasing chain mobility with an increase in temperature.
- On further heating, polymer chains, being organic in nature and very often having unsaturation present in the backbone, become susceptible to degradation, depolymerization, and cyclization reaction, resulting in drastic changes in mass, enthalpy, and other properties. This process depends on atmosphere, additives used in polymer composites, and experimental conditions. Finally, ash may be induced due to degradation depending on the nature of the polymer.

Since the entire phenomenon described above can be studied only by using different techniques of thermal analysis, it is impossible to characterize the polymer or polymer composite completely without these techniques.

The variety of techniques (listed in Table 2.8) depends upon the variety of physical properties that can be measured for a particular material. The result of measurement of these properties is a thermal analysis curve, and the characteristic features of this curve

TABLE 2.8
Main Thermal Analysis Techniques

Property	Technique	Abbreviation
Mass	Thermogravimetry Derivative thermogravimetry	DTG TG
Temperature	Differential thermal analysis	DTA
Enthalpy	Differential scanning calorimetry	DSC
Mechanical properties	Thermomechanical analysis (thermomechanometry)	TMA
Dynamic mechanical properties	Dynamic mechanical analysis	DMA
Optical properties	Thermoptometry or thermomicroscopy	EGA
Mass/functional groups detection after pyrolysis	Evolved gas analysis	GC-IR TG-IR GC-MS
Magnetic properties	Thermomagnetometry	TM
Electrical properties	Thermoelectrometry	
Acoustic properties	Thermosonimetry and thermoacoustimetry	TS
Evolution of radioactive gas	Emanation thermal analysis	ETA
Evolution of particles	Thermoparticulate analysis	TPA

(discontinuities in peaks, changes of slope, etc.) are related to thermal events in the sample (Table 2.9).

2.3.3 Some Important Technical Terms Related to Thermal Analysis

Highly volatile matter: Moisture, oil, plasticizer, emulsifier (e.g., in styrene butadiene rubbers), curatives (sulfur, accelerator), antioxidant, anti-ozonant, and other low boiling components (approximately 300°C or lower).

Medium volatile matter: Materials that decompose at temperature ranges between 200 and 750°C. The materials in this category include processing oil, processing aid, elastomer, resin (used as curing agent), etc.

Combustible material: Material that is oxidizable in nature. Carbon black is an example of such a material.

Ash: Nonvolatile residue in an oxidizing atmosphere. These are metallic oxide, inert reinforcing material (e.g., silica), etc.

Mass loss plateau: A region of a thermogravimetric curve with a relatively constant mass.

Derivative: The first derivative (mathematical) of any curve with respect to temperature or time.

Specific heat capacity (Cp): The energy necessary to provide a unit temperature increase to a unit mass of material; it is expressed as J/(kg K).

Inflection temperature: The point on the thermal curve corresponding to the peak of the first derivative of the thermal curve.

Differential scanning calorimetry (DSC): A technique in which the difference in energy input into a substance and a reference material is measured as a function of temperature.

Differential thermal analysis (DTA): A technique in which the temperature difference between a substance and a reference material is measured as a function of temperature. A differential scanning calorimeter and a differential thermal analyzer are used to determine the transition temperatures of the materials. Differential scanning

TABLE 2.9
Some Major Thermal Events

Thermal Analysis Technique	Thermal Events	Instrument Used
Thermogravimetric analysis	i) Composition of polymers and polymer compounds	TGA
	ii) Decomposition characteristics	
	iii) Polymer blend ratio	
Differential scanning calorimeter	i) Specific heat and T_g,	DSC
	ii) Melting point (T_m) and heat of fusion	
	iii) Curing characteristics	
	iv) Decomposition characteristics	
Dynamic mechanical analysis	i) Stress relaxation analysis	DMA
	ii) Creep analysis	
	iii) Dynamic modulus	
	iv) Loss factor	
Thermomechanical analysis	i) Expansion and contraction	TMA
	ii) Modulus	
	iii) Glass to rubber transition temperature (T_g)	
	iv) Softening point	
	v) Gel point	
	vi) Viscosity at low temperature	
	vii) Crosslink density	
	viii) Shrinkage forces and percent shrinkages of films and fibers	
	ix) Coefficient of thermal expansion (CTE)	
	x) Testing of coatings on metals, films, optical fibers, and electrical wires	
	xi) Composite delamination temperature	
	xii) Melting temperature	
	xiii) Dimensional compatibilities of two or more different materials	
	TMA is widely used in the analysis of fibers. It measures the following:	
	i) Coefficient of linear thermal expansion (CTE)	
	ii) Thermal shrinkage (ST)	
	iii) Shrinkage force (SF)—usually reported in units of stress (i.e., cN/dtex)	
	iv) T_g and melting temperature (T_m) of fiber	
	v) Kinetics of shrinkage and shrinkage force phenomena	

calorimetry (DSC) is used to determine the heat or enthalpy of transition. Power compensation differential scanning calorimetry and heat flux differential scanning calorimetry are commonly used for measurement by DSC.

Dielectric thermal analysis (DETA): Dielectric constant (permittivity or capacitance) and dielectric loss (conductance) of a substance under an oscillating electric field are measured as a function of temperature or time.

Dynamic mechanical analysis (DMA): The modulus (visco-elastic response) of a substance under oscillatory load is measured as a function of temperature, time, strain, or frequency of oscillation.

Thermal analysis (TA): A group of techniques in which the physical property of a substance is measured as a function of temperature or time while the substance is subjected to a controlled-temperature program.

Thermally stipulated current (TSC) analysis: In TSC analysis, the current generated when dipoles change their alignment in a substance is measured as a function of temperature or time.

Thermomechanical analysis (TMA): Deformation of a substance under nonoscillatory load is measured as a function of temperature or time.

Thermophotometry: An optical characteristic of a substance is measured as a function of temperature or time.

Torsional braid analysis (TBA): A dynamic mechanical analysis in which the material is supported on a braid and the specimen is examined in torsion.

Note: While measuring thermal properties, it is possible that some materials degrade near the transition region. From the analysis spectrum, the clear differentiation between degradation and transition needs to be carefully observed. The glass transition is manifested as a step change in specific heat capacity. For amorphous and semi-crystalline materials, determination of the glass transition temperature may lead to important information about their thermal history, processing conditions, stability, progress of the chemical reaction, and mechanical and electrical behavior.

2.3.4 Principle of Differential Scanning Calorimetry (DSC) Operation

DSC can be used for calorimetric measurement—characterization and analysis of thermal properties of materials (crystallization, curing, etc.).

DSC responds to an enthalpy change accompanying a physical or chemical event in a material. Depending on the method of measurement employed, two types of DSC are categorized:

1. Power compensation
2. Heat flux

The difference of enthalpy between sample and reference pans is measured, keeping both at the same heating program. Blank pan, in general, is used as reference in DSC. At temperatures below 500°C, samples are usually contained in aluminum sample pans. Above 500°C, or for samples that react with aluminum pans, gold or graphite pans are used. To study the reaction at high pressure, a high pressure sample kit is also available.

The theory of operation of the PerkinElmer DSC (Waltham, Massachusetts) is based on the power compensated "null-balance" principle. According to this principle, energy absorbed or evolved by a sample is compensated by adding or subtracting an equivalent amount of electrical energy to a heater located in the sample holder. A platinum resistance heater and a thermometer are used to measure the temperature and energy. With the continuous adjustment of power, keep the sample holder temperature identical to that of the reference holder. This principle helps to capture the varying electrical signals equivalent to the varying thermal behavior of the sample. The ultimate measurement is directly in energy unit (milliwatts).

Note: Aluminum sample pans should not be used above 600°C. Since aluminum melts at 660°C, the pans will alloy with and destroy the sample holders. Always encapsulate indium, tin, and zinc standard in aluminum or graphite pans; these metals will alloy with gold, copper, or platinum pans. High-quality (99.95) purge gas is required, and a high-quality filter dryer is necessary.

According to the heat flux principle, the differential temperature (ΔT) between the sample and reference is converted to differential heat flow in a way that is analogous to current flow in Ohm's law. Heat flow causes a temperature difference ΔT. The temperature difference is measured as the voltage difference ΔU between the sample and reference junctions. The voltage is adjusted for thermocouple response S and is proportional to heat flow. In conventional heat flux DSC design, the magnitude of the thermal constant in the system varies as a function of temperature.

For degradation studies or for runs on sample that liberates gas during transition or reaction, the recommended flow rates of the purge gas are 40 to 50 cc/min. The flow rates will be realized by an inlet purge gas pressure of 40 to 50 psi. For liquid nitrogen, the sub-ambient operation flow rates of purge gas are 20 to 30 cc/min. A typical DSC curve is shown in Figure 2.13.

2.3.5 Application of DSC

2.3.5.1 Glass Transition Temperature, T_g

Polymers have different applications depending on their structure. Though they have different thermal characteristics, overall features could be described by a hypothetical DSC trace (Figure 2.13). The main thermal transitions are given in Table 2.10.

A transition at low temperature appears as a shift in the baseline. The temperature at which it occurs is called the glass transition temperature (T_g) in polymer science. This is due to a change in the specific heat of the polymer as a result of the onset of chain segment motion in amorphous regions of polymers.

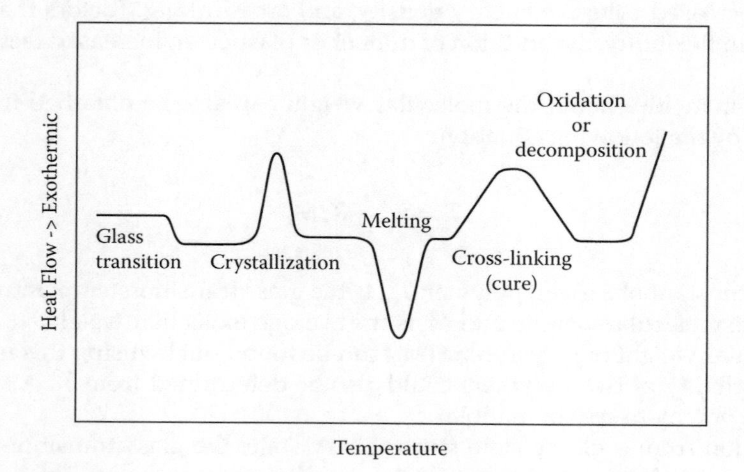

FIGURE 2.13
Typical DSC thermogram of rubber compound.

TABLE 2.10

Main Thermal Transition in DSC for Polymeric Compound

Temperature Rise	Transition or Reaction	Heat Effect
↓	Glassy transition, T_g	—
	Crystallization	Exo
	Melting, T_m	Endo
	Viscous flow	Endo
	Modification	Exo
	Crosslinking	Exo
	Oxidation	Exo
	Degradation	Endo
	Depolymerization	Endo
	Destruction	Endo

Next, an exotherm appears for the crystallization of polymers followed by an endotherm which is due to the melting of the polymer crystal. These peaks are used for the characterization of crystalline polymers. This also depends on the nature of the thermal history of the samples. The melting peak position and shape are both significant factors. A sharp peak indicates a highly crystalline and highly oriented material, while a broad peak indicates an imperfectly crystalline material.

Oxidation is indicated next by an exotherm, and finally polymers degrade at a high temperature. Both of these portions of the trace give an idea about the stability of polymers or reactions occurring due to the presence of oxygen.

A polymer having T_g lower than the room temperature and having an amorphous nature behaves as rubber. The T_g of fibers and plastics is above room temperature. There is a strong effect of structure on the values of T_g. T_g increases with the increase in bulkiness of the group on the backbone. Side chain also affects the value of T_g. All of these results could be related fundamentally to the free volume of a polymer. Factors that favor an increase in T_g are main chain rigidity, increased polarity, bulky or rigid side chain, increased molecular weight, increased cohesive energy density, and crosslinking. Factors that decrease T_g are main chain flexibility, the addition of diluent or plasticizer, increased tacticity, branching, etc.

An approximate idea about the molecular weight can also be obtained if T_g is known. This is given by the following equation:

$$T_g = T_{g\alpha} - K/M_n \quad (2.37)$$

where K is a constant of a given polymer, $T_{g\alpha}$ is the glass transition temperature for a polymer of infinite molecular weight, and M_n is the average molecular weight.

The molecular weight of polystyrene (M_n) can be found out by using this method.

The compatibility of two polymers could also be determined from T_g. A single T_g indicates that the polymers are compatible.

The transition from a glassy state to a rubbery state, the glass transition, is accompanied by a change in heat capacity, but there is no change in enthalpy. Thus, the transition appears on the DSC curve as a discontinuity in the baseline at the glass transition temperature, T_g.

T_g determination can be used as a tool for:

1. Polymer identification
2. Structure and uniformity
3. Product uniformity
4. Plasticizer efficiency
5. Degree of curing
6. Quality control

In the case of a random co-polymer, a single T_g, which is intermediate between the values of two homopolymers, is obtained, whereas two T_g values, characteristics of component homopolymers, would be expected for the block co-polymer. Compatibility of two polymers in a blend can be judged by measurement of T_g.

2.3.5.2 Melting and Crystallization

Crystallization and melting are important for semi-crystalline and crystalline polymers (Figure 2.13). DSC has been used for studying the crystallization kinetics of polyvinylidine fluoride. The time dependence of crystallization could be obtained at higher rates with sufficient accuracy. Crystallization of natural rubber has also been studied. The rate of crystallization increases as the temperature drops below T_m. At some temperature, a maximum rate will occur which may or may not be realizable experimentally. The rate again decreases at lower temperature and reaches zero at T_g.

The degree of crystallinity also could be measured from DSC trace, after dividing the measured heat of fusion ΔH_f of the sample by ΔH_{fo}, the heat of fusion for 100% crystalline polymer. However, for polymers that have more than one crystalline structure, this method is a little complex.

Crystallinities of Nylon-6, polyethylene terephthalate (PET), etc., have been determined by this method. The amount of crystallinity and T_m depend on processing conditions and heat history.

Peaks due to crystallinity and melting are important for fiber characterization. Nylon-6 and PET tire cord are characterized by these peaks.

As the temperature is slowly increased, an exotherm for crystallization of polymer appears, followed by an endotherm, due to the melting of polycrystallites (i.e., melting point, T_m) of crystalline/semi-crystalline polymers. T_g, T_m, and crystallinity of representative polymers are given in Table 2.11.

TABLE 2.11

Phase Transition Data for Polymers by DSC

Polymer	T_g (°C)	T_m (°C)	Crystallinity (%)
Polybutadiene	–105	–30	—
Nylon-6	60	215	40
Nylon-6,6	65	255	42
Polyethylene	–120	122	32

DSC can also be used for determination of the degree of crystallization and crystallization kinetics. Crystallization and melting are important for semi-crystalline and crystalline polymers (e.g., fibers). Polyethylene, polypropylene, and polyethylene terephthalate are well-known polymers for crystallization study. The degree of crystallization can also be measured from DSC trace for fusion by the following equation:

$$\text{Degree of crystallinity} = \frac{\text{Enthalpy change for the sample} \times 100}{\text{Enthalpy change for 100\% crystalline polymer}} \tag{2.38}$$

The crystallinity of Nylon-6 and PET has also been determined by this method. Peaks due to crystallinity and melting are important for fiber characterization. The purity of the polymer additives can also be judged from T_m and the degree of crystallinity.

2.3.5.3 Curing or Vulcanization

The state of curing or vulcanization of rubber can be found out by TG/DSC. Many curing reactions also could be studied using DSC. Common examples are rubber or epoxy resins, which are cured with different chemical species. Cure exotherm could be used for this purpose.

The activity of catalysts and the effect of fillers or other additives on curing can also be measured by thermal analysis.

Curing/crosslinking/vulcanization can be studied by DSC because of its exothermic nature of reaction. The appearance of peak strongly depends on the polymer to be cured and curing agents, including accelerators/activators.

The curing of phenolic, epoxy resins, polyurethane, and different rubbers can be studied by isothermal or non-isothermal DSC in a N_2 atmosphere simulating the exact curing conditions (i.e., following a specific time-temperature program). Typical curing data for natural rubber (NR) vulcanization and the effect of the type of crosslinking system are illustrated in Table 2.12.

The extent of curing of semi-cured materials can also be estimated by the following equation:

$$\text{Extent of curing (\%)} = \frac{\text{Enthalpy of curing for semi-cured material} \times 100}{\text{Enthalpy of curing for 100\% uncured material}} \tag{2.39}$$

2.3.5.4 Oxidation and Degradation

The oxidation or degradation of a polymer indicates its thermal stability, when a polymer is heated in air/oxygen. It combines with oxygen and gives oxidative degradation products

TABLE 2.12

DSC Study of Compounded NR in Nitrogen Atmosphere

Sample	Vulcanization System	Onset (°C)	Peak (°C)	ΔH (J/g)
NR-compound	Conventional (CV)	178	199	26
NR-compound	Semi-efficient (SEV)	186	190	14
NR-compound	Efficient (EV)	192	195	06

(noted by exothermic peak). Sometimes it breaks into its components (noted by endothermic peak). DSC has been used to determine the thermal stability of polyvinyl chloride (PVC).

The evaluation of an antioxidant in rubber can be correlated by the amount of heat change or enthalpy during degradation. The smaller the heat change, the more effective is the antioxidant. Relative effectiveness of different antioxidants can also be measured by estimating the shifting in oxidation exotherm to a higher temperature.

Wire and cables for electrical applications are coated with many polymeric materials including PVC, polyethylene, and different rubbers. Degradation of polymer sheathing on cable is a major concern in the choice of suitable polymers. An inhibitor is always added to suppress degradation. The effect of the inhibitor on polymer properties could be monitored by DSC using elevated temperature and oxygen atmosphere.

At higher temperatures, a polymer decomposes (degrade) or oxidizes, depending upon the surrounding atmosphere. In the presence of inert atmosphere, energy is required to rupture the chemical bonds, and the overall pyrolytic process is endothermic. However, in some cases, an exothermic transition prior to final decomposition is obtained which indicates bond-forming reactions like crosslinking, cyclization, etc. The relative thermal stability of different polymers can be obtained from the position of the peak in DSC curve.

Oxidative degradation of polymers occurs in oxygen atmosphere by chain scission involving peroxide formation followed by distinct gelation, which is characterized by an exothermic peak in the DSC curve with no or marginal weight loss. This ultimately leads to autocatalytic oxidation, which is very fast and is characterized by a very strong and erratic exotherm in the DSC curve. Degradation data by DSC are listed in Table 2.13. For polymers, the following can be useful:

- To assess thermal and thermoxidative stability
- To assess the effect of additives like antioxidants
- To elucidate the mechanism and kinetics of oxidation
- To predict lifetime in service

2.3.5.5 Specific Heat or Heat Capacity

Thermal properties like specific heat of polymers can be measured from a DSC run at any temperature (where no physical/chemical transition occurs) by the following equation:

$$C_p = \frac{1}{M}\frac{dH/dt}{dT/dt} \tag{2.40}$$

where dH/dt is the energy difference between the sample and blank at temperature T; dT/dt is the heating rate of the DSC run; and M is the mass of the polymer.

TABLE 2.13

Degradation Data for Polymers by DSC

	In Nitrogen		In Oxygen	
Polymer	**Peak (°C)**	**ΔH (J/g)**	**Peak (°C)**	**ΔH (J/g)**
Natural rubber	—	—	218	1.45
Styrene butadiene rubber	387	1.05	214	0.60
Polybutadiene rubber	377	1.95	200	0.91

2.3.6 Principle of Thermogravimetric Application

The standard test method for composition analysis by TG describes a general technique to determine the quantity of four arbitrarily defined components: low boiling, high volatile matter; medium boiling matter; oxidizable or combustible material; and non-combustible or ash left after oxidative decomposition. The definition of each component is based on its relative volatility or non-volatility which was mentioned above. The success of the methods depends on each component having a different thermal stability range in inert and in oxidizing atmosphere.

The analysis is performed by first tearing the electrovalence, introducing and weighing the specimen, and establishing the inert atmosphere. The desired program is then initiated, and the specimen mass is continuously monitored by a recording device (Figures 2.14 and 2.15). The mass loss profile may be expressed in either milligram or percent of the original sample mass. After the medium volatile matter, mass loss plateau is established, usually at 600°C or above, and the atmosphere is changed from inert to oxidative. The analysis is complete when a mass loss plateau, corresponding to the residual sample mass, is established. Smaller sample size provides better resolution of the fraction and is more advantageous with newer instruments that have better sensitivity.

Analysis of water and process solvent could best be accomplished by holding the sample at isothermal conditions under an inert atmosphere or possibly under reduced pressure, as a preliminary step in the TG procedure. Most of the oils and plasticizers have ranges of volatilization, rather than a discrete volatilization point, because they are chemically blends of various molecular weights and volatilities.

DTG curves for elastomers have also been used as fingerprints to identify many single elastomers and blends. Quantitative determination of the ratio of the elastomers has also been calculated from DTG.

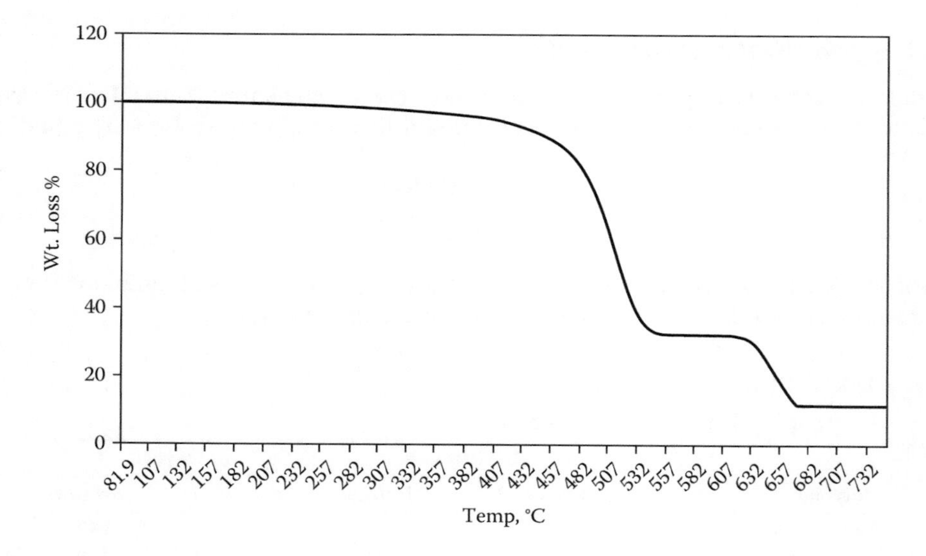

FIGURE 2.14
Typical TGA thermogram of rubber compound.

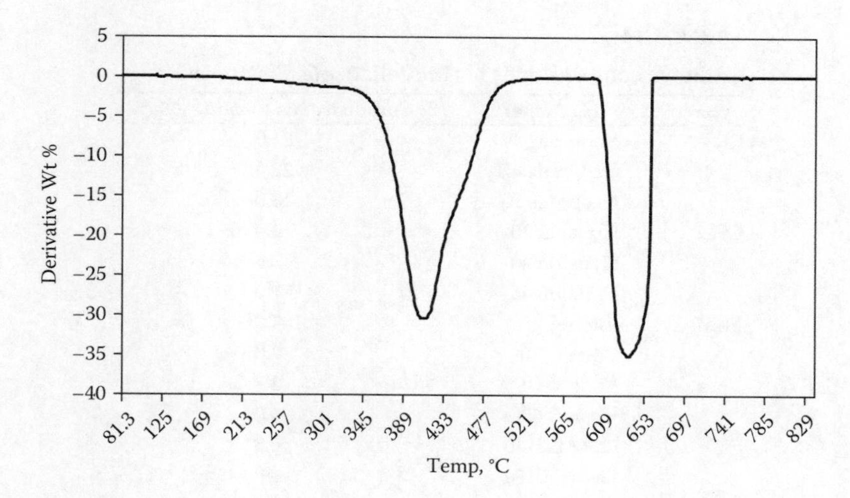

FIGURE 2.15
Typical DTGA thermogram of rubber compound.

2.3.6.1 Interference

Several factors can contribute to interference with a sample TG-DTG procedure. One factor is the overlap of the process oil, resin, etc., with a polymer decomposition region. The second and by far the greatest complication is observed for elastomers with a heteroatom in the monomer, like acrylonitrile butadiene rubber (NBR), chloroprene rubber (CR), polyvinyl chloride (PVC), chlorosulfonated polyethylene rubber (CSM) (e.g., DuPont™ Hypalon®), polyacrylate (ACM), fluoroelastomer (FKM), epichlorohydrine rubber (CO), epichlorohydrine-ethylene oxide copolymer (ECO), etc. These polymers leave a char (carbon residue) after degradation in nitrogen (Table 2.14), making it difficult to estimate the elastomer or carbon black included in the recipe.

The literature contains several references that clearly illustrate that adjusting (slowing) the heating rate during weight change improves resolution. Polymers containing heteroatom leave a carbonaceous residue (char) after degradation in nitrogen, which is oxidized along with the carbon black included in the recipe, when the environment is changed to air or oxygen. This gives a higher estimation of carbon black and lower polymer content value than are actually present. It may be observed that carbonaceous residue amounts estimated by different workers for the same polymers are somewhat different. This is attributed to the following factors: different chlorine content of different poly(chloroprene rubber), temperature of determination, run time (prolonged heating), and rate of heating.

It has been observed that there are at least three stages of the degradation process for chloroprene rubber. Sulfur curing does not increase the carbon residue but gives rise to an additional DTG peak at 190°C. The presence of calcium carbonate increases carbon residue. The amount of carbon residue increases with acrylonitrile content of NBR. So when acrylonitrile content is plotted against carbon residue, a correction factor can be determined. In order to determine the acrylonitrile content, estimation of T_g (glass transition temperature) is required.

The most promising approach to separate polymer, char, and carbon black oxidation peaks in a TG-DTG experiment is in the use of lean oxygen gas. Further work needs to be carried out with different lean gas compositions (different nitrogen/oxygen or helium/oxygen ratio)

TABLE 2.14

Carbonaceous Residues for Miscellaneous Elastomers

Type	Elastomer	% Carbonaceous Residue (550°C)
CR	Neoprene W	21.0
	Neoprene GT	22.0
	Neoprene AJ	23.0
CSM	Hypalon 20	2.0
	Hypalon 40	3.5
	Hypalon 45	2.0
FKM	Viton A	4.0
	Viton C-10	7.0
	Viton E-60	3.7
	Viton E-60C	4.0
	Fluorel 2140	5.5
	Fluorel 2160	8.0
CO	Hydrine 100	13.0
ECO	Hydrine 200	8.0
ACM	Hycar 4041	7.5
	Hycar 4042	6.0
	Hycar 4043	5.0

on various formulations. The imposition of lower heating rate, isothermal oxidation, and/or use of reduced pressure, along with the slow feed of lean oxygen would help improve resolution.

To overcome the problem of overlap of process oils and polymer, the following exercise can be practiced:

1. Extract the sample to remove oil, excess curatives, etc., prior to TG analysis. This provides a reasonable estimate of oil/plasticizer content if corrections are made for various low molecular weight polymeric and non-polymeric materials (excess curatives, antioxidant fragment, etc.) removed with oil.
2. Establish a correction curve based on a reference temperature for a given polymer compound. This procedure is somewhat lengthy and requires knowledge of the polymer and oil types as well as curatives.
3. Analyze isothermal curves below the polymer decomposition temperature.
4. Use reduced pressure to aid in removing oil at a lower temperature where polymer decomposition is not significant.

2.3.6.2 Application of Thermogravimetric Analysis in Polymer

The TGA method is used mainly:

- To determine percent volatiles
- For polymer stability
- For analysis of additives in polymers
- For kinetics of degradation

The useful life of a polymer could be predicted from the dynamic TG using the Arrhenius equation and activation energy. Since there are myriad of complications, lifetime prediction is never accurate, but it gives approximate data.

TGA is the oldest among all of the thermal procedures. It monitors continuous weight loss of sample as a function of temperature with the help of thermobalance. This is a combination of suitable electronic microbalance with a furnace associated temperature programmer.

Typical TG and DTG curves for polymers are shown in Figures 2.14 and 2.15. As the temperature is gradually increased from ambient to a higher temperature at a fixed heating rate, the low molecular weight additives such as monomer, stabilizer, plasticizer/extender, moisture, etc., volatilize away giving rise to weight loss at 300°C. On further heating, polymer chain starts to degrade, involving rupture of chemical bonds. The nature of TG curve depends upon the surrounding atmosphere of the sample under testing. In an inert atmosphere, degradation due to chain cleavage occurs, whereas in the presence of oxygen, oxidative degradation takes place through peroxide (free radical) formation by oxygen absorption.

2.3.6.3 Decomposition

The DTG curve in nitrogen atmosphere allows fingerprinting of elastomer type in a single polymer/blend of polymers/polymer composites, since each polymer has a characteristic temperature T_{max}. It depends upon the nature of the polymer as well as the heating rate and surrounding atmosphere. T_{max} values of some typical elastomers in different atmospheres are shown in Table 2.15.

Along with T_{max} values, the onset of decomposition and decomposition range (the temperature range in which major decomposition occurs in between 250 and 500°C in nitrogen atmosphere) is found to be very useful to characterize the polymeric material. In order to avoid the effect of heating rate on these measurable parameters, most often two parameters, T_o and T_{max} (onset temperature and T_{max}) are calculated from the data obtained from at least three TG runs at different rates.

In general, the decomposition of polymer in oxygen (oxidation) occurs at a comparatively lower temperature (Table 2.15). This process, compared to that in nitrogen atmosphere is fast, erratic, and generally affected by sample size and additives. Most often, multiple oxidation steps (i.e., T_{max}) are observed in the TG/DTG curve. These often lead to inconclusive results in polymer characterization. But with sophisticated instruments having greater control over the sample and experimental parameters, one can get useful information regarding the characterization of polymers.

TABLE 2.15

Typical T_{max} Values of Different Polymers (at 40°C/min Heating Rate)

Elastomer	T_{max} (°C) in N_2	T_{max} (°C) in O_2
Natural rubber	411	380
Polybutadiene rubber	500	440
Styrene-butadiene rubber	488	443
Isobutylene isoprene rubber	430	389
Ethylene propylene diene rubber	498	353
Nitrile rubber	491	442
Polychloroprene rubber	403	396
Acrylic rubber	441	384

2.3.6.4 Stability and Kinetics

TGA method is the quickest method in estimating the thermal stability of polymeric material under inert as well as thermo-oxidation environment. TGA provides the mechanism of degradation at lower and higher temperatures. The kinetic parameters as well as the estimation of lifetime in service or polymer degradation can also be obtained from a TGA decomposition thermogram. The effect of curing agent, filler, and other additives on a polymer can be studied using this method.

2.3.6.5 Composition

TGA is also used for quantitative estimation of different polymers and additives in a polymeric composite. The oil amount in oil extended rubbers like OE-SBR and OE-EPDM can be determined as oil is volatile and such rubbers are thermally stable in the lower temperature zone (<350°C). With the help of an established calibration curve and knowing the DTG peak temperature, identification of oil type is also possible.

TGA can be used to identify carbon blacks since particle sizes of blacks affect oxidative TG trace. TGA is a good tool to investigate the reinforcing cord's material in a cord rubber composite (e.g., tires, conveyor belts, etc.).

Composition in four major fractions, namely polymer, volatile, black filler, and non-black filler/mineral oxide additives, can be determined by TG curve from ambient temperature in nitrogen atmosphere, followed by changeover of atmosphere to oxygen in between 600 and 650°C (Figure 2.16). Under nitrogen atmosphere, low temperature volatilizates like oil, plasticizers, etc., volatilize at somewhat lower temperature, followed by decomposition of the polymer at higher temperature. The carbon black content can be estimated by burning the sample after changeover in air or oxygen atmosphere. The ash left over contains mineral fillers and ZnO.

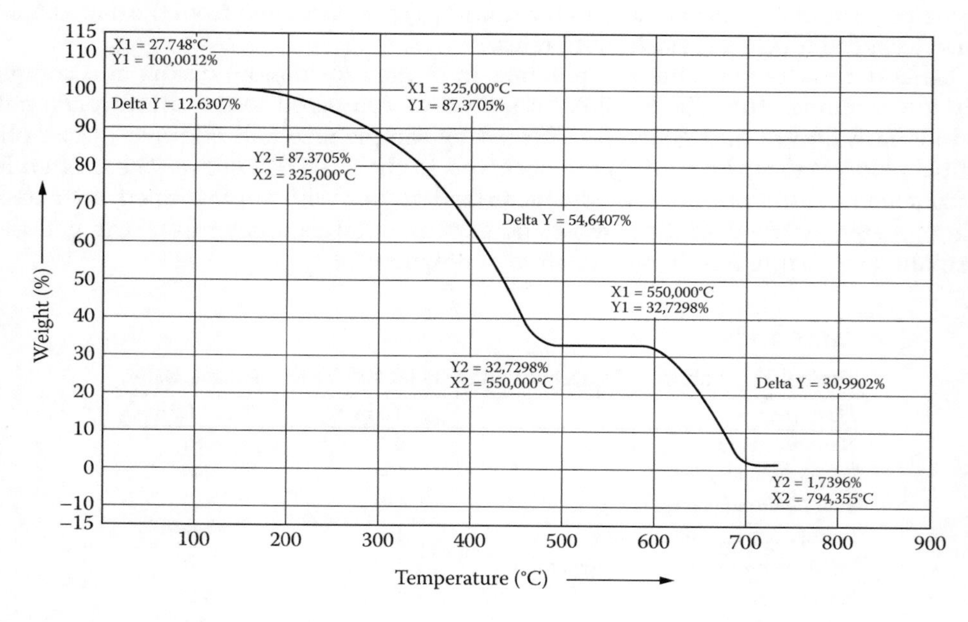

FIGURE 2.16
Four major fractions in a TGA curve.

2.3.7 Differential Thermal Analysis (DTA)

DTA is the simplest and most widely used thermal analysis technique. The difference in temperature ΔT between the sample and a reference material is recorded while both are subjected to the same heating program. Application of DTA in polymer field can be well covered by DSC technique.

2.3.8 Thermomechanical Analysis (TMA) and Thermodilatometry (TD)

Measurement of the expansion of solids and liquids as a function of temperature is known as thermodilatometry. It is often included under the general heading of thermomechanical analysis (TMA).

TMA consists of four techniques: penetration, extension, flexure, and torsional measurement. Penetration (the expansion or contraction of a sample) is measured as a function of temperature under compression, whereas the extension (the expansion or contraction of a sample) is measured as a function of temperature under tension.

The following two types of experiments are possible:

1. Measurement of dimension with temperature at a fixed load
2. Measurement of dimension with load at a fixed temperature

Major applications of TMA measurement in polymeric fields are as follows:

- Expansion coefficient as a function of temperature
- Glass transition temperature (T_g)
- Stiffness (modulus) as a function of temperature (curing)
- Softening temperature (penetration), viscous flow

2.3.9 Dynamic Mechanical Analysis (DMA)

Changes in Young's modulus indicate changes in rigidity and hence strength of the sample. Damping measurements (i.e., measurement of storage and loss-modulus) give information about glass transitions, changes in crystallinity, the occurrence of crosslinking, and some features of polymer chains. This information is widely used in studies of vibration dissipation, impact resistance, and noise abatement.

DMA is also used to study the behavior of rubbers like SBR and BR used in tire manufacturing. It gives an idea about different isomers and additives used in various formulations.

Figures 2.17a and b exhibit the dynamic mechanical spectra of hydrogenated nitrile rubber filled at various loadings of carbon black.

2.3.10 Evolved Gas Detection (EGD) and Evolved Gas Analysis (EGA)

Different hyphenated techniques, such as a combination of basic techniques of TA and either detection or evaluation of gas(es) from the sample (EGD) or detection and quantification of the gas(es) evolved (EGA), are becoming more useful in polymeric studies because of improved instrumentation like TG-IR. These techniques have the advantage of the measurement being continuous and hence able to readily relate to thermal analysis curves.

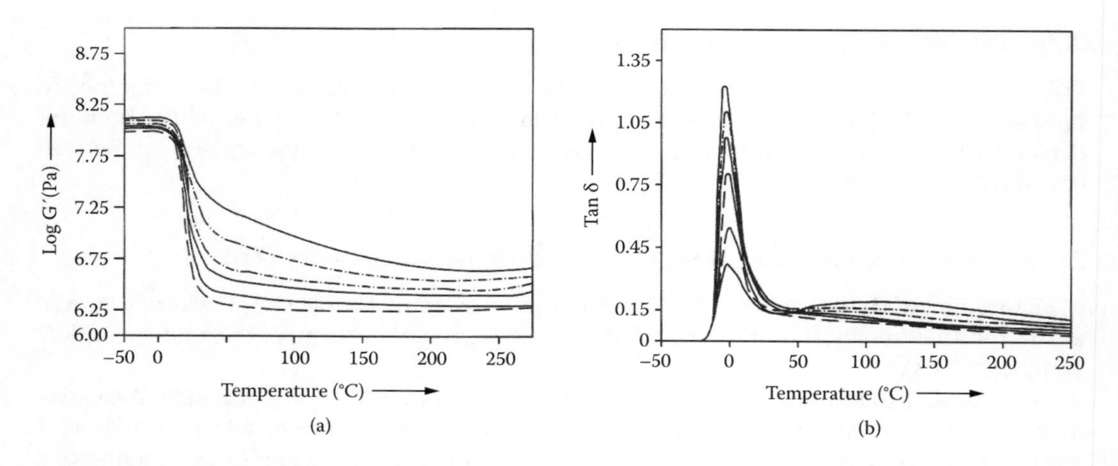

FIGURE 2.17
(a) Storage modulus versus temperature plot. (b) Tan δ versus temperature plot for hydrogenated nitrile rubber at 10, 20, 30, 40, 50 phr of carbon black loading.

EGD is used mainly to distinguish between phase transitions and endothermic decompositions. In the polymer-related fields, while degrading a polymer, it is useful to have information about the gaseous products to elucidate the reaction mechanism, kinetics, service life, etc.

These are a few thermal analysis techniques that are used extensively to characterize polymeric materials. The need for reliable instrumental techniques for polymeric chemists, quality control laboratories, where testing of polymeric material is performed, as well as in research and development laboratories where new polymers, co-polymers, and polymer products are designed and evaluated is well known. Thermal analysis has taken the lead and major share in this field.

2.4 Chromatography

2.4.1 Introduction

Chromatography is basically a technique of separation. This separation is either physical or chemical. In chromatography the two immiscible phases are brought together: one phase is stationary, and the other is mobile. A sample introduced into a mobile phase is carried along through a column containing a distributed stationary phase. Components in the sample undergo repeated interactions (partitions) between the mobile phase and the stationary phase.

Every chromatographic separation is based on differences in the rates of migration of the sample components through the column. The different sample components spend different times in the stationary phase depending on its molecular characteristics, whereas the time spent in the mobile phase is identical for all components. The mobile phase, which alone affects transport through the column, can be a gas, a liquid, or a supercritical fluid. Classification of the chromatographic process is based on the types of mobile and stationary phases used.

Various types of chromatography include:

- Gas chromatography
- Liquid chromatography
- Bonded-phase chromatography
- Ion-pair chromatography

If a solid with a large active surface is employed as the stationary phase and a gas as the mobile phase, the chromatography principle is called gas-adsorption chromatography or gas-solid chromatography (GSC). With a liquid mobile phase, the method is liquid-adsorption chromatography or liquid-solid chromatography (LSC). When the separation involves predominantly partitioning between two immiscible liquid phases, the process is called liquid-liquid chromatography (LLC). Two other liquid chromatographic methods differ somewhat in their modes of action. In ion-exchange chromatography (IEC), ionic components of the samples are separated by selective exchange with counter-ions of the stationary phase. The use of exclusion packing as the stationary phase brings about a classification of molecules based largely on molecular geometry and size. Exclusion chromatography (EC) is referred to as gel-permeation chromatography by polymer chemists and as gel filtration by biochemists.

When a solid is used as the stationary phase, one generally speaks of adsorption chromatography; if a liquid is coated on an inactive solid support, it is called partition chromatography. No clear differentiation can be made between the two, especially in adsorption chromatography where precoated adsorbents are frequently used and there is a continuous transition from pure adsorption to a more or less distinct partition.

Gas chromatographic methods are appropriate for the separation of volatile substances. Many substances that are nonvolatile at ordinary pressures can be converted simply and quantitatively into volatile derivatives that can be separated by gas chromatography.

Liquid chromatographic methods are utilized mainly for the separation of substances, which decompose on vaporization.

The chromatographic behavior of a solute can be described in numerous ways like retention volume, V_R (or corresponding retention time, t_R), and the partition ratio k.

2.4.1.1 Retention Behavior

Retention behavior reflects the distribution of a solute between the mobile and stationary phases. The volume of mobile phase necessary to convey a solute from the point of injection, through the column, and to the detector is defined as the retention volume V_R. It may be obtained directly from the corresponding retention time t_R on the chromatogram by multiplying the latter by volumetric flow rate F_c, expressed as the volume of the mobile phase per unit time:

$$V_R = t_R F_c \tag{2.41}$$

2.4.1.2 Partition Ratio

The partition ratio (capacity ratio) k relates the equilibrium distribution of the sample within the column to the thermodynamic properties of the column and to the temperature. For a given set of operating parameters, k measures the time spent by the sample in the stationary phase relative to the time spent in the mobile phase. It is defined as the ratio of moles of a solute in the stationary phase to the moles in the mobile phase.

A suitable chromatography column may be selected on the basis of three criteria:

1. The attainable resolution
2. The speed of analysis
3. The load capacity of the column

2.4.1.3 Resolution

The resolution R of two sample bands is defined in terms of the distance between the two peak maxima, expressed as the difference in the two retention times, t_{R2} and t_{R1}, and the arithmetic mean of the two bandwidths, W_1 and W_2, and is expressed as

$$R = \frac{2(t_{R2} - t_{R1})}{W_1 + W_2} \quad (2.42)$$

In chromatography one should go for optimum resolution (i.e., the peaks should be separated from each other as far as necessary).

2.4.1.4 Speed of Analysis

A separation is optimum if it completes in the shortest time. High resolution is undesirable because it can only be achieved at the expense of analysis time. For every analysis there is an optimum system of stationary and mobile phases with an optimum temperature which offers the greatest selectivity for the desired separation and thus provides a large relative retention. The column properties and packing exert considerable influence on the analysis time. However, a certain length is necessary to attain the required plate number. It is important to remember that the plates do not really exist; they are imaginary and are used to help in conceptualizing the working principle of the column. Plates also indicate column efficiency, either by stating the number of theoretical plates in a column, N (the more plates the better), or by stating the plate height—the *height equivalent to a theoretical plate* (HETP) (the smaller the better).

If the length of the column is L, then the HETP is

$$\text{HETP} = L/N \quad (2.43)$$

The number of theoretical plates that a real column possesses can be found by examining a chromatographic peak after elution:

$$N = \frac{5.55 t_R^2}{w_{1/2}^2} \quad (2.44)$$

where N is the number of theoretical plates, t_R is the time between sample injection and an analyte peak reaching a detector at the end of the column which is termed as the retention time, and $w_{1/2}$ is the peak width at half-height. Since the plate number is inversely proportional to the square of the particle size, smaller particles should always be used.

For a given separation, the required plate number can be obtained with a long column packed with larger particles as well as with a shorter one containing smaller particles. However, the speed of analysis is always greater with shorter columns packed with smaller particles. Another advantage of small particles is in the higher detection sensitivity they permit, which results from the sample components migrating as sharper zones, hence reaching the detector less diluted and giving rise to higher peaks.

The temperature affects the analysis time only indirectly. The viscosity of the mobile phase decreases with rising temperature, resulting in increasing eluent velocities for a constant pressure drop. Since the rate of diffusion also increases with rising temperature, sharper peaks are obtained at higher temperatures.

The chromatographic techniques principally used in the rubber industry are as follows:

1. Thin-layer chromatography (TLC)
2. Liquid-solid column chromatography (LSC)
3. Gas-liquid chromatography (GLC)
4. High-performance liquid chromatography (HPLC)

These techniques may be variously used as analytical tools to establish the complexity of mixtures and purity of samples, and as preparative tools for separating mixtures into individual components. The selection of a particular technique is to some extent a matter of experience.

2.4.2 Thin-Layer Chromatography (TLC)

In this technique, glass plates coated with layers of solid stationary phase are used. Some of the binding agents like calcium sulfate are also used for better adhesion of the stationary phase with the glass plate. The most commonly used stationary phases are silica gel, alumina, kieselguhr, and cellulose powder. Some of these are also available in fluorescent grade (containing zinc sulfide). This helps in detecting the resolved components of the mixture while viewing the plates under ultraviolet light (as shown in Figure 2.18).

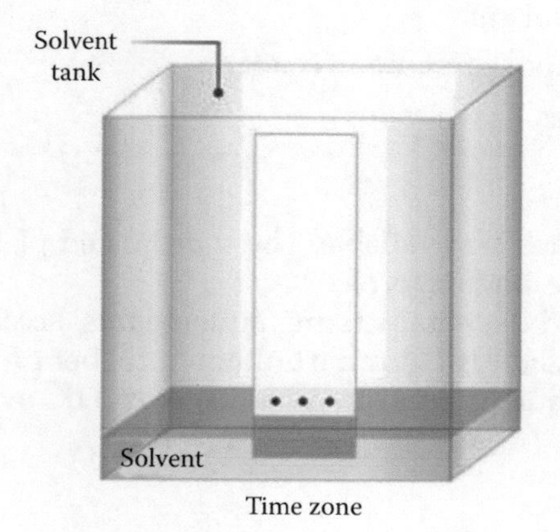

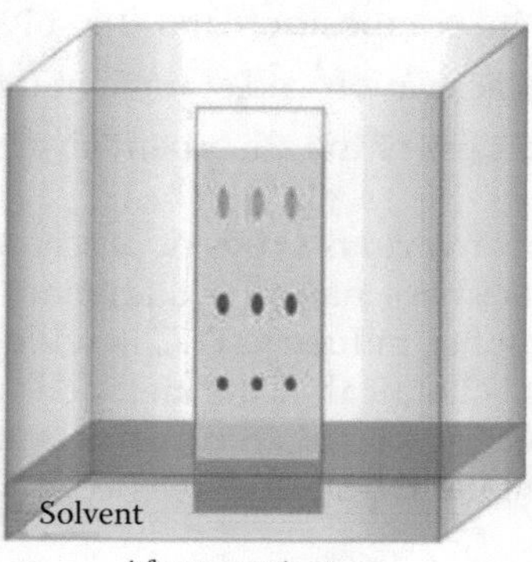

FIGURE 2.18
Typical instrumental diagram of thin-layer chromatography.

TABLE 2.16

Different Mobile Phases Used to Separate Antioxidants on Silica Gel

Sl Number	Mobile Phase
1.	90 parts *n*-heptane and 10 parts ethyl acetate by volume
2.	95 parts toluene and 5 parts ethyl acetate by volume
3.	75 parts cyclohexane and 25 parts diethylamine by volume
4.	50 parts toluene and 50 parts *n*-heptane by volume
5.	100 parts toluene, 5–10 parts acetone, and 0.1–0.2 part ammonium hydroxide
6.	50 parts toluene, 15 parts heptane, 10 parts chloroform, 10 parts ethyl ether

The various steps involved in thin-layer chromatography are preparation of plates, sample application, development of plates, and evaluation of the developed plates. In developing the plates, the selection of the solvent system is very important. Some commonly used mobile phases for antioxidant separation are shown in Table 2.16. If the chromatographic behavior of the substance under investigation is unknown, the most satisfactory developing solvent must be ascertained by preliminary trial runs using micro-plates. Solvents that cause all the components to remain near the spot of origin or to move near the solvent front are clearly unsatisfactory.

Different methods of evaluation of developed plates include viewing the plates under UV light, using iodine vapor, and applying inorganic adsorbent (also known as spotting reagents). Selectivity and sensitivity of a wide range of spotting reagents may be found in the literature on thin-layer chromatography.

The movement of any substance relative to the solvent front in a given chromatographic system is constant if the experimental conditions are reproducible. This is the characteristic of a sample. This constant is known as the R_f value and is defined as:

$$R_f = \frac{\text{Distance traveled by solute}}{\text{Distance traveled by the solvent front}} \tag{2.45}$$

Reproducibility of the R_f value is rarely achieved in practice due to minor changes in a number of variables, such as:

- Particle size of different batches of adsorbent
- Solvent composition and the solvent vapor pressure in the tank
- Prior activation and storage of the plates
- Thickness of the adsorbent layer

A number of various types of precoated plates are available. These precoated plates vary in plate thickness, nature of stationary phase, plate size, etc.

In the rubber industry, TLC is widely used in characterizing antioxidants, accelerators, and special chemicals. This method is also used to determine unknown rubber processing ingredients for formula reconstruction and quality control. Advantages of TLC over other instrumental analysis are:

- Inexpensive process
- Minimum instrumentation
- Specific coloration for specific materials, etc.

Most of the work done using TLC for identification of rubber processing ingredients is focused on the determination of antidegradants using silica gel adsorbent. ASTM D 3156 is a widely used procedure for the identification of antioxidants. Researchers are using fluorescent silica gel as a solid stationary phase.

There are several situations that lead to various anomalies in the development. One of the most frequent anomalies in TLC is "tailing." There are two major reasons for tailing. The first is improper cleaning of the sample, so that the interfering substances affect the development. The second is too much sample applied to the layer. Because of the tailing, the R_f values calculated are not accurate and will appear higher than those of a standard of the same material which is not influenced by the effect of the tail.

2.4.3 Gas Chromatography (GC)

Gas chromatography is a process by which a mixture of samples is separated into its constituents by a moving gas phase passing over a stationary sorbent, as shown in Figure 2.19. Gas chromatography is divided into two major categories: gas-liquid chromatography (GLC) with which separation occurs by partitioning a sample between a mobile gas phase and a thin layer of non-volatile liquid coated on an inert support, and gas-solid chromatography (GSC) that employs a solid of large surface area as the stationary phase.

A gas chromatograph consists of several basic modules joined together to:

1. Give a constant flow of mobile phase, which is carrier gas
2. Allow the introduction of sample vapors into the flowing mobile phase
3. Have the sufficient length of a stationary phase
4. Maintain the column at the appropriate temperature
5. Detect the sample components as they elute from the column
6. Provide a readable signal proportional in magnitude to the amount of each component

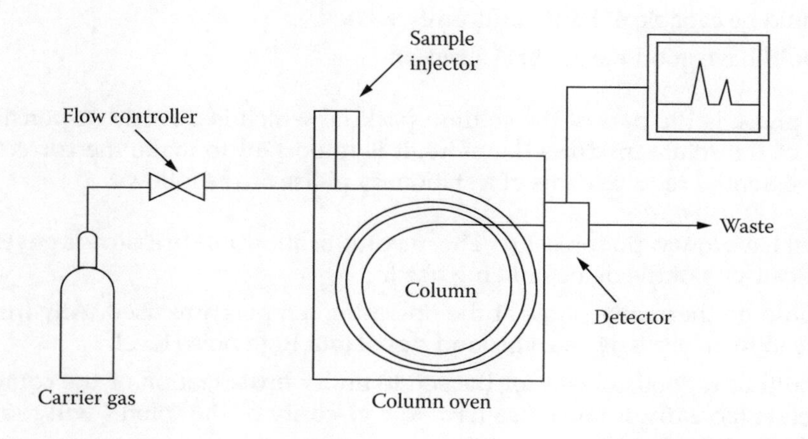

FIGURE 2.19
Typical instrumental diagram of gas chromatography.

2.4.3.1 Carrier Gas

The mobile phase used is either helium, nitrogen, hydrogen, or argon—the selection of gas depends on factors such as availability, purity required, consumption, and the type of detector employed. Helium gas is preferred when thermal conductivity detectors are employed because of its high thermal conductivity relative to that of the vapors of most organic compounds. The operating efficiency of the apparatus is highly dependent on the maintenance of a constant flow of carrier gas.

2.4.3.2 Sample Injection

Various applicators are available for introducing the sample, but the major applicator is a hypodermic syringe, a self-sealing silicone rubber septum, as common applications involve liquid samples that are injected onto a glass liner within a metal block. The temperature of the sample port is kept high so that the liquid is rapidly vaporized but without decomposing or fractionating the sample. Generally, the sample port temperature is around the boiling point of the least volatile component.

Many samples are, however, unsuitable for direct injection into a gas chromatograph because of their polarity, low volatility, or thermal instability.

2.4.3.3 The Column

Two basic types of columns are generally used: the packed column and the open tubular or capillary column. The actual separation of sample components is affected in the column where the nature of the solid support, type and amount of liquid phase, method of packing, length, and temperature are important factors in obtaining the desired resolution.

The internal column diameter should be at least eight times the diameter of the support particles. The essential requirements for an ideal support are:

1. It must be chemically inert to solutes passing through the column.
2. It must have a large surface area to volume ratio.
3. It must have a low resistance to gas flow.
4. It should be capable of being uniformly wetted.
5. It should have good mechanical strength.

Stationary phase is the part of the column packing which is directly responsible for the separation of the solute mixture; therefore, it is important to make the correct choice of phase. The essential requirements of a stationary phase are as follows:

1. It must have low vapor pressure. The maximum allowable pressure is governed by the sensitivity of the detector being used.
2. It should be thermally stable at the operating temperature used. Any instability will lead to an excessive background signal and high noise level.
3. It should be a good solvent for the solute under investigation or the components will elute too early. It must also have low viscosity or the solutes will remain too long in the liquid phase.
4. It must be chemically inert to the solute, column material, etc.

The temperature range of the stationary phase must also be considered. Every phase has a maximum recommended temperature (MRT) which is defined as the temperature at which the phase has a vapor pressure of 0.1 mm of mercury. Columns should not be used at temperatures above the MRT.

These columns require conditioning because the liquids used as stationary phases are frequently mixtures of polymers which have varied molecular weights. Due to this spread of molecular weight, such polymers exert a considerable vapor pressure even at low temperature. The finite vapor pressure of these liquids results in a small amount of stationary phase being swept by the carrier gas into the detector. This removal of column stationary phase by the carrier gas is called column bleed. If a highly sensitive detector like a flame ionization detector is used, a large background current will be generated due to the column liquid burning in the flame. By conditioning the column, low molecular weight impurities and residual solvents are removed. Conditioning of the column can be achieved by heating the column overnight at 20 to 30°C below the MRT of the stationary phase with a low flow rate of carrier gas passing through the column. It is important to disconnect the column from the detector during conditioning to avoid the risk of contamination.

Capillary columns may be prepared from stainless steel or glass, but these have been almost completely replaced by fused silica. Fused silica is an inert material that is extremely flexible, robust, and has a low thermal mass; it is thus ideal for a capillary column. Capillary columns have an internal diameter of 1 mm or less.

The main disadvantage with the capillary column is the sample capacity of the column, which depends on film thickness. With thin-film and narrow-bore columns, maximum column efficiencies are achieved but the sample capacity is low. With wide-bore, thick-film columns, the sample capacity is increased but at the cost of column efficiency.

A large selection of columns is therefore available for gas chromatography. Packed columns are relatively cheap, simple to use, and do not require specialized injectors. However, the separating power of packed columns is limited. Capillary columns have a high separation capability but are more expensive and require a higher level of competence.

2.4.3.4 Detectors

A detector located at the exit of the separation column detects the presence of the individual components as they leave the column. The choice of detector will depend on factors such as the concentration level to be measured and the nature of the separated components. Detectors usually translate the column's separation process into an electrical signal to be used for qualitative and quantitative measurements by means of recorders, electronic integrators, etc. Some of the important properties of a detector in gas chromatography are:

- Sensitivity
- Linearity
- Stability
- Universal or selective response

The following are different detectors in gas chromatography:

- Thermal conductivity detector (TCD)
- Flame ionization detector (FID)

- Thermionic emission detector (TED)
- Electron capture detector (ECD)
- Flame photometric detector (FPD)
- Photoionization detector (PID)

2.4.3.4.1 *Thermal Conductivity Detectors*

Detectors in this group, including hot wire and thermistor types, are designed to continuously measure the thermal conductivity of column effluent. The principle of operation is simple. The carrier gas is passed into a cell, the walls of which are maintained at a constant temperature, with the use of an electrical filament or thermistor. When a constant current is applied to the hot element, the rate of production of heat is constant. This heat must be dissipated. Such dissipation is largely conduction of heat through the carrier gas. The appearance of vapor in the gas changes its thermal conductivity so that different temperature gradients are necessary to maintain the required rate of dissipation, thus causing the hot element to change temperature.

The sensitivity of the detector is affected by the current through the wires, the relative thermal conductivity of the carrier gas and the sample compartments, the carrier gas flow rate, and the relative temperatures of the cell block and the hot elements. The heavier the carrier gas and the lower the thermal conductivity, the less is the sensitivity.

The advantages of thermal conductivity detectors are that they are non-destructive (i.e., the sample can be collected for further investigation if required), and they will respond to all compounds eluting from the column. Their disadvantages are that an expensive thermostated oven is required to house the detectors, and compared with the various ionization detectors, they are relatively insensitive.

2.4.3.4.2 *Flame Ionization Detector*

The flame ionization detector is the most popular detector in use today. This popularity is due to its high sensitivity, the wide sample concentration range over which its response is linear, and its robust design. It is a mass sensitive detector—that is, its response is proportional to the total mass of compound entering the detector per unit time.

A good flame ionization detector should have the following characteristics:

- The change in current through the detector should be as large as possible for a given mass of sample (i.e., it should have high sensitivity).
- The change in current through the detector should be proportional to the mass of sample over as wide a range of sample sizes as possible (i.e., it should have a wide linear dynamic range).
- The signal generated for the same masses of different compounds should be equal (i.e., the relative response factor should be unity).
- The signals generated when no sample is passing through the detector should be minimal (i.e., low noise level).
- The detector should not respond to changes in the carrier gas flow rate, and because of their small dead volume, they should have a fast response speed.

The advantages of a flame ionization detector are high sensitivity, fast response, and wide dynamic range. In addition, it is insensitive to changes in temperature, is robust, and can be dismantled easily for cleaning. The disadvantages are that it is somewhat specific in response (i.e., the detector will not respond to inorganic compounds including the permanent gases, and organic molecules containing a single carbon atom which is part of a carbonyl group, e.g., formic acid) and that it is a destructive detector in that the sample cannot be collected for further study.

2.4.3.4.3 *Electron Capture Detector*

The electron capture detector (ECD) was the first truly selective detector for gas chromatography. The operation of the ECD is based on the ability of certain substances to react with free electrons. Most ionization detectors are based on measurement of the increase in current, which occurs when a more readily ionized molecule appears in the gas stream.

Compared with the FID, however, the ECD is more specialized and tends to be chosen for its selectivity which can simplify chromatograms. The ECD requires careful attention to obtain reliable results. Cleanliness is essential, and the carrier gases must be pure and dry.

Other detectors that have not been mentioned here are also available for chromatographic analysis.

2.4.3.5 Carrier Gas Flow

During an isothermal run, control of the carrier gas flow can be maintained by setting up a constant pressure drop across the column and maintaining the pressure drop by means of a pressure controller. During a temperature programmed run, as the column is heated, the molecular interactions in the carrier gas increase, causing the viscosity to increase and the flow rate to decrease. The flow rate drop during a temperature program can have two effects: the column efficiency may deteriorate rapidly and labile components on the column may be lost. The problem can be overcome by including a flow controller in the carrier gas line. The flow controller increases the pressure drop across the column in proportion to any decrease in flow rate, consequently maintaining a constant flow rate throughout the column regardless of temperature.

2.4.3.6 Injector System

A gas chromatographic system is closed off from the external atmosphere so that any sample introduction must be achieved while maintaining the internal gas pressure. One of the easiest means of doing this is to use a rubber septum that can be pierced by the needle of a micro-syringe, although some injectors employ pneumatically operated seals as an alternative. Some of the common injectors are:

- Packed column injector
- Split/splitless injector
- On-column capillary injector
- Programmable temperature vaporizer

2.4.3.6.1 *Split/Splitless Injector*

This is a very versatile injection system able to cope with wide concentrations from trace amounts upward. In the split mode the evaporated sample is homogeneously mixed with the carrier gas, and then the mixture is split into two portions, the smaller of which is directed onto the column. This allows concentrated samples to be chromatographed without overloading the capillary column. The split ratio is defined as the ratio of inlet flow to column flow. The major part of the split sample is vented to the atmosphere using a needle valve to vary the split ratio.

Some samples contain trace quantities to be analyzed and thus require the whole sample to go onto the column to ensure adequate sensitivity. The splitless mode was developed to allow this feature. This splitless mode is ideal for low concentrations of components.

2.4.3.7 Sample Handling Techniques

Different types of samples are analyzed by gas chromatography. Depending on sample characteristics, the handling of the sample plays an important role in gas chromatography. Some of the sample handling techniques are:

- Gas sampling
- Pyrolysis
- Headspace analysis

2.4.3.7.1 Gas Sampling

Gas sampling requires:

- Constant temperature (high enough to prevent any condensation of the components in the mixture)
- Constant pressure
- Gas-tight seals on sample transfer lines and the sampling device
- Fixed amount of sample injected into the GC

2.4.3.7.2 Pyrolysis

Pyrolysis is a technique applied mainly to solid samples, usually polymers. The pyrolysis system involves the thermal decomposition of the sample—the degradation products enter the gas chromatograph where they are separated and identified.

There are two types of pyrolyzer in general use:

- Filament type: this system is capable of operating in the range of 500 to 1000°C.
- Curie point: in this case, the sample is coated onto the probe.

2.4.3.7.3 Headspace Analysis

In headspace analysis, a sample of the vapor produced by a liquid or solid material is obtained and introduced into the gas chromatograph. The sample is placed in a sealed container and thermostated for a period sufficient to establish equilibrium in the headspace above the sample. When the headspace sample is injected, a chromatogram is obtained with peaks corresponding to the components.

2.4.4 High-Performance Liquid Chromatography (HPLC)

High-performance liquid chromatography separates macromolecules and ionic species, labile natural products, polymeric materials, and a wide variety of other high molecular weight polyfunctional groups. Chromatographic separation in HPLC takes place due to

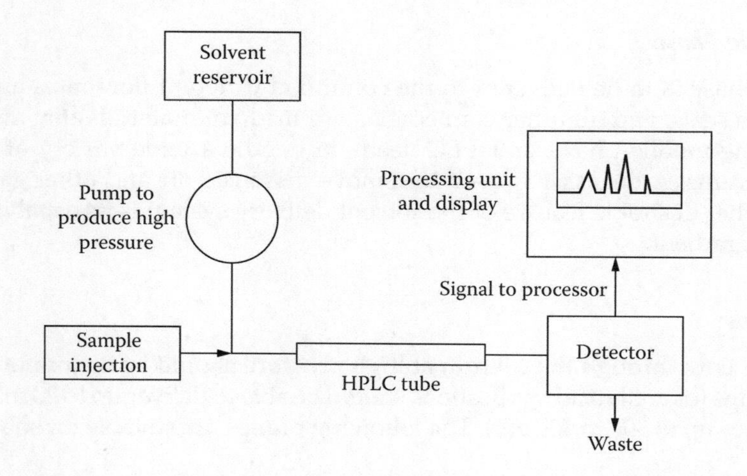

FIGURE 2.20
Typical instrumental diagram of HPLC.

specific interactions between sample molecules with the stationary and mobile phases. These interactions are not present in the mobile phase of gas chromatography. There are varieties of stationary phases for HPLC, which allows a greater variety of these selective interactions and more possibilities for separation. Sample recovery in HPLC is easy. Recovery is usually quantitative, and separated sample components are readily isolated from the mobile-phase solvent. A typical instrumentation diagram is shown in Figure 2.20.

The general instrumentation for HPLC incorporates the following components:

- There is a solvent reservoir for the mobile phase.
- The mobile phase must be delivered to the column by some type of pump. To obtain separations either based on short analysis time or under optimum pressure, a wide range of pressure and flows is desirable. The pumping system must be pulse-free or have a pulse damper to avoid generating baseline instability in the detector.
- Sampling valves or loops are used to inject the sample into the flowing mobile phase at the head of the separation column. The sample should be dissolved in a portion of the mobile phase to eliminate an unnecessary solvent peak.
- Ahead of the separation column there may be a guard column or an in-line filter to prevent contamination of the main column by small particulates.
- To measure column inlet pressure, a pressure gauge is inserted in front of the separation column.
- The separation column contains the packing needed to accomplish the desired HPLC separation. These may be silicas for adsorption chromatography, bonded phases for liquid-liquid chromatography, ion-exchange functional groups bonded to the stationary support for ion-exchange chromatography, gels of specific porosity for exclusion chromatography, or some other unique packing for a particular separation method.
- A detector with some type of data handling device completes the basic instrumentation.

2.4.4.1 Mobile Phase

The mobile phase is to be delivered to the column at different flow rates and pressures. The pump, its seals, and all other connections are made of materials that are chemically resistant to the mobile phase, as HPLC needs to handle a wide variety of organic and inorganic solvents. A degasser is used to remove dissolved air and other gases from the solvent. Another desirable feature of the solvent delivery system is its capability to generate a solvent gradient.

2.4.4.2 Pumps

Mobile phase flow through the column at high pressure should be continuous and pulse-free. The pumps for analytical applications should be able to deliver up to 20 mL/min of eluent at pressures up to 300 to 400 atm. The following pumps are suitable for solvent delivery:

1. Single-stroke piston pumps with constant eluent flow
2. Reciprocating piston pumps and diaphragm pumps with a pulsating flow and constant stroke frequency
3. Reciprocating piston pumps with variable stroke frequency
4. Gas-driven displacement pumps

The pumps are designed for either constant pressure or constant flow operation.

2.4.4.2.1 Single-Stroke Piston Pumps

In these pumps the piston is driven at a slow, constant rate, and the mobile phase is delivered continuously. The advantages of these types of pumps are the absence of valves and the delivery of a constant, pulse-free flow of solvent. The disadvantage is the necessity of more frequent interruptions to refill at the higher flow rates. Pumps of this type are relatively expensive.

2.4.4.2.2 Reciprocating Piston Pumps and Diaphragm Pumps

These pumps deliver a continuous but pulsating flow. Diaphragm pumps are recommended for chromatographic purposes because the parts that are contacted by the eluent can be readily made from inert materials such as stainless steel. In such pumps the piston movement is transmitted by means of hydraulic fluid to the diaphragm and then to the eluent. Since the piston seals contact only hydraulic fluid, the sealing problems are reduced, and the reliability is increased. The disadvantage of this type of pump is in the dependence of the delivered amount on back pressure caused by the dead volume and the valves. Their output decreases with increasing pressure.

2.4.4.2.3 Reciprocating Piston Pumps with Variable Stroke Frequency

These pumps are particularly well suited for installation in a gradient system.

2.4.4.2.4 Gas-Driven Displacement Pumps

Gas-driven pumps, in their simplest form, consist of a plastic vessel within a pressurized chamber that is connected to a gas cylinder. In an even simpler method, pressure from a gas cylinder is used to drive the eluent contained in a long tube. These pumps were used at the beginning of the development of HPLC.

2.4.4.3 Sample Introduction

Introduction of sample is either by syringe injection through the septum of an injection port into the eluent stream or by a sample loop from which it is swept into the system by the eluent. The sample should reach the column without any appreciable mixing with the eluent. Furthermore, the pressure and flow equilibria should not be disturbed during sample introduction.

2.4.4.4 The Column

Columns are generally constructed with heavy-wall, glass-lined metal tubing to withstand high pressure up to 680 atm and the chemical action of the mobile phase. The interior of the tubing must be smooth with a uniform bore diameter. Straight columns are preferred and are operated in the vertical position.

The properties of the most important tubing materials may be described as follows:

- Glass: smooth inner wall, transparent, inert, usable up to about 70 atm
- Stainless steel: relatively corrosion resistant, can be passivated, no pressure limitations
- Tantalum: smooth inner wall, largely inert, hard, difficult to shape
- Copper: easy to work, subject to corrosion

Column end fittings and connectors must be designed with zero void volume. Most column lengths range from 10 to 30 cm.

The method employed for column packing depends on the particle size of the stationary phase. Many HPLC separations are done on columns with an internal diameter of 4 to 5 mm. Such columns provide a good compromise between efficiency, sample capacity, and the amount of packing and solvent required. Although a decrease in response is associated with an increase in column diameter, there are benefits of using the wider-diameter radial compression columns. A decrease in the overall operating pressure allows decreasing analysis time by increasing solvent flow. Decreasing the internal diameter of the column by a factor of two increases the signal of a sample component by a factor of four the square of the change in diameter. This is the case with the narrow-bore columns. Narrow-bore columns also reduce solvent consumption and as a result save the solvent cost.

To prolong the life of analytical columns, guard columns are often inserted ahead of the analytical column where they act as both chemical and physical filters. Guard columns are relatively short, usually 5 cm, and contain a stationary phase similar to that in the analytical column. They protect the analytical column from particulate contamination that may arise from the contaminated mobile phase. A guard column extends the lifetime of the expensive separation column.

2.4.4.5 Detectors

The composition of the column effluent is continuously monitored by a detector. The two most frequently used detectors in LC are the UV and differential refractometers. UV detectors are the most sensitive for samples having relatively high absorption coefficients at appropriate wavelengths. Differential refractometers are very sensitive to temperature and pressure fluctuations. The following criteria are used for the characterization and description of detectors: The noise level governs the lowest detection limit. A chromatographic peak can only be recognized as such if its height is at least twice that of the highest noise peak. A drift in the baseline is undesirable.

In considering the sensitivity, distinction must be made between the absolute and the relative sensitivity of a detector. The sensitivity is one of the most important characteristics of the detector.

2.4.4.5.1 Ultraviolet (UV) Detectors

With low susceptibility to temperature and flow rate fluctuations, UV detectors are widely used. Most of the instruments operate with a single wavelength of 253.7 nm. In some instruments a band at 280 nm is also used. The UV cells should have an optical path length of 5 to 10 mm and very small volume. The instruments may be single or double beam. The disadvantage of UV detectors is in their specificity. Only molecules that absorb in the UV region near the wavelength of the detector can be monitored. The sensitivity of UV detectors depends strongly on the molar absorption coefficients of the sample components. Because of its sensitivity and selectivity, a UV detector is applicable to gradient elution only if the eluent has no UV absorption in the region of the wavelength used.

2.4.4.5.2 Differential Refractometer

A differential refractometer measures the bulk refractive index of a sample-eluent system. In order to obtain adequate sample response, the refractive index of the mobile phase must be compensated by means of a differential technique. Any substance with a refractive index that differs sufficiently from that of the eluent can be detected. However, as this type of detector detects every change in eluent composition, it cannot be used for gradient elution unless solvents are chosen with identical refractive indices. Another disadvantage is in the strong temperature dependence of the refractive index. To attain adequate sensitivity, the temperature of the eluent and measuring cells must be held constant at ±0.001°C.

Other detectors for HPLC are as follows:

- Micro-adsorption detector
- Transport detector (FID)
- Fluorescence detector
- Electrochemical detector
- Conductivity detector
- Capacity detector
- Radioactivity detector
- Mass detector
- Infrared detector

2.4.4.6 HPLC Separation

Most of the LC separations are based on adsorption effects. Such separations are governed by the interaction of adsorbent, solute, and eluent. A linear adsorption isotherm is essential for reproducible chromatographic work.

2.4.4.6.1 Effect of Water on Separation

The oxide adsorbents such as alumina and silica gel are good drying agents for nonpolar organic solvents. The absorbed water exerts a considerable effect on the chromatographic properties. The addition of small amounts of water or other polar moderators to an adsorbent or eluent reduces the retention volumes to the extent that nonpolar compounds are no longer retained.

2.4.4.6.2 Effect of Eluent on Separation

The choice of the proper solvent frequently affects the success of a separation more than the selection of the stationary phase. Depending on the properties of the eluent, on a given adsorbent a sample may be excessively retained or not at all, or its retention time may fall into the desired level. The eluent should not interact irreversibly with either the sample or adsorbent. In some critical cases, eluent viscosity can be another criteria for eluent selection, because the smaller the viscosity, the lower is the required pressure drop to achieve a given flow velocity.

2.4.4.6.3 Effect of Sample Structure

The molecular structure of the sample determines the elution order to a greater extent compared to the properties of the solid stationary phase and the eluent. Knowledge of the composition of a sample and the structure of its components simplifies the choice of system and enables predictions to be made about the elution order.

2.4.5 Gel Permeation Chromatography (GPC)

Gel permeation chromatography (GPC) is also known as size exclusion chromatography (SEC) or gel filtration chromatography (GFC). GPC separates dissolved molecules on the basis of their size by pumping them through specialized columns containing a microporous packing material. Typical instrumentation of GPC is shown in Figure 2.21.

GPC/SEC is used to characterize natural and synthetic polymers, biopolymers, proteins, or nanoparticles.

When GPC/SEC separation is coupled with light scattering, viscometer, and concentration detectors together (triple detection), it will provide a distribution of absolute molecular

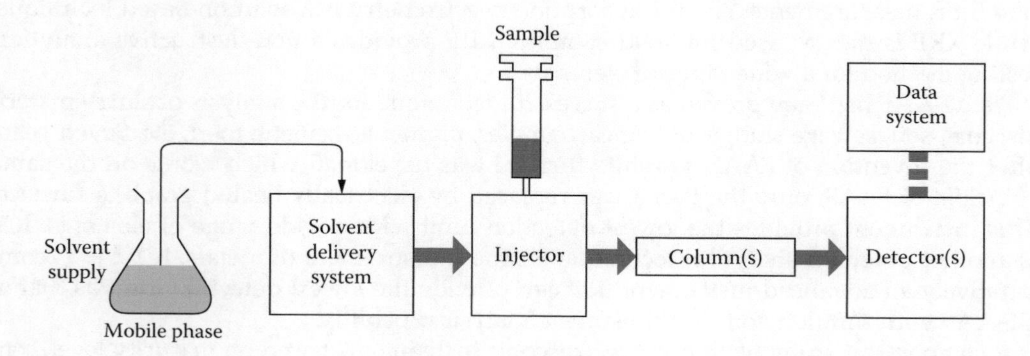

FIGURE 2.21
Typical instrumental diagram of GPC.

weight, molecular size, and intrinsic viscosity as well as information on macromolecular structure, conformation aggregation, and branching.

Gel permeation and size exclusion chromatography (GPC/SEC) is a versatile analytical technique available for understanding and predicting polymer performance. It is used to characterize the complete molecular weight distribution of a polymer, which is one of its most fundamental characteristics.

This is important, since it affects many of the characteristic physical properties of a polymer. Molecular weight and its distribution affect the end-use properties of a polymer which include tensile strength, toughness, brittleness, adhesive strength, impact strength, elastic modulus, drawability, coefficient of friction, and stress-crack resistance to name a few.

Where a polymer's end-use application requires precision performance or endurance, the need for polymer characterization is particularly acute. Because GPC fulfils these needs better than any other single technique, it has become an extremely valuable tool for materials characterization in the polymer industry.

2.5 Atomic Absorption Spectroscopy (AAS)

2.5.1 Introduction

The determination of metal content is an important parameter for rubber-based industries, because it gives a guideline in the fields of research, quality control, and root cause analysis of many problems. There are a variety of instruments available to determine the metal content in a given sample, such as x-ray fluorescence spectrophotometer (XRF), flame atomic absorption spectrophotometer (FAAS), graphite furnace atomic absorption spectrophotometer (GF-AAS), hydride atomic absorption spectrophotometer (Hydride-AAS), atomic emission spectrophotometer (AES), inductively coupled plasma-atomic emission spectrophotometer (ICP-AES), inductively coupled plasma-mass spectrophotometer (ICP-MS), direct current plasma spectrophotometer (DCPS), anodic stripping Voltameter (ASV), potentiometric titrator (auto-titrator), UV visible spectrophotometer (UV-Visible), etc. Of the above, XRF is considered as the best alternative technique to atomic absorption spectroscopic techniques like Hydride-AAS, GF-AAS, and ICP-MS in terms of low detection limit measurements. Atomic absorption spectrometry is a solution-based technique, while XRF is mostly used for solid samples. XRF provides a non-destructive analytical testing method for a wide range of elements.

Flame-AAS has been proven as a successful technique for the analysis of slurry materials, coal, soil, sewage sludge, biological samples, pigments, cement, food, etc. Seven years after the invention of FAAS, graphite furnace was developed which works on the same principle of FAAS, only the flame was replaced by electrically heated graphite furnace. This instrument provides the lowest detection limit with a wide range of elements. ICP is another potential analytical technique for the measurement of metals. ICP-MS is comparatively an advanced instrument that can provide the lowest detection limit as that of GF-AAS with simultaneous multi-element analysis capability.

A comparative study of atomic spectroscopic instruments based on maturity level, sensitivity, and interference is briefly presented in this text. Considering the maturity level, flame-AAS, GF-AAS, Hydride-AAS, and ICP are well-established techniques, whereas ICP-MS is a comparatively new and growing technique. As for sensitivity, DCP (refractory metals), ICP-MS, Hydride-AAS (hydride-forming metals), and GF-AAS (except refractory metals) are excellent as compared to Flame-AAS and ICP-AES. As for interference problems, Flame-AAS, Hydride AAS, and ICP are well understood. In GF-AAS, interference can be controlled by a stabilized temperature platform furnace (STPF); in ICP, spectral interference; and in ICP-MS, mass overlaps as interference.

In this section, important applications of atomic spectrometric instruments in rubber-based industries, with special emphasis on atomic absorption spectrophotometer (AAS) are discussed.

2.5.2 Analysis Principle of Atomic Absorption Spectrometric Instruments

The atomic spectrometric instruments can be broadly categorized based on their working principles: atomic absorption spectroscopy, atomic emission spectroscopy, and atomic fluorescence spectroscopy.

2.5.2.1 Atomic Absorption Spectrophotometer (AAS)

An atom is made up of a nucleus and is surrounded by electrons. Electrons normally occupy a stable orbital configuration called the ground state. If energy of the correct magnitude is applied to an atom, energy will be absorbed, and its outermost electrons can be promoted to a higher energy level which is in a less stable state called the excited state. The amount of radiation absorbed is proportional to the number of free atoms present, given by the Lambert-Beer law.

$$\text{Absorbance, } (A) = \log I_O/I_t = \text{K.C.L.} \quad (2.46)$$

where I_O is the intensity of incident radiation emitted by the light source; I_t is the intensity of transmitted radiation; C is the concentration of the sample solution; K is a constant called the absorption coefficient, and L is the path length.

This equation can be deduced in atomic absorption analysis as Absorbance $(A) = K \times C$, where K is a constant and it follows a first-order equation. The instrument is calibrated by standards and sample concentration is interpolated from the calibration curve.

The basic instrumentation for a flame atomic absorption spectrophotometer is as follows: A primary light source (different for different metals) which emits a characteristic wavelength radiation by either hollow cathode lamp (HCL) or electrode less discharge lamp (EDL). The sample is introduced into the system with air by the Venturi principle, using a nebulizer. The solution that enters from the nebulizer is mixed with fuel and oxidant and forms small droplets of size less than 10 μm (aerosol). This is generally done with a flow spoiler and/or with impact bead. The sample aerosol is then passed through an atomization source that is either an air-acetylene or nitrous oxide–acetylene flame that can produce maximum temperatures of 2400 and 2800°C, respectively. In this situation, the aerosol converts to an atomic vapor cloud that is now ready to absorb the characteristic incident radiation from the light source. A part of incident radiation is absorbed by the atomic cloud of the element, and the transmitted light is detected by a mono-chromator and detector assembly.

In the case of GF-AAS, an electrically heated graphite tube replaces the flame. Sample is introduced directly into the tube, which is then heated in a programmed series of steps to remove the solvent and major matrix components, and then the remaining sample is atomized. All the analytes are atomized, and atoms are retained within the tube for an extended period; as a result, sensitivity and detection limit are significantly improved.

Hydride-AAS is basically used for elements that can form volatile hydrides, such as arsenic (As), bismuth (Bi), antimony (Sb), selenium (Se), tellurium (Te), tin (Sn), and mercury (Hg), and it is normally carried out with a flow injection analysis system (FIAS) in conjunction with AAS. Here sodium borohydride solution is used as a reducing agent that liberates nascent hydrogen in acidic medium and reacts with the element to form a volatile hydride. This hydride is carried to the AAS atomizer using argon as the carrier gas.

2.5.2.2 Atomic Emission Spectroscopy

In emission spectrometry, thermal/electrical energy in the form of flame, spark, or plasma source is used to excite free atoms/ions from ground state to an excited state of higher energy levels (unstable configuration). The atoms/ions returning to the more stable configuration will subsequently emit characteristic radiation that can be isolated by a monochromator. The emitted radiation is proportional to the concentration of atoms/ions present. For emission analysis, the temperature of excitation source has to be high in order to excite large quantities of free atoms/ions.

In the case of direct current plasma (DCP), the sample is aspirated into a premix spray chamber through a nebulizer using argon as the transport gas. The sample aerosol in the stream of argon is directed at a set of electrodes across which a high electrical potential is applied. The resulting electrical discharge between the electrodes supplies enough energy to ionize the argon into a plasma of positively charged argon ions and free electrons. The thermal energy of plasma, in turn, atomizes sample constituents and creates an excited state of atoms that emit their characteristic emission spectra. The DCP is proven to be a better technique for alkali metals and alkaline earth metals than ICP-AES, and the cost of the instrument is lower than ICP-AES. But the disadvantage is that the electrodes that form the DC arc are continually eroded during operation. Due to a highly resistant skin effect, the sample does not penetrate the hottest part of plasma but instead is deflected around it, so chemical and ionization interference remain in the system.

Inductively coupled plasma (ICP) is similar to DCP. Here, the sample aerosol is directed up the central tube of the ICP torch. The torch consists of concentric tubes with independent argon streams flowing through each. The top of the torch is centered within a radio frequency induction coil that is a source of energy in the system. The ICP sample experiences temperatures between 6000 K and 10,000 K. These temperatures allow complete atomization of the element to minimize the effects of chemical interference. The detection limit attainable with ICP-AES is comparable to AAS for many elements. Elements like lead (Pb) and cadmium (Cd) can be better analyzed with AAS than ICP-AES. Because of the dynamic and wide range of ICP-AES, the multiple dilution problems of AAS can be reduced or eliminated. Physical and spectral interference are more prominent in ICP-AES.

ICP-Mass spectrometry is one of a growing number of hyphenated techniques where output of one technique is input of another. The ions generated by ICP are directed to the mass spectrophotometer that separates the ions according to their mass-to-charge ratio. ICP-MS combines multi-element capabilities and the broad linear range of ICP emission with the exceptionally low detection limit of GF-AAS. Moreover, it is considered to be one of the few analytical techniques that permit the quantification of elemental isotopic concentration.

2.5.2.3 Atomic Fluorescence Spectroscopy

In AFS, the free atoms are excited to higher energy levels by characteristic radiation as in atomic absorption spectrophotometers; however, the detector is placed at a right angle to the radiation source to observe fluorescence. There are two types of fluorescence commonly observed:

1. Resonance fluorescence: In this case the emitted radiation has the same wavelength as the excitation radiation.
2. Stepwise line fluorescence: In this case the emitted radiation is at different wavelength to the excitation radiation.

The inductively coupled plasma-atomic fluorescence spectrophotometer (ICP-AFS) is one of the successful instruments in this category and is a good choice for selected trace element analysis, notably zinc, cadmium, and alkali metals. The detection limits of important spectrometric techniques are given in Table 2.17. All detection limits are given in microgram per liter and were determined using element standards in dilute aqueous solution. All detection limits are based on a 98% confidence level.

2.5.3 Various Applications of Atomic Absorption Spectrometric Instruments in Rubber Industries

2.5.3.1 Analysis of Natural Rubber, Synthetic Rubbers, and Recycled Materials

Trace amounts of transition metal ions that undergo one electron transfer reactions, like copper (Cu), iron (Fe), manganese (Mn), cobalt (Co) ions, etc., have been reported to be rubber poisoning metals. These metals catalyze the aging reaction, and as a result, life of

TABLE 2.17

Detection Limit (Microgram per Liter) of Various Atomic Spectrometric Techniques

Elements	Flame AAS (microgram per liter)	Hg/Hydride-AAS (microgram per liter)	GF-AAS (microgram per liter)	ICP Emission (microgram per liter)	ICP-MS (microgram per liter)
Al	45.00	—	0.30	6.00	0.0060
As	150.00	0.030	0.50	30.00	0.0060
Ba	15.00	—	0.90	0.15	0.0020
Ca	1.50	—	0.03	0.15	2.0 (44_{Ca})
Cd	0.80	—	0.02	1.50	0.003
Co	9.00	—	0.40	3.00	0.0009
Cr	3.00	—	0.08	3.00	0.0200
Cu	1.50	—	0.25	1.50	0.0030
Fe	5.00	—	0.30	1.50	0.4 (54_{Fe})
Hg	300.00	0.009	1.50	30.00	0.0040
K	3.00	—	0.02	75.00	1.0000
Mg	0.15	—	0.01	0.15	0.0070
Mn	1.50	—	0.09	0.60	0.0020
Mo	45.00	—	0.20	7.50	0.0030
Na	0.30	—	0.05	6.00	0.0500
Ni	6.00	—	0.80	6.00	0.005 (60_{Ni})
Pb	15.00	—	0.15	30.00	0.0010
Si	90.00	—	2.50	5.00	0.7000
Sn	150.00	0.2	0.50	60.00	0.0020
Zn	1.50	—	0.30	1.50	0.0030

Source: Courtesy of PerkinElmer, Waltham, Massachusetts.

a product decreases. The metal impurities may come in rubber compound from different sources including rubber chemicals, synthetic rubbers that are polymerized by using some catalysts, and natural rubber that comes from nature. The metal catalyst is reported as an active hydroperoxide decomposer in both its higher and lower oxidation states.

In the overall reaction, two molecules of hydroperoxide decompose to peroxy and alkoxy radicals:

$$ROOH + M^{n+} \longrightarrow ROO^{\cdot} + M^{(n-1)+} + H^{\cdot}$$

$$ROOH + M^{(n-1)+} \longrightarrow RO^{\cdot} + M^{n+} + {}^{\cdot}OH$$

$$2ROOH \longrightarrow ROO^{\cdot} + RO^{\cdot} + H_2O$$

Copper has an effect on a gum compound of oil extended styrene-butadiene rubber, and cobalt affects the stability of divinyl rubber. The metal poisoning is more destructive in polymers containing an unsaturated backbone. Copper, as a poisoning metal, drastically reduces the effective service life in NR-based formulations.

The raw natural or synthetic rubbers are kept in a muffle furnace at a temperature of 550°C. The inorganic residue (ash) thus obtained is dissolved in either hydrochloric acid or nitric acid and is quantitatively transferred to a volumetric flask. The solution is now ready for AAS analysis.

Microwave digestion is also used for sample solution preparation because of the advantages including the following: contamination is minimized, volatiles are retained, microwave heat goes to the core of the sample, the result is reproducible, and the sample is super heated. Various decomposition reagents are used for microwave digestion. Nitric acid (HNO_3) is used for the oxidation of organic matter, and it also makes metal in soluble format. It is used in combination with hydrogen peroxide (H_2O_2), sulfuric acid (H_2SO_4), hydrochloric acid (HCl), and hydrofluoric acid (HF). H_2SO_4 is used for dehydration and charring of organic material in combination with H_2O_2, HNO_3, perchloric acid ($HClO_4$), and HF. HF is particularly used for dissolution of silicates; it is also used in combination with HNO_3, HCl, phosphoric acid (H_3PO_4), fluoro-boric acid (HBF_4), and boric acid (H_3BO_3). The solutions prepared above are now ready for analysis by using atomic spectrophotometers. The AAS analysis conditions used for some typical metals, which are known as poisoning metals, are given in Table 2.18.

The poisoning metal analysis is a very important test parameter for refined or recycled materials. Analysis results of metal content of virgin reclaim, crumb rubber, and virgin oil and recovered oils determined through AAS-3300 equipment (from PerkinElmer, Waltham, Massachusetts) are given in Table 2.19. The upper limit of recycled material in a rubber compound formulation is also obtained.

TABLE 2.18

Typical AAS Analysis Condition of Some Metals in Rubber Industry

Parameters	Cu	Mn	Fe	Co
Wavelength (nm)	324.8	279.5	248.3	240.7
Slit width (nm)	0.7	0.2	0.2	0.2
Flame used	Air-acetylene oxidizing	Air-acetylene oxidizing	Air-acetylene oxidizing	Air-acetylene oxidizing
Technique	Background corrected (deuterium arc)	Background corrected (deuterium arc)	Background corrected (deuterium arc)	Background corrected (deuterium arc)

TABLE 2.19

Comparative Analysis Data of Various Virgin and Recycled Materials in AAS

Sample ID	Cu (ppm)	Mn (ppm)	Fe (ppm)
Regular ISNR-20	3.0	2.9	5.0
40-mesh crumb rubber	8.1	5.9	960.5
80-mesh crumb rubber	13.0	1.3	282.0
100-mesh crumb rubber	8.8	25.2	3250.0
Super-fine reclaim rubber	86.7	64.5	1300.0
White reclaim rubber	60.3	51.2	5700.0
Regular aromatic oil	1.13	0.08	2.7
Recycled dust seal oil (centrifuged for 24 h)	33.9	3.8	4.5

2.5.3.2 Steel Wire Characterization

In the case of tires, with the advancement of radial technology, the multi-filament brass-plated steel cord is being used as an important reinforcing material for tire carcass and belt reinforcement. The steel cord is coated with brass for better bonding with rubber. Sulfur vulcanizable rubbers during vulcanization form a direct bond to brass, zinc (Zn), palladium (Pd), nickel (Ni), and nickel/zinc or zinc/cobalt (Co). Brass plating on the surface of steel cord reacts with the sulfur in rubber compounds during the curing process of tire, forming a copper sulfide (Cu_xS) film of brass. The desired chemical reaction taking place at rubber/brass-plated steel cord interphase may be considered as mixed ionic and covalent chemical reactions.

A steel wire sample of 0.2 to 0.4 g is accurately weighed, treated with chloroform toluene mixture (50:50) to remove any organic contaminants, and dried in oven at 105°C. After drying, the sample is cooled in desiccators, and the coating is stripped with 5 mL concentrated nitric acid to 200 mL volumetric flask and made up to mark. The solution is now ready for analysis by AAS. The typical AAS analysis condition and characterization data of steel wire samples are shown in Table 2.20. The typical analysis data of two steel wire samples used in the tire industry are given in Table 2.21.

The bead wire of 0.965 mm and 1.60 mm is widely used in the tire industries. The function of bead wires is to hold the tire with the rim of the vehicle. Bead wire is made of steel, and the wire is normally coated with copper and tin. The bead wire generally contains 97 to 99% copper and 1 to 3% tin. The coating weight is typically found in the range 0.15 to 0.65 g/kg of sample. If a 10 g sample is used for analysis and stripped coating is made up to

TABLE 2.20

Typical AAS Analysis Condition

Parameters	Copper	Zinc
Wavelength	324.8 nm	213.9
Slit width	0.7 nm	0.7
Flame type	Air-acetylene oxidizing	Air-acetylene oxidizing
Technique	Background corrected (deuterium arc)	Background corrected (deuterium arc)
Read time	5 s	5 s (read delay 2 s)

TABLE 2.21

Analysis of Coating of Steel Wires Used in Tire Industries

Sample ID	Parameters	Cu (%)	Zinc (%)
2 + 2 × 0.25 style steel cord	Median	63.4	36.6
	Range	61.3–68.1	31.9–38.7
2 × 0.30 style steel cord	Median	63.9	36.1
	Range	62.8–64.7	35.1–37.2

Note: Data are on the basis of 10 samples.

100 mL, only 0.15 to 2 mg/L tin will be available in solution. The low level of tin is analyzed by GF-AAS, AAS-FIAS or ICP-AES and ICP-MS.

The coating analysis method described in this study is briefly described here. A 10 g bead wire sample is treated with concentrated HNO_3, and the coating is stripped out to 100 mL volumetric flask. Diluted sample is analyzed for copper in AAS, and sample solution is diluted with FIAS carrier solution (saturated boric acid with 1% HCl) and analyzed with FIAS-AAS. Characterization with the AAS-FIAS technique provides good repeatability and reproducibility of bead wire characterization. The AAS-FIAS analysis condition used in the above study and characterization data of bead wire samples are shown in Tables 2.22 and 2.23, respectively.

2.5.3.3 Analysis of Rubber Compounding Ingredients

The estimation of metals is essential for process control, product acceptance, and research activities. The sample is dissolved in mineral acid and analyzed by AAS equipment.

The AAS analysis condition and most commonly found metal impurities in various rubber compounding ingredients are provided in Tables 2.24 and 2.25, respectively.

TABLE 2.22

AAS-FIAS Instrumental Analysis Parameters of AAS

Parameters	Copper	Tin
Wavelength	324.8 nm	286.3 nm
Slit width	0.7 nm	0.7 nm
Technique	Background-corrected atomic absorption (impact bead also used)	Atomic absorption
Data processing	Time average	Peak height with 37 point smoothening
Cell temperature	—	900
Read time	5 s	15 s
Flame type	C_2H_2:Air	—

TABLE 2.23

Repeatability of Bead Wire Characterization by FIAS-AAS

Parameters	Tin (%)	Copper (%)	Plating Weight (g/kg)
Mean (five samples)	1.1	98.9	0.48
Standard deviation (five samples)	0.07	0.07	0.003

TABLE 2.24

Typical AAS Analysis Condition

Parameters	Mn	Cd	Pb	Fe	Ni
Wavelength	279.5	228.8	283.3	248.3	232.0
Slit width	0.2	0.7	0.7	0.2	0.2
Flame type	Air-acetylene oxidizing	Air-acetylene oxidizing	Air-acetylene oxidizing	Air-acetylene oxidizing	Air-acetylene oxidizing
Technique	Background corrected (deuterium arc)	Background corrected (deuterium arc)	Background corrected (deuterium arc)	Background corrected (deuterium arc)	Background corrected (deuterium arc)

TABLE 2.25

Commonly Found Metal Impurities in Rubber Compounding Ingredients

Clay Mn	ZnO (Rubber Grade)		Stearic Acid	VP Latex	
(ppm)	Cd (ppm)	Pb (ppm)	Fe (ppm)	Fe (ppm)	Ni (ppm)
≤40 (3.65)	≤170 (10.00)	≤1500 (210)	≤110 (5.67)	≤23.98 (8.3)	≤10.04 (1.17)

The values shown in parentheses are medians of 10, 20, 5, and 15 of different clay, ZnO, stearic acid, and VP latex samples, respectively.

2.5.3.4 Reverse Engineering/Benchmarking

AAS is widely used for the analysis of rubber vulcanizate. This is especially used to find out the type of vulcanizing system, amount of inorganic fillers, and amount of special additives used in an unknown compound.

For example, in a sulfur cure system, zinc oxide is used as an activator. In a metal oxide curing system, ZnO is used as a crosslinking agent along with magnesium oxide which acts as an acid scavenger. Aluminum hydroxide, aluminum silicate, antimony trioxide, barium ferrite, barium sulfate, calcium carbonate, calcium oxide, calcium hydroxide, calcium sulfate, lead monoxide, lithopone, magnesium carbonate, magnesium oxide, titanium oxide, and zinc carbonate are used in rubber compounding to achieve various properties for finished products.

An analysis summary of some rubber-based products by AAS is described in Tables 2.26 and 2.27. It would be useful in the reverse engineering of any rubber product formulation.

TABLE 2.26

Typical AAS Analysis Condition

Parameters	Zn	Ca	Mg	Al
Wavelength	213.9	422.7	285.2	309.3
Slit width	0.7	0.7	0.7	0.7
Flame type	Air-acetylene oxidizing	Air-acetylene oxidizing	Air-acetylene oxidizing	Nitrous oxide-acetylene reducing
Technique	Background corrected (deuterium arc)	Background corrected (deuterium arc)	Background corrected (deuterium arc)	Background corrected (deuterium arc)

TABLE 2.27

Analysis Summary of Different Rubber Products through AAS

Rubber Products	Zn (%)	Ca (%)	Mg (%)	Al (%)
Pharmaceutical stopper (IIR based)	57.7	11.7	2.4	Nil
V-Belt (EPDM based)	71.1	—	—	—
V-Belt (CR based)	26.2	—	26.4	—
White sidewall of tire (SBR based)	20.6	—	—	66.2

Metal content analysis also may be helpful in the analysis of root causes of failure of market-returned products.

2.6 Microscopy and Image Analysis

2.6.1 Introduction

As discussed in other chapters in this book, various chemical and analytical techniques are used in rubber compound analysis for formula reconstruction. Transmission electron microscopy (TEM) is primarily used for determining the type of carbon black used in the unknown rubber compound, and characterizing the morphology of any elastomer blend present. Actual elastomer types and blend ratios are analyzed by infrared spectroscopy. TEM can basically confirm the presence of any blend of elastomers and interdispersion of elastomers on a microscopic scale.

In this section we include general TEM procedures used by the rubber industry for carbon black identification and to determine the blend morphology. We also provide two examples of compound carbon black analysis and corresponding data and TEM images of known polymer blend samples prepared in the laboratory.

2.6.2 Identification of Carbon Black Type

2.6.2.1 Carbon Black Identification by Transmission Electron Microscopy (TEM)

TEM is used to determine the average particle size of carbon black in a compound which is then compared with that of the known carbon black standard.

2.6.2.1.1 Sample Preparation and TEM Technique

Rubber compound containing carbon black is pyrolyzed under vacuum at 800 to 900°C to remove all compounding ingredients except carbon black and inorganic materials. The pyrolyzed carbon black is then analyzed under a TEM, using a technique based on ASTM D 3849, and representative areas are transmitted to an image measurement system through an online camera interfaced with TEM. In this case, we used TEM Phillips model CM12 interfaced with a wide-angle TV camera. The images were captured on a computer using a frame-grabber for subsequent image processing. The individual particle sizes were measured. Data were transferred to the Excel spreadsheet of a Windows®-based PC, and corresponding particle size histogram and average particle size were obtained.

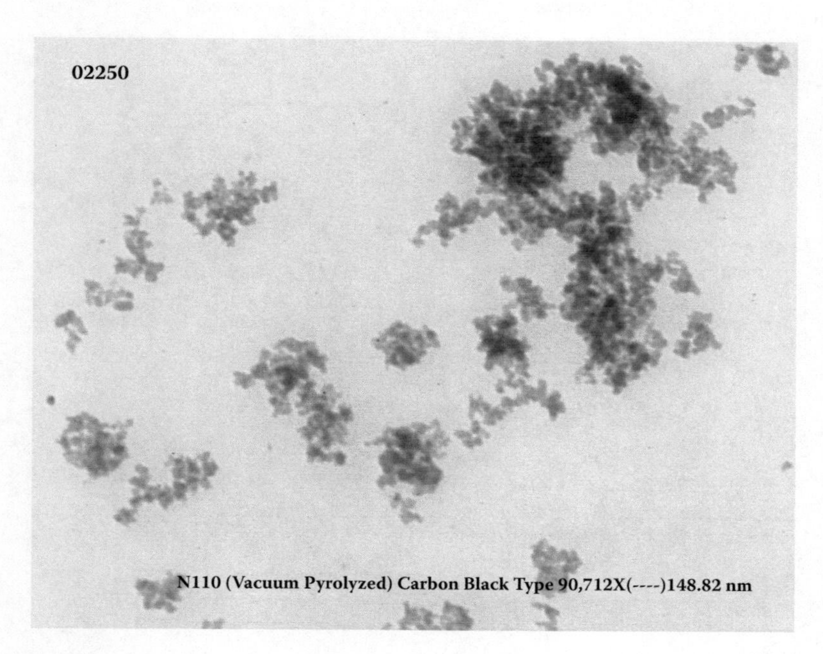

FIGURE 2.22
TEM image of carbon black N-110.

2.6.2.1.2 Image Analysis Results—Examples

Two compounds were mixed with carbon blacks and cured using suitable curing agents. Both compounds were based on 100% SBR. One compound contained 50 phr of N-110 carbon black, and the second compound contained 50 phr of N-330 carbon black.

Figure 2.22 shows a TEM image of a compound containing N-110 carbon black. A particle size histogram is shown in Figure 2.23. Its average particle size was 17.17 nm which falls in the ASTM standard range for the N-100 family of carbon blacks. Thus, this black was determined to be N-110.

Figure 2.24 shows a TEM image of N-330 carbon black. Its histogram is shown in Figure 2.25. Table 2.28 lists its average particle size of 33.03 nm which is in the particle size range of the ASTM standard of 26 to 35 nm for the N-300 carbon black family. Hence this

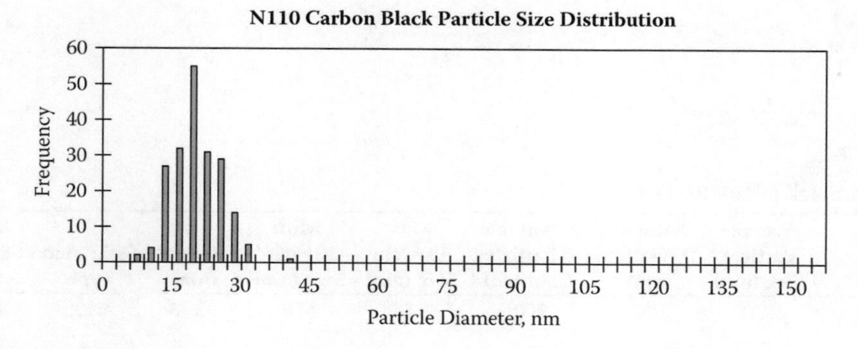

FIGURE 2.23
Particle size histogram of N-110.

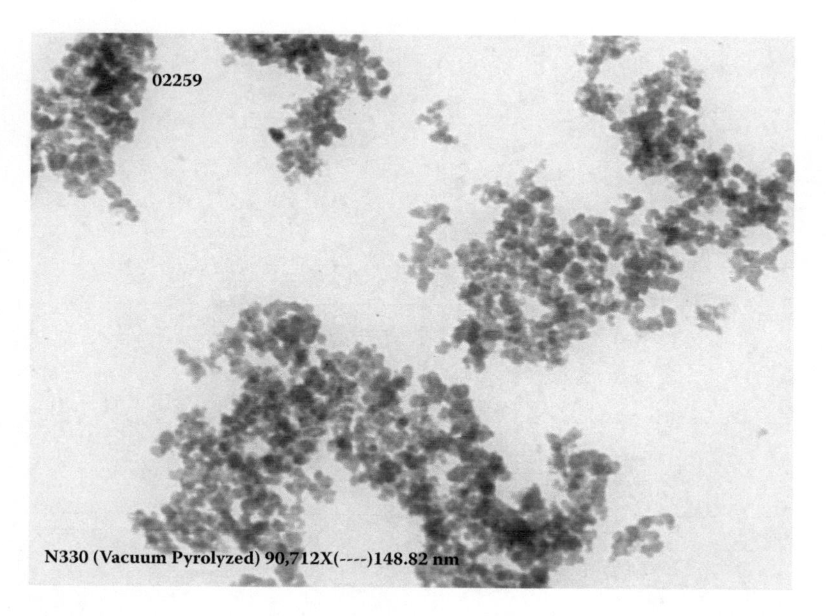

FIGURE 2.24
TEM image of N-330.

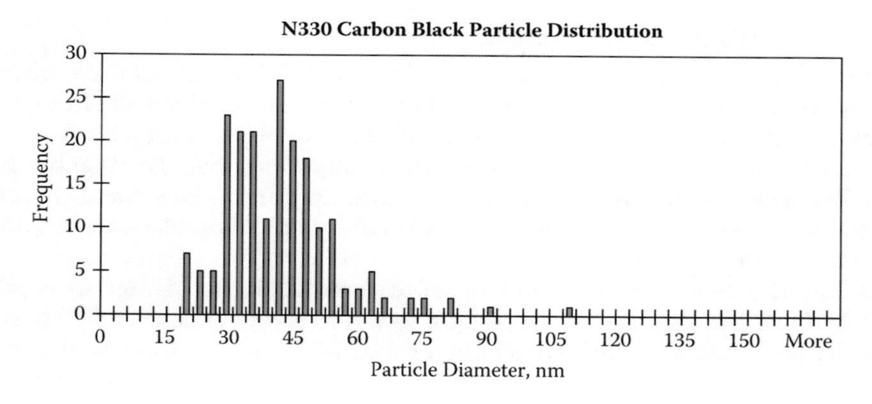

FIGURE 2.25
Histogram of N-330.

TABLE 2.28

Carbon Black Diameter Data

Sample	Average Particle Size (nm)	Standard Deviation (nm)	n (Number of Particles Counted)	Max Particle Size (nm)	Min Particle Size (nm)	Range (nm)	Estimated Type	ASTM Standard (nm)
Cabot N110	17.17	5.24	200	38.50	4.78	33.73	N100	11–19
Cabot N330	33.03	8.97	200	73.05	6.75	66.30	N300	26–35

TABLE 2.29
Elastomer Blends

Compound Number	Blends
Compound A	70/30 Natural rubber/EPDM
Compound B	70/30 Natural rubber/polybutadiene

Note: Each blend contained 40 phr of N-650 carbon black.

black was determined to be of N-330. For more examples of carbon black determination, see Coz and Baranwal in the Bibliography.

2.6.3 Elastomer Blend Morphology by TEM

2.6.3.1 Microtomy, Staining, and TEM Analysis

Two compounds (compounds A and B listed in Table 2.29) were mixed in the Banbury with 40 phr of N-650 carbon black and suitable curing agents. Samples were cured for TEM analysis. Samples of cured compounds containing polymer blends and carbon black were microtomed at temperatures below T_g of rubber (–120°C). The resulting sections were placed on uncoated copper grids and then exposed to osmium tetroxide (OsO_4) for staining. Images of stained and unstained sections were obtained to determine the effect of staining on samples. In general, for a blend of two elastomers, one with higher unsaturation would be stained more.

2.6.3.2 TEM Image Analysis—Examples

In our examples we used two cured elastomer blend compounds each containing 40 phr N-650 carbon black and appropriate curing agents. Elastomer blend ratios are given in Table 2.29. Cured samples were microtomed with an RMC Powertome XL with an RXL cryo attachment at a temperature lower than the glass transition temperature of polybutadiene (–120°C). The resulting microtomed sections were approximately 1000 angstroms thick. The sections were collected on a dry diamond knife and placed onto uncoated 400 mesh copper grids. Sections were then exposed to OsO_4 for 10 minutes.

The samples were analyzed with a Phillips CM12 TEM interfaced with a Gatan Model 673 TV camera and a Scion Image frame grabber. Photomicrographs of both stained and non-stained sections were taken to determine the effect of staining on the samples.

According to the literature, the polymer phases would accept the OsO_4 stain in the following order: polybutadiene rubber, natural rubber, and EPDM rubber. Therefore, the polybutadiene rubber would be viewed as being darker than the natural rubber, which would be darker than the EPDM rubber for stained samples. The carbon black would tend to go into the polymer phases in the same order—that is, polybutadiene rubber, natural rubber, and EPDM rubber—unless it is compounded in such a way that it would be forced into the different polymer phases prior to blending them together.

2.6.4 TEM Observations

TEM photomicrographs of unstained sections of compound A (a blend of NR/EPDM) show that the carbon black is located in the natural rubber phase and little or no carbon black is seen in the EPDM (white) phase (see Figures 2.26 through 2.28).

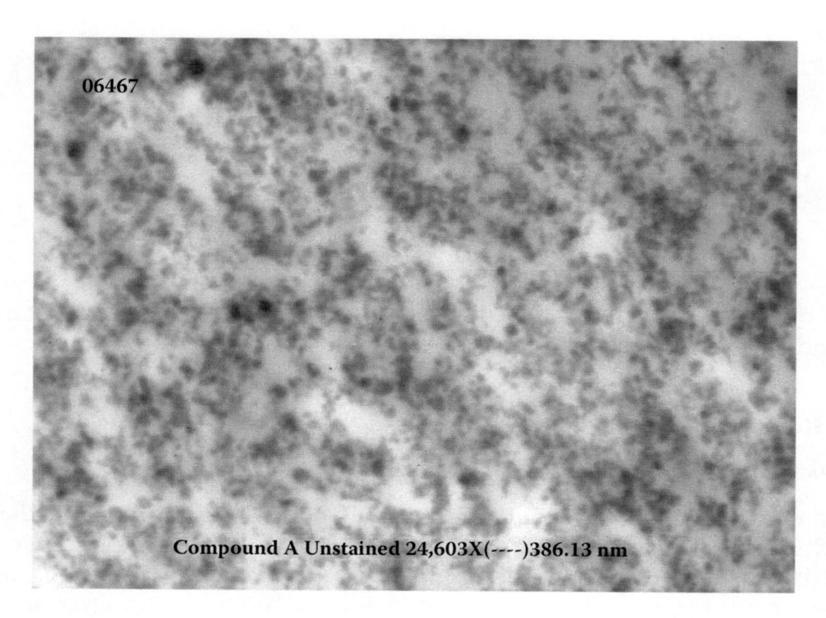

FIGURE 2.26
Compound A unstained TEM photomicrograph.

TEM photomicrographs of OsO_4 stained thin sections for compound A show that the EPDM is the dispersed unstained (white) phase and the natural rubber is the continuous stained phase (see Figures 2.29 through 2.30).

In the compound B (NR/polybutadiene) sample shown in Figures 2.31 through 2.36, the carbon black would tend to go into the polybutadiene phase rather than into the natural

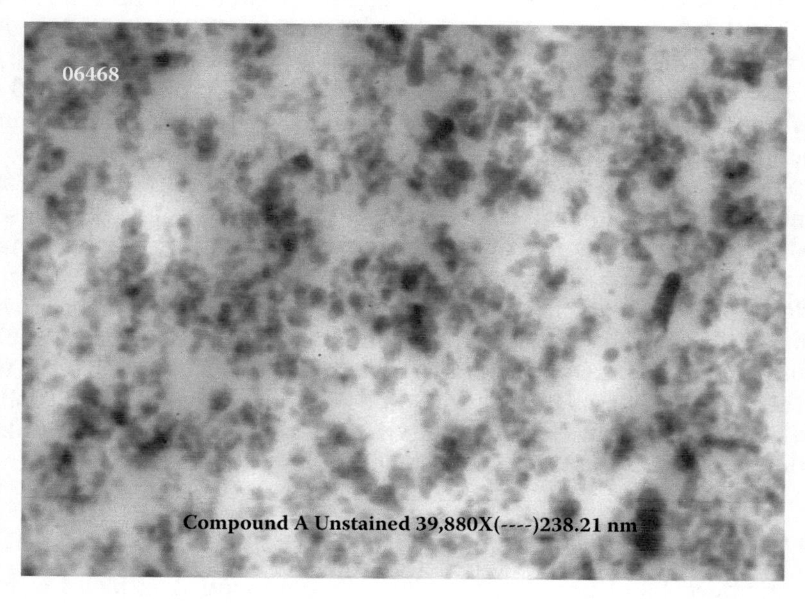

FIGURE 2.27
Compound A unstained TEM photomicrograph.

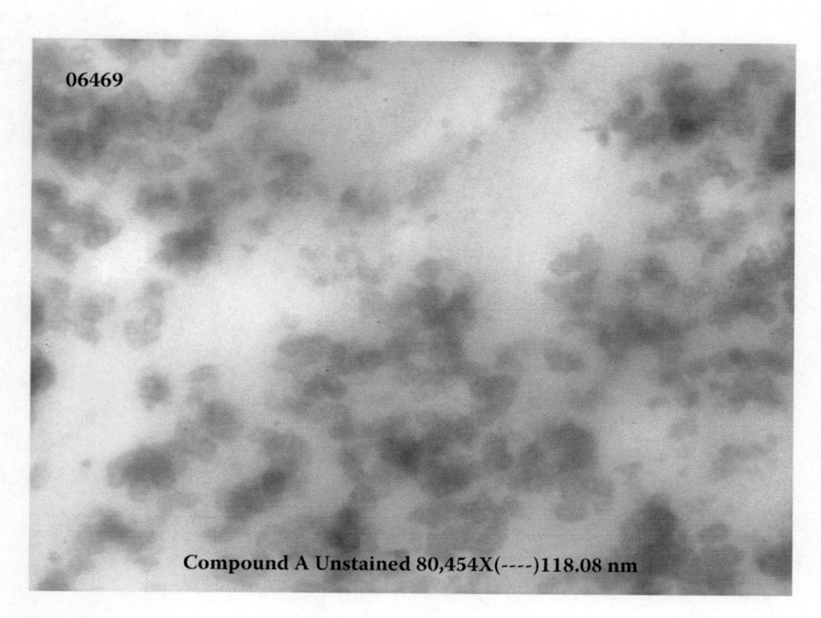

FIGURE 2.28
Compound A unstained TEM photomicrograph.

rubber phase, but some carbon black would be present in the natural rubber. The lighter areas seen in the unstained sections would likely be natural rubber (see Figure 2.33).

TEM photographs of OsO_4 stained thin sections for compound B show that the natural rubber and polybutadiene rubber are mostly co-continuous stained phases. Since both the natural rubber and the polybutadiene rubber accept the osmium stain, the lighter areas

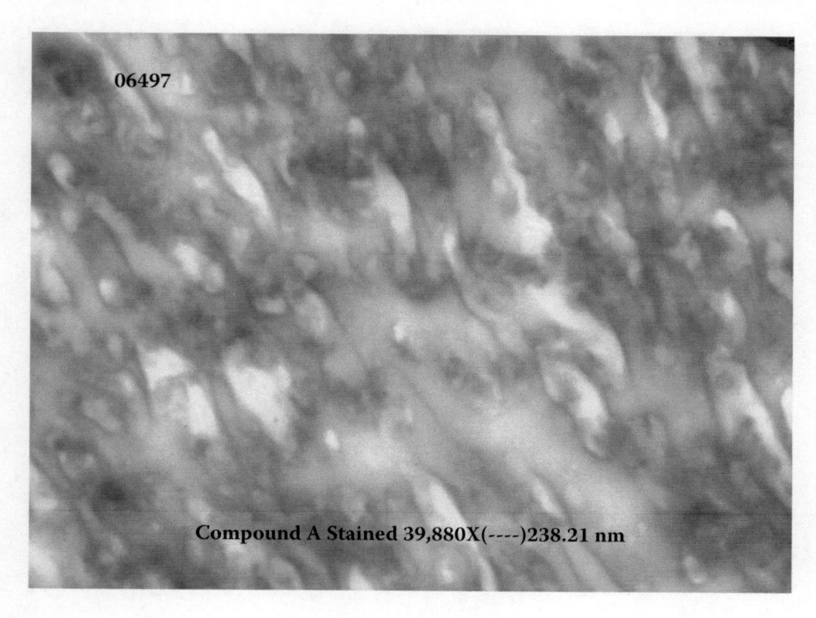

FIGURE 2.29
Compound A OsO_4 stained TEM photomicrograph.

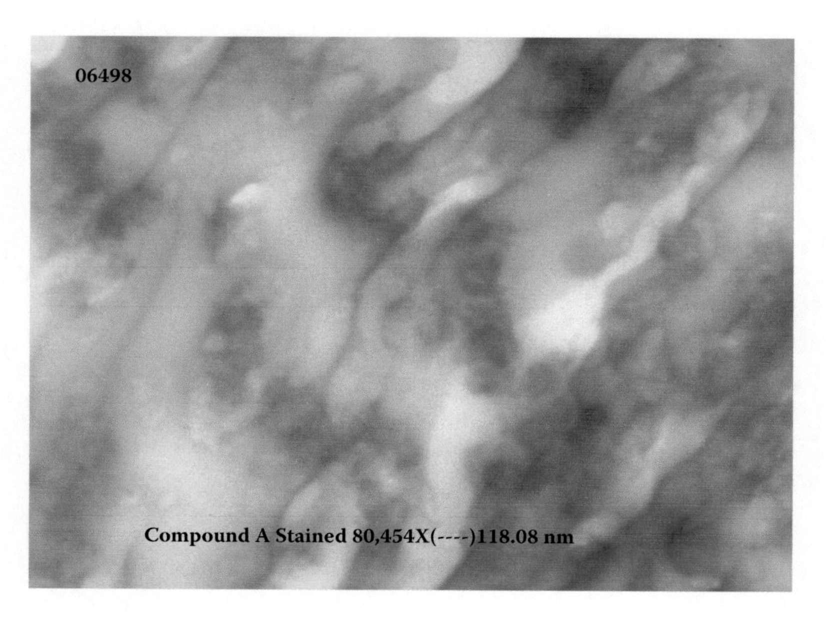

FIGURE 2.30
Compound A OsO_4 stained TEM photomicrograph.

of staining would be composed of mainly natural rubber, and the darkest would be composed mostly of polybutadiene rubber (see Figures 2.34 through 2.36).

Thus, unstained sample images showed the elastomer phase with higher unsaturation contained more carbon black, which was observed earlier by Hess et al. Similarly, stained sample images showed that the elastomer phase with higher unsaturation was stained more (darker phase).

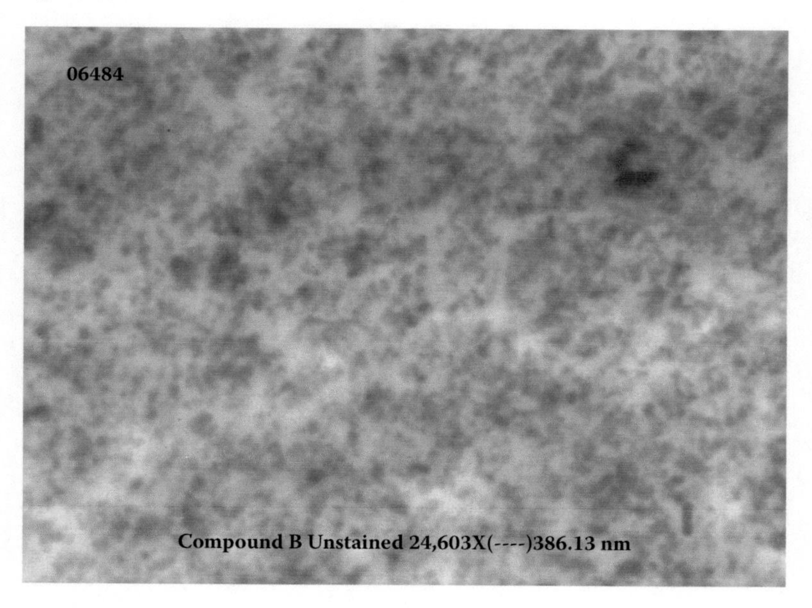

FIGURE 2.31
Compound B unstained TEM photomicrograph.

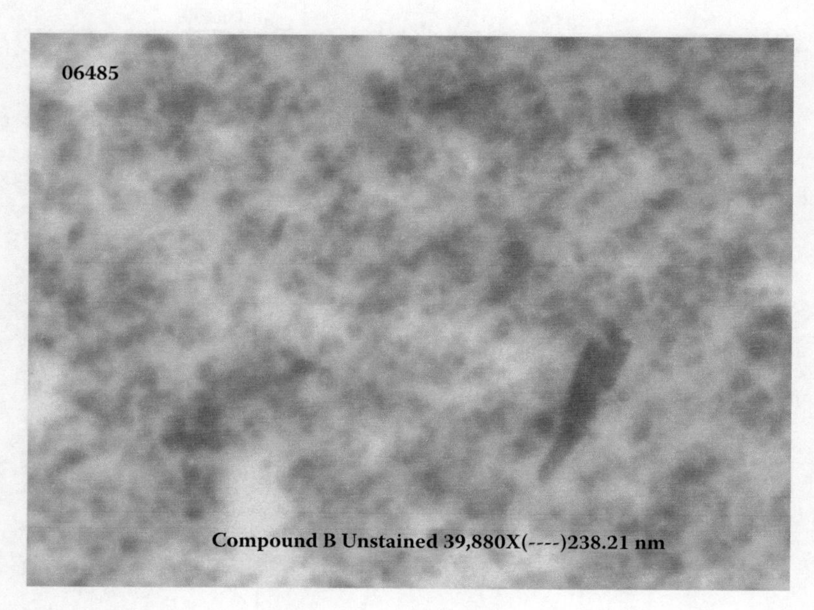

FIGURE 2.32
Compound B unstained TEM photomicrograph.

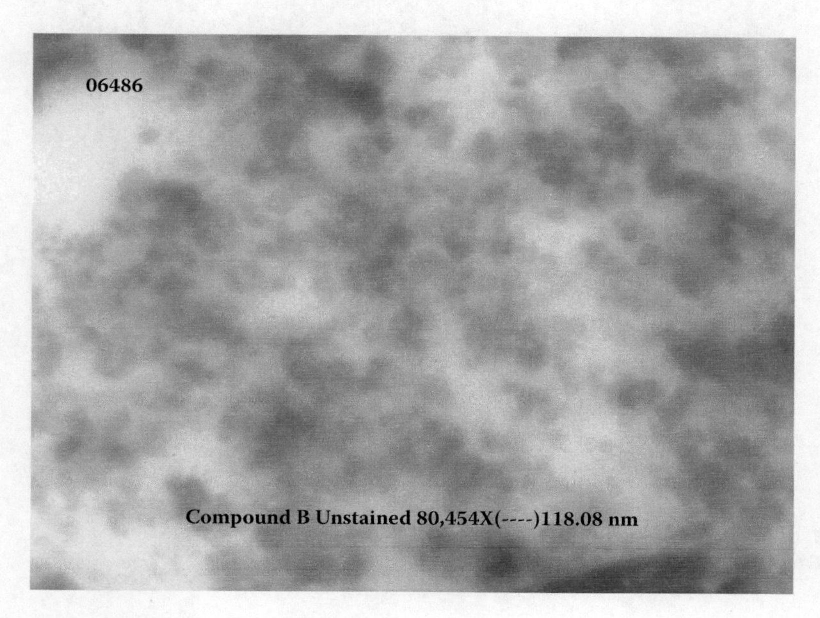

FIGURE 2.33
Compound B unstained TEM photomicrograph.

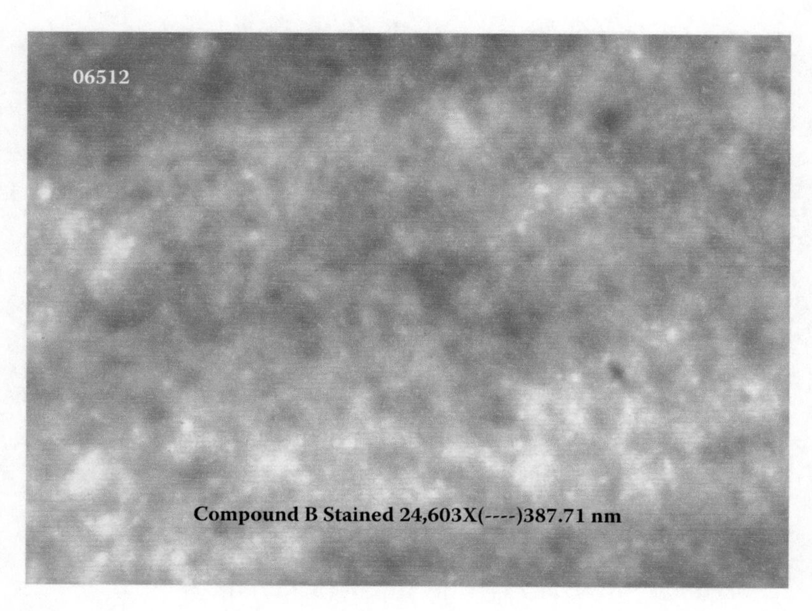

FIGURE 2.34
Compound B OsO_4 stained TEM photomicrograph.

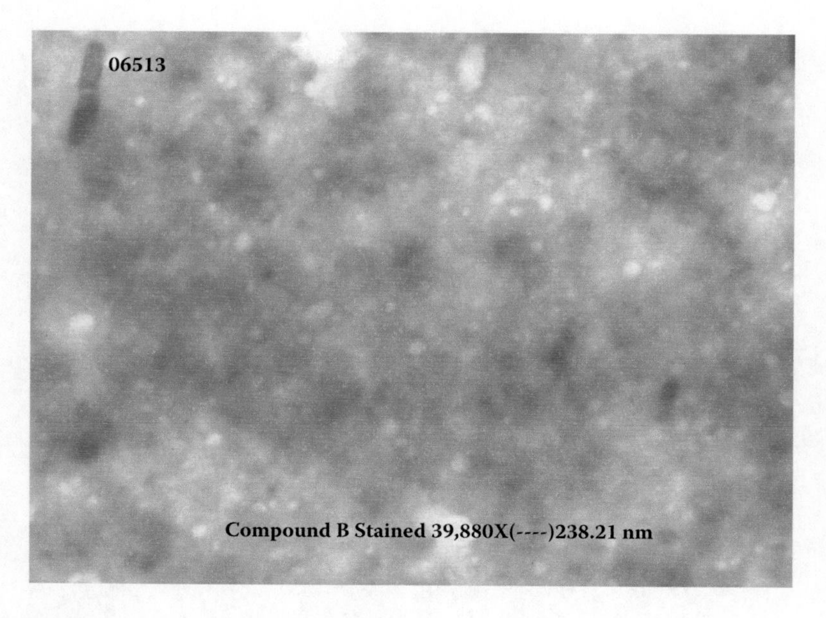

FIGURE 2.35
Compound B OsO_4 stained TEM photomicrograph.

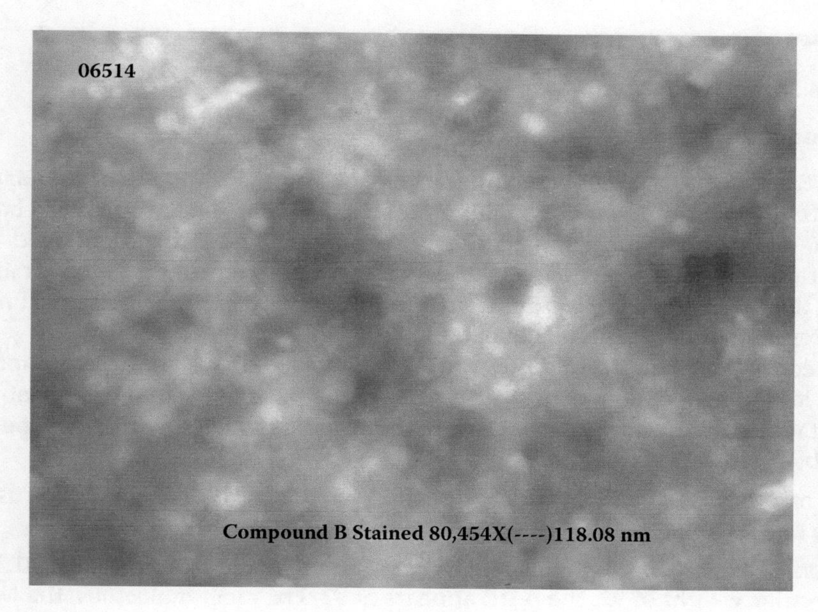

FIGURE 2.36
Compound B OsO_4 stained TEM photomicrograph.

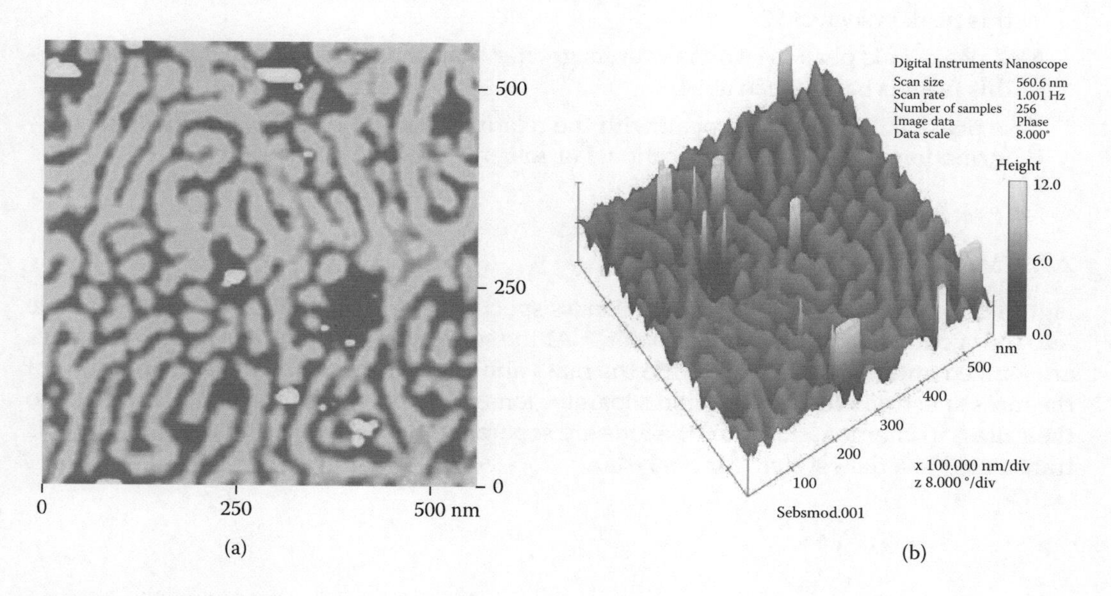

FIGURE 2.37
Tapping mode phase morphology of the nanocomposites. (a) SEBS-Clay at four parts. (b) Positions of clay layers of the same in the three-dimensional image.

These examples show that TEM technique is useful in confirming the presence of elastomer phases, especially if they have different unsaturation contents.

AFM in conjunction with TEM/3D-TEM is also used for understanding dispersion. Figure 2.37 shows a representative micrograph of a nanocomposite using clay.

2.7 Mass Spectrometry

2.7.1 Introduction

Mass Spectrometry is a powerful technique for identifying unknown molecular structure. It is essentially a technique for "weighing" molecules. Mass spectrometry is based upon the motion of a charged particle, called an ion, in an electric or magnetic field. The mass to charge ratio (m/z or m/e) of the ion affects this motion. Since the charge of an electron is known, the mass to charge ratio is a measurement of an ion's mass. Typical mass spectrometry research focuses on the formation of gas phase ions.

Mass spectrometry is used for determining the molecular weight of a compound. Mass spectrum identifies a cation radical or cation. If the particle of interest is not positively charged, it will not be seen. Some of the terms that are associated with mass spectrometry are described here.

m/z or m/e: Mass to charge ratio. For general purposes, the charge will be assumed to be one so that the value of m/z or m/e corresponds to the mass.

Mass ion peak: M+ is the molecular ion for the molecule. For toluene which has a molecular weight of 92, the peak appears at 92. For most molecules, the M+ peak is seen on the mass spectrum.

M+1 peak: This peak is one mass unit greater than the mass ion peak. For toluene, this peak is seen at 93.

M+2 peak: This peak is two mass units greater than the mass ion peak. For toluene, this peak is barely seen at 94.

Base peak: This peak is the peak with the relative intensity of 100%, which is due to formation of the most stable cation. For toluene, the base peak is at 91.

2.7.2 Method

Figure 2.38 shows a block diagram of a mass spectrometer. The inlet transfers the sample into the vacuum of the mass spectrometer. At the source region, neutral sample molecules are ionized and then accelerated into the mass analyzer. The mass analyzer is the heart of the mass spectrometer. This section separates ions, either in space or in time, according to their mass to charge ratio. After the ions are separated, they are detected and the signal is transferred to a data system for analysis.

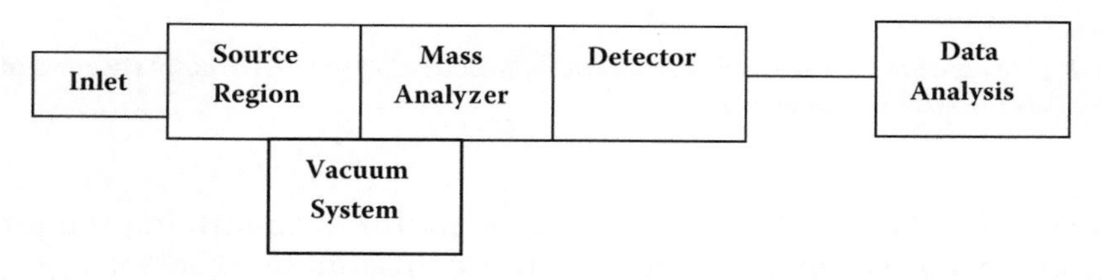

FIGURE 2.38
Block diagram of a mass spectrometer.

There are several methods to generate the cation for mass spectrometer, including the following:

EI: Electron ionization with a stream of high energy electrons (70 electron volts).

CI: Chemical Ionization is caused by a carrier gas (usually methane) reacting with electrons to make primary ions ($CH4^+$ and $CH3^+$). These ions then react with a test sample. This may be used when the mass ion peak does not show up with EI.

FAB: Fast atom bombardment is accomplished by using Xenon atoms.

ESI: Electrospray ionization concerns using a voltage across a spray coming out of a liquid chromatograph machine.

Soft Laser Desorption: This technique utilizes a laser source to form a cation from large molecules. An example of this technique is MALDI (Matrix Assisted Laser Desorption/Ionization).

Some of the common mass spectrometry systems are:

MS: The first type is to use just an MS to generate a mass spectrum. This is an older technique and has been overtaken by other methods.

GC/MS: This type stands for gas chromatography/mass spectroscopy. The GC separates compounds while the MS determines the mass spectrum for each component.

LC/MS: This type stands for liquid chromatography/mass spectroscopy. It is a newer method and uses an HPLC to separate compounds before they are analyzed.

MS/MS: This type stands for having two mass spectral machines in tandem. The first MS separates the molecule into parent ions. The parent ions are then collapsed to form daughter ions in the second machine.

FT-MS: This type stands for Fourier Transform-Ion Cyclotron Resonance Mass Spectrum.

2.7.3 Interpretation of Mass Spectrum Data

2.7.3.1 The Molecular Ion

The molecular ion in an EI spectrum may be identified using the following rules:

1. It must be the ion of highest m/z or m/e in the spectrum, apart from isotope peaks resulting from its molecular formula. Conventionally in mass spectrometry, the molecular weight is calculated based on the masses of the most abundant isotopes of all the elements in the compound. For instance, for chlorothiophene we calculate the molecular weight as

$$^{12}C_4\,^{1}H_3\,^{32}S_1\,^{35}Cl_1 = 118 \text{ Da}$$

   The possibility of higher m/z or m/e peaks due to impurities should also be considered.

2. In an EI spectrum, the molecular ion is formed by loss of an electron; therefore, one electron is unpaired, making it a radical species. Such ions are known as odd-electron (OE) ions. If the molecular formula of an ion is known (or suspected), the "rings and double bonds" formula may be used to determine whether it is an even- or odd-electron species. This formula may be generalized as follows:

For a formula $C_x H_y N_z O_n$, the number of rings plus double bonds is given by $x - y/2 + z/2 + 1$. This calculation will yield a whole number for an odd-electron ion and a number ending in 1/2 for an even-electron ion. Other atoms in the formula are counted as equivalent to whichever of C, H, N or O they correspond to in valence. For instance, Si is equivalent to C, halogens are equivalent to H, P is equivalent to N.

3. The ion must obey the "nitrogen rule." This states that for compounds containing most of the elements common in organic chemistry, for a molecule to have an odd numbered molecular weight it must contain an odd number of nitrogen atoms. If a molecule contains an even number of nitrogen atoms or no nitrogen atoms at all, then it will have an even-numbered molecular weight. This fact arises from the fact that nitrogen has an even mass and an odd valency.
4. The assumed molecular ion must be capable of producing the high mass fragments in the spectrum by plausible losses of neutral fragments.

2.7.3.2 Isotopes

Many elements have more than one stable isotope. The effect of naturally-occurring isotopes is to produce peaks in the mass spectrum other than the "main" peak (the peak due to the most abundant isotopes). For example, in the spectrum of chlorothiophene, the peaks at m/z at 118 and 120 represent the contributions from the two isotopes of chlorine, ^{35}Cl and ^{37}Cl.

Elements that have isotopes at "A+2" values are chlorine, bromine, oxygen (very weak, only 0.2%), silicon and sulfur. Of these, chlorine and bromine are readily recognizable. Elements that have isotopes at "A+1" values are carbon and nitrogen.

2.7.3.3 Accurate mass measurements

Using the appropriate instrument and techniques, the masses of ions may be determined with accuracy of the order of <10 ppm. This information can be used to determine the elemental composition of any ion due to the fact that the exact masses of the isotopes of element are, with the exception of the carbon 12 standard, not whole numbers. e.g. the "nominal" mass of acetone (C_3H_6O) is $3 \times 12 + 6 \times 1 + 1 \times 16 = 58$.

The exact mass of this elemental combination, taking the mass of ^{12}C as 12.000000, ^{1}H as 1.0078246 and ^{16}O as 15.994915 is 58.04186. Using accurate mass measurement, this can be readily distinguished from the mass of butane (C_4H_{10}), which is 58.07825.

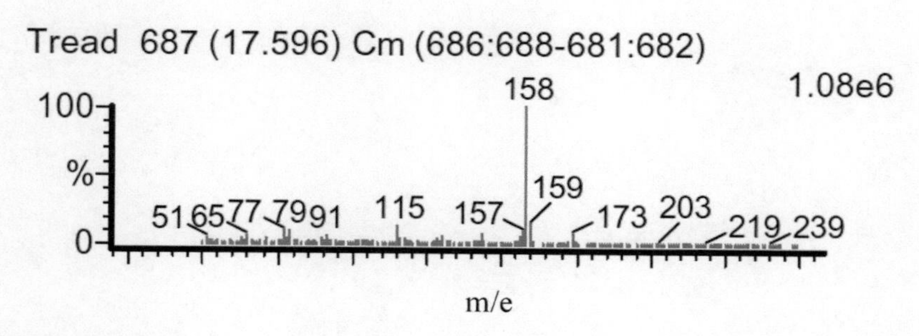

FIGURE 2.39
Mass spectrum of an antidegradant.

2.7.3.4 The General Appearance of the Spectrum

The intensity of the molecular ion usually parallels the chemical stability of the molecule, and compounds with high unsaturation (numbers of rings and double bonds) show the most abundant molecular ions. For aromatic compounds containing an alkyl chain, the abundance of the molecular ion will decrease with increasing length of the alkyl chain. A representative mass spectrum is shown in Figure 2.39.

3

Principal Physical Test Methods

3.1 Introduction

Rubber has a wide range of unique properties and it is necessary to utilize them in a variety of products and applications, like tires, hoses, conveyor belts, bridge bearings, shoe soles, cables, mounts, gaskets, seals, rocket insulation, etc.

Rubber differs considerably from other engineering materials—for example, it is a highly deformable material, exhibiting virtually complete recovery, and it is nearly incompressible with a bulk modulus some thousand times greater than its shear modulus.

Why is it necessary to carry out many complicated tests on rubber-like materials? The simplest answer is that rubber cannot be described by the mathematical laws applicable to ideal materials such as Hookean solids and Newtonian fluids, and is a material that possesses complex properties that are not related to each other.

Rubber differs from metals in that it does not obey Hooke's law except at very small extensions (up to 50%), and there is no elastic limit that can be specified for "safe loads." The stress/strain curve for vulcanized rubber is "S" shaped, and this shape can be modified by the temperature of the test, speed of testing, dimensions of sample, and its previous history. There is a large hysteresis effect when rubber is deformed. This is largely due to the Gough-Joule effect that is evidenced by the fact that rubber heats upon stretching and cools on retraction. The evolution of heat utilizes some of the energy input, and the retraction stress/strain curve does not follow the same course as the extension curve.

One of the chief differences between rubber and other engineering materials is in the large elastic strains to which the former can be subjected without rupture. The engineer wishing to make use of this capacity of strain encounters two difficulties:

1. The normal methods of engineering calculations fail.
2. The material properties, on which normal engineering design is founded, cannot be measured easily or directly under conditions of considerable strain.

The reasons for failure of the rules of normal engineering design are that most of these rules assume that change of shape under load is small and that deformation at the loading or gripping points is also small. Neither assumption is true for rubber. The result is that the average engineer cannot forecast the mechanical behavior of a sample of rubber, even if he has been supplied with the best available data regarding its fundamental physical properties of stiffness and viscosity.

The basis for collaboration between a conventional engineer and a rubber technologist lies in the understanding of each other's problems. For instance, the rubber technologist expects that the engineer should be familiar to some degree with the nature of rubber-like materials, the general effects of compounding ingredients, and the difficulties of fabrication and mold design. The engineer expects the rubber technologist to have some knowledge of basic applied mechanics in order for the engineering factors of the problem to be understood. The rubber technologist has often been found to prefer to experiment with rubbers of different compositions, when it becomes clear that the original mechanical design requires modification.

The general engineer's first difficulty arises because with metals he has become used to a small strain proportional to stress. Young's modulus or the modulus of rigidity (stress/strain) is readily determined and available for use in design calculations, within the safe working range of metal. For rubber in either extension or compression, the stress/strain relationship is variable. By *modulus,* the rubber technologist means the stress at a given elongation (e.g., 100%, 300%). *Flexure* to an engineer means bending, but to some rubber technologists it means any form of straining. *Aging* to an engineer means stress relieving before final machining; to the rubber technologist *aging* means deterioration with age. *Resilience* to an engineer means the energy stored per unit volume; in rubber it means the rebound property or coefficient of restitution. It is important that these and other similar differences are appreciated.

With such terms as hardness, fatigue, and creep, we are at least on fairly common ground. Hardness with a metal is assessed by measurement of the permanent indentation produced by a hardened ball or diamond at a given load. Rubber hardness is resistance to elastic indentation. Creep, which is change of strain at constant stress, normally occurs with rubber but is not noticeable with metals except at elevated temperatures.

Fatigue is probably the most important factor in the dynamic application of rubber or a metal. With a metal, stress reversals do no serious harm; in rubber design they should be avoided by pre-loading. Poisson's ratio applies to both metals and rubber. It is important that Poisson's ratio of 0.5 makes rubber virtually incompressible.

The conventional engineer should know that rubber has the following advantages: strength; good energy absorption; the ability to undergo large deformations and recover (160,000 times more elastic than steel in shear); good electrical resistance properties; resistance to fatigue, abrasion, and corrosion; and moldability.

The disadvantages are that it can be attacked by oils and greases; depending on the structure, it is susceptible to aging which is accelerated by exposure to heat or light; it can be attacked by the ozone; and it is a poor conductor of heat. The first three of the above can be more or less avoided by using a suitable type of synthetic rubber.

From the above considerations it is clear that the elastic behavior of rubbers differs fundamentally from that of metals. Deformation of a metal involves changes in the inter-atomic distances, and very large forces are needed to change these distances, hence very high elastic modulus characteristics of a metal. The forces involved are so great that before the deformation reaches a few percent, other actions come into place involving slippage between adjacent crystals. In other words, a metal shows a yield point above which the deformation increases much more rapidly than the stress, so that the stress/strain curve turns away from the stress axis (Figure 3.1). Moreover, the deformation above this point is irreversible. With rubber, on the other hand, the stress/strain curve (Figure 3.2) bends the other way, as explained above, and there is no yield point. The rubber recovers to almost its original form from any point on the stress/strain curve. Moreover, the deformation of rubber does not involve any straining of the inter-atomic bonds, and hence the force required is lower than that of a metal.

It is essential that selected tests in some way relate to the functions a rubber has to perform. In other words, we should be able to assume that the results are related to

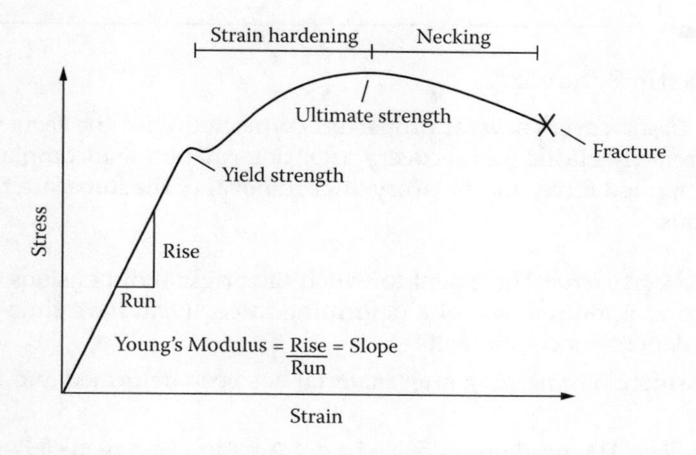

FIGURE 3.1
Typical stress/strain curve of metal.

performance at least to the extent that if the properties tested remain constant (or above predetermined values), its performance will be satisfactory. This is the crux of the whole question of laboratory tests, because the relation between basic properties and performance is only imperfectly understood, so that the selection of the most suitable tests calls for careful judgement.

Laboratory tests can be divided into three main groups:

1. Simple destruction tests and those involving non-recurrent cycles of loading (e.g., tensile strength, compression and shear modulus, bond strength, tear resistance, plasticity, hardness, resilience, permanent set and creep, low-temperature flexibility, and swelling)
2. Service tests (e.g., fatigue, heat buildup, abrasion, flexing, and aging)
3. Development tests (e.g., dynamic stiffness and dynamic resilience, serviceability tests)

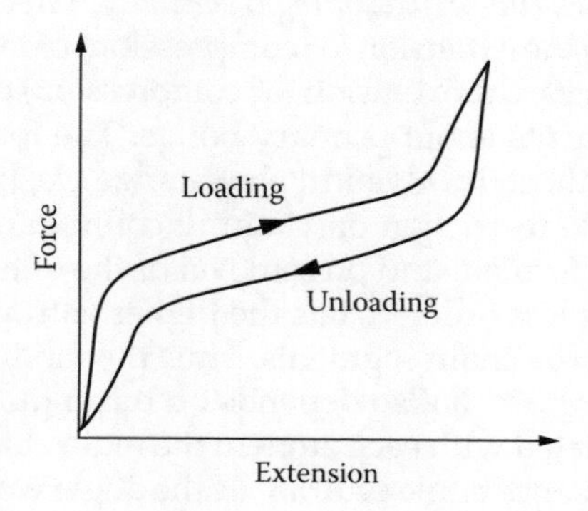

FIGURE 3.2
Typical stress/strain curve of rubber.

3.2 Visco-Elastic Behavior

The term *visco-elastic* covers several properties connected with the facts that most polymers are not perfectly elastic (i.e., recovery after deformation is incomplete) and that the response to an applied force, and recovery after removal of the force, are time dependent, not instantaneous.

Recovery (elastic recovery): The extent to which the original dimensions of a material are recovered upon removal of a deforming force. It can have time-independent and time-dependent components.

Set: The deformation remaining after material has been deformed and then allowed to recover.

Creep or cold flow: The gradual increase in deformation in a material subjected to a constant force. It also indicates the dimensional stability of a rubber product.

Stress relaxation: The proportion of the applied energy returned during a cycle of deformation and recovery. It is measured as a decrease in retractive force of a cured rubber product held at constant deformation. It depends on the rate of deformation (i.e., the cycle time).

Hysteresis: The energy lost during a given cycle of deformation and recovery.

Heat buildup: The heat generated in a succession of cycles of deformation and recovery, due to the conversion of hysteresis into thermal energy. It is usually measured by the rise in the temperature of the test specimen. The deformation of a polymeric material is a complex process involving the following phenomena:

Short-range (high-modulus) elasticity: This involves the stretching (or compression) of bonds between atoms and also of bonds between molecules and deformation of bond angles. It is important in rigid materials (e.g., polymers that are crystalline or below their glass temperature). It extends over only a very short range of deformation, and the modulus (Young's) is high and only slightly affected by temperature, Hooke's law is obeyed, and the strain is in phase with the stress (i.e., produced, or disappearing, instantaneously on applying, or removing, the deforming force). This kind of elasticity can be linked to (is, in fact, nearly identical to) that of a helical steel spring.

Long-range (low-modulus, rubber-like, or high) elasticity: This property, unique to high polymers, is due to the extension (or compression) of randomly coiled or kinked long-chain molecules, the extension (or compression) occurring largely by rotation of chain segments about primary bonds. The forces are many thousandfold smaller than those involved in short-range elasticity; hence a rubber—in which there is little restriction on chain uncoiling and coiling—can be easily and extensively deformed, and on removal of the deforming force the original state (which, being less ordered, has the higher entropy) is quickly restored by thermal motion of the chain segments. Thus the modulus is low and markedly affected by temperature. It also depends on the angular disposition of the primary bonds associated with each atom in the main chain. For carbon chains the bonds are at the valence angle of 109.5° (if the angle were 180°, the chains would have only a rod-like configuration).

Long-range deformation and recovery from it are however delayed in action (i.e., time dependent), so that the strain is more or less out of phase with the stress, causing hysteresis and heat buildup. This kind of elasticity can be linked to that of a system consisting of a helical spring and an oil dash-pot in parallel.

Non-elastic (viscous or plastic) deformation: Stress applied to a linear polymer, or to any one in which the molecules are not firmly and permanently crosslinked, may bring about relative displacement of the molecules which is irrecoverable. Creep, cold flow, and stress relaxation, also set, at least in part are associated with such rheological behavior, which can be linked to that of a dash-pot (movement in such being irrecoverable in the absence of a spring).

Short-range (high-modulus) elasticity, though often marked by the effects of high elasticity, is a component of the deformation of any polymeric material. It is in part responsible for the behavior of bulked textile yarns; and though it may also be delayed in action in a complex system, it is important in the crease recovery, smooth-drying, and crease retention of textile fabrics.

Long-range (low-modulus or high) elasticity of a very evident kind is particularly manifest in unfilled rubbers that have been lightly vulcanized (crosslinked) to prevent non-elastic flow. These, at normal or moderately elevated temperatures, show recovery that is substantially complete and instantaneous ("snap"). However, below the glass transition temperature such materials become rigid (i.e., lose their long-range elasticity).

3.3 Elastic Modulus

Elastic modulus is the ratio of stress to strain, measured within the range where deformation is reversible and proportional to the stress. Young's modulus is the ratio of the tensile (or compressive) stress to the extension (or compression) strain, i.e.,

$$E = (F/a)/(e/L) = FL/ae \quad (3.1)$$

where F/a is the ratio of the tensile (or compressive) force to the initial cross-sectional area and e/L is the ratio of the length increase (or decrease) to the initial length. For materials subjected to shear deformation (i.e., relative displacement of two parallel to them), the shear or rigidity modulus, G, applies. This is represented by Figure 3.3.

$$G = \text{shear stress/shear strain} = \text{(force/area of one face)/(displacement/distance between faces)} \quad (3.2)$$

It is less than Young's modulus, being related to the latter; thus, $E/2G = (1 + \text{Poission's ratio})$; for rubber, Poisson's ratio for small strains is 0.5 and G is thus about one-third of E.

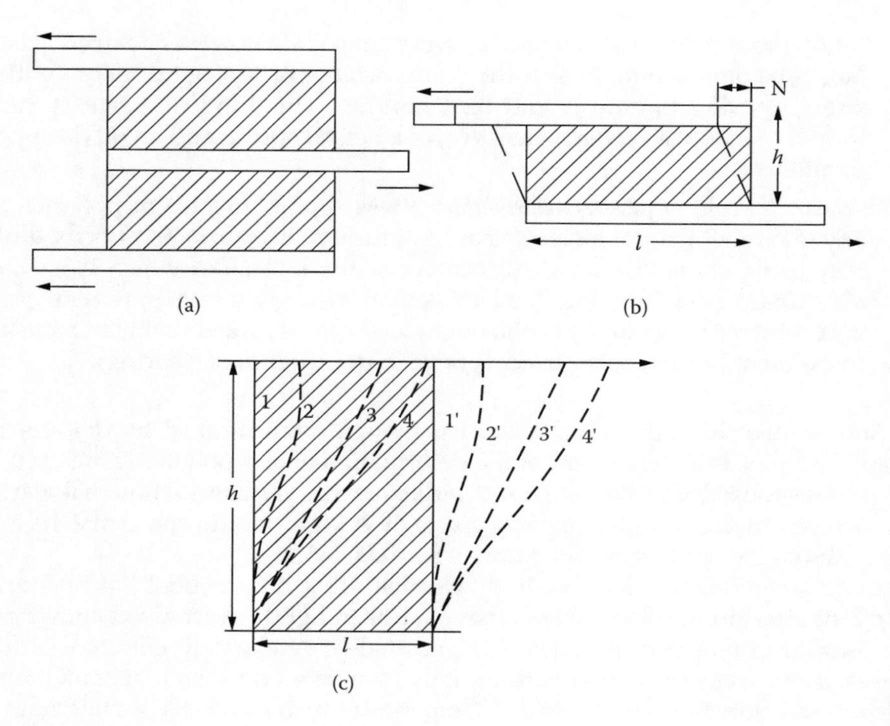

FIGURE 3.3
Typical shear specimen.

Elastic modulus (Young's or shear) has the same dimensions as stress and is listed in kgf/mm² or mP$_a$; for fibers it can be expressed in gf/denier. Poisson's ratio is lateral strain divided by longitudinal strain in a material subjected to stress; thus, in a stretched tensile test piece it equals (the proportional decrease in width or thickness)/(the proportional increase in length).

The value of Young's modulus indicates the resistance of a material to reversible longitudinal deformation. It can be considered as the theoretical stress required to double the length of a specimen, but this is not realized in practice because either the material breaks short (e.g., glass, most materials and hard plastics), or the stress/strain relationship is not linear. For most plastics materials, Young's modulus is less than 1/10 that of metals, while for rubbers it is only 1/10,000 or less; some fibers, however, have a modulus approaching that of metals. For materials in the rubber-like (high-elastic) state, elastic modulus increases with an increase in the degree of crosslinking. Young's modulus can be derived from measurements of extension under load or of the bending of a rod or beam or of deformation under compressive load. The strain should be kept small so as to maintain a linear stress-strain relationship. This applies particularly to compression, where the linear range is especially small if the shape factor (ratio of cross-sectional dimensions to height) is large, and if the end faces of the test piece cannot slip freely over the compressing surfaces (the test piece then "barrels" and in shear gives a bigger linear range), so that measurements of shear modulus are often advantageous. The measurement of complex modulus is also useful, because it can be analyzed into an in-phase or storage modulus and an out-of-phase or loss modulus, corresponding respectively to components of the stress in phase and 90 degree out of phase with the applied strain; the out-of-phase modulus determines the energy loss in cyclic deformations. The ratio of out-of-phase modulus to in-phase modulus is the loss tangent (tan δ, δ being the loss angle). A related quantity is the internal friction, equal to the out-of-phase modulus divided by the angular frequency of the deformation cycles. Both the in-phase and out-of-phase moduli, especially the latter, increase with an increasing rate of deformation. Hence, modulus measured at high deformation rates, notably under "dynamic" conditions (e.g., impact vibration, rapidly repeated deformation cycles), is higher than the "static" modulus measured under equilibrium conditions or by slow deformation.

Bulk modulus is the ratio of the change in external pressure to the change in column for reversible conditions, and it indicates the resistance of a substance to volume compression. It is high in organic polymers; materials appear in about the same order as for Young's modulus, but the values are higher, sometimes greatly so. Thus, the bulk modulus for soft rubbers is similar to that for water and the less compressible organic liquids (i.e., initially 200 kgf/mm^2, as compared with Young's modulus 0.2 kgf/mm^2). The rubber technologist's "modulus" is the stress (calculated on the initial cross section) at a stated elongation, usually a multiple of 100%. It is not a modulus in the strict sense and is better called "stress at X% elongation," or simply "X% stress value."

Hardness as measured on vulcanized rubber is essentially a function of elastic modulus.

The preparation of test pieces is as follows:

- A test piece of proper dimension must be prepared prior to each test.
- Direct molding can be possible from a mixed compound.
- Specimens are needed to be cut, sliced, or buffed from the finished products.

3.3.1 Effect of Mixing and Molding

Processing variables can affect to a great extent the results obtained on the final product; different physical tests are carried out in order to detect the results of these variables, such as state of cure and level of dispersion.

Mixing is carried out in open two-roll mills and/or internal mixers, like Banbury or Intermix types following the standard methods.

The conditions and time of storage between mixing and vulcanization can affect the properties; hence it is necessary to store the material in a dark and dry atmosphere. The preferred conditioning time is 24 h.

To get a better idea from the laboratory test results with the full-sized factory equipment, the tightest possible control on equipment, times, temperatures, and procedures is necessary.

3.3.2 Effect of Cutting/Die Cut from Sheet

The accuracy of the final test result depends considerably on the accuracy with which the test piece was prepared. The first requirement is that the test piece should be dimensionally accurate.

It is essential that cutters be sharp and free from nicks or unevenness in the cutting edge as that can produce flaws in the test piece resulting in premature failure. Blunt knives lower tensile strength on ring test pieces by 5-10%.

It is normal to restrict stamping to sheets no thicker than 4 mm as the "dishing" effect becomes more severe as the thickness increases.

Rotary cutters can be used to produce discs or rings from thin sheets and are necessary for sheets above about 4 mm thick to prevent distortion.

A lubricant that has no effect on the rubber can be applied during cutting the sample, particularly when using a rotating cutter.

3.3.3 Test Pieces from Finished Products

It is desirable to test wherever possible on the actual finished product rather than on specially prepared test pieces that may have been produced under rather different conditions. To obtain a test piece from a finished product, it is necessary to cut a large block and then reduce the thickness and remove irregularities by using the buffing/slitting machine. Some examples of test pieces are shown in Figure 3.4.

The particular disadvantage of buffing is that heat is generated which may cause significant degradation of the rubber surface. The effect of buffing on tensile properties (drop) on soft rubbers is 10-15%, whereas for a tire tread type the drop is about 5-10%.

A discussion of all the test methods available to the rubber compounder is not intended. Here we discuss only a few of the most common tests, their significance, and equipment used, considering the limited operating budget available to small/medium-scale manufacturers. Table 3.1 presents processability test information, and Table 3.2 presents information for vulcanizate testing. The samples are cured at optimum cure time determined from a rheometer (see Figure 3.5).

3.4 Some Special Features of General Physical Tests

3.4.1 Tensile Stress-Strain

Tensile strength is the maximum tensile stress reached in stretching a test piece, usually a flat dumbbell shape, to its breaking point. By convention, the force required is expressed as force per unit area of the original cross section of the test length.

Elongation, or strain, is the extension between benchmarks produced by a tensile force applied to the test piece and is expressed as a percentage of the original distance between the marks. Elongation at break, or ultimate elongation, is the elongation at the moment of rupture.

Tensile stress, or modulus, is the stress required to produce a certain elongation. Thus if a stress of 7 MPa produces an elongation of 200%, the rubber is said to have a 200% modulus of 7 MPa.

Unlike metals, the stress in a rubber is not directly proportional to strain, and modulus is therefore the stress at a certain strain. It is neither a ratio nor a constant but merely the coordinates of a point on the stress/strain curve.

The theoretical strength of polymers, calculated from bond energies, is some 10 times higher than tensile strength found in practice. The discrepancy is ascribed to randomness in the molecular configuration and to minute faults, voids, and inhomogeneities at which

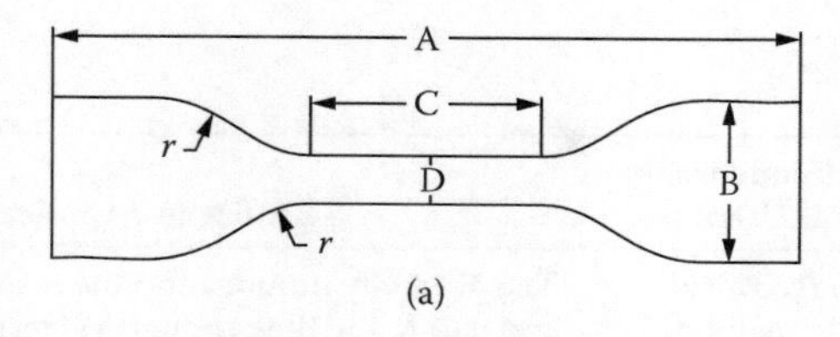

Dimension in mm	Type 1	Type 1A	Type 2	Type 3	Type 4
A Overall length	115	100	75	50	35
B Width of ends	25 ± 1.0	25 ± 0.5	12.5 ± 1	8.5 ± 0.5	6 ± 0.5
C Length of narrow portion	33 ± 2	20 + 2/−0	25 ± 1	16 ± 1	12 ± 0.5
D Width of narrow portion	6 + 0.4/−0	5 ± 0.1	4 ± 0.1	4 ± 0.1	2 ± 0.1
E Transition radius outside	14 ± 1	11 ± 1	8 ± 0.5	7.5 ± 0.5	3 ± 0.1
F Transition radius inside	25 ± 2	25 ± 2	12.5 ± 1	10 ± 0.5	3 ± 0.1
Gauge length	25 ± 0.5	20 ± 0.5	20 ± 0.5	10 ± 0.5	10 ± 0.5

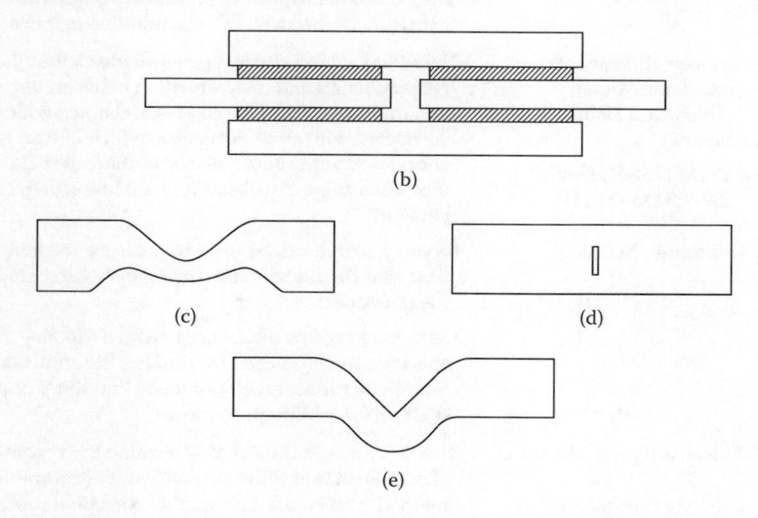

FIGURE 3.4
Some test pieces for the mechanical testing of rubbers: (a) tensile test piece, (b) quadruple shear test piece, (c) crescent tear test piece, (d) delft tear test piece, and (e) angle tear test piece.

concentration of stress initiates breakdown. A typical stress/strain curve of a vulcanized rubber is shown in Figure 3.6.

In practice, the tensile properties—strength, elongation, stress at fixed elongation—depend on many factors: type of polymer, degree of polymerization, degree of crystallinity, molecular orientation of crosslinks, the reinforcing effect of fine particle fillers, the presence of plasticizers, strain rate during testing, and whether the test uses constant strain rate or constant rate of stress and size of test piece (since the chance of occurrence of a weak spot increases with the volume of material tested).

It is of interest to note the order that different types of substances assume when arranged in comparison tables according to strength. In this respect the appearance of vulcanized rubbers among the weak materials, when some of them are quite strong, is apt to be

TABLE 3.1

Processability Tests

Sl Number	Type of Test (Standards) [Equipment Used]	Significance/Application of the Test
1	Plasticity and recovery (parallel plate method) (ASTM D 926) [Wallace Plastimeter]	The plasticity number and the recovery of the test specimen are related to the flow properties, respectively, and may be useful in predicting the processability characteristics such as ease of forming and extrusion characteristics.
2	Extrudability of unvulcanized compounds (ASTM D 2230) [Brabender Plasti-coder]	One objective in the extrusion of rubber compounds is to obtain a smooth extrusion that closely reproduces the contours of the extrusion die. This test method provides for a subjective determination of this under controlled conditions. As the rating is subjective, it does not lend itself readily to incorporation as a specification requirement. It does not measure other aspects of extrudability such as rate of extrusion or die swell in a quantitative manner.
3	Mooney viscosity, stress relaxation, Mooney scorch, and Delta Mooney (ASTM D 1646, SS-ISO 289, ASTM D 3346) [Monsanto MV2000]	The Mooney viscosity test gives an idea about the processing characteristics and the correlation between viscosity values and molecular mass. Visco-elastic characteristics can be correlated with stress relaxation which in turn has a correlation with rubber structure characteristics such as molecular mass distribution, chain branching, and gel content. Mooney scorch can be used to measure the incipient cure time and the rate of cure during very early stages of vulcanization. Delta Mooney (the difference between Mooney viscosity at two specified times) is used for ranking the emulsion styrene-butadiene rubber (SBR) polymers that differ appreciably in their processability properties.
4	Vulcanizing characteristics (ASTM D 2084, ASTM D 5289, SS-ISO 3417, SS-ISO 6502) [Monsanto MDR2000]	This test method is used to determine the vulcanization characteristics of (vulcanizable) rubber compounds. This test method may be used for quality control in rubber manufacturing processes, for research and development testing of raw rubber compounded in an evaluation formulation, and for evaluating various raw materials used in preparing (vulcanizable) rubber compounds. The test piece in a rotorless cure meter (ASTM D 5289) approaches the test temperature in a shorter time, and there is a better temperature distribution in the test piece due to the elimination of the unheated rotor found in oscillating disk cure meters (ASTM 2084). A typical rheometer curve is shown in Figure 3.5.

misleading. It arises because the strength is normally measured with respect to the initial cross-sectional area. If it were given in terms of the cross section at break, the order of the substances would be changed, since the strength of relatively inextensible materials would not be altered.

The characteristics of rubbers and metals, in general, lie at extreme and opposite limits; thus, whereas a vulcanized rubber cord is extended by 50% of its length for a load of 0.05 kgf/mm^2, a load of 50 kgf/mm^2, extends steel by only 0.25% of its initial length.

TABLE 3.2

Physical Tests of Vulcanized Rubber

Sl Number	Type of Test (Standards) [Equipment Used]	Significance/Application of the Test
1	Specific gravity (IS 3400, part-IX)	It is a very quick test for checking any weighing error during mixing. This test determines batch-to-batch variation (i.e., quality control checking; also used for development purposes).
2	Hardness (ASTM D 2240, ASTM D 1415, SS-ISO 7619, SS-ISO 48) [Shore-A durometer, Wallace dead load hardness tester (IRHD)]	This test method is based on the penetration of a specific type of indentor when forced into the material under specified conditions. It is also defined as the modulus at low indention and also gives the idea about degree of curing. This is an empirical test for the control process. Although no simple relationship exists between indentation hardness and any fundamental property, but for substantially elastic isotropic materials like well-vulcanized natural rubbers, the hardness in International Rubber Hardness Degrees (IRHD) bears a known relation to Young's modulus.
3	Stress-strain including tear (ASTM D 412, ASTM D 624, SS-ISO 1798, SS-ISO 37, SS-ISO 34 -1) [Universal Testing Machine]	These tests are used to determine the modulus, tensile strength, elongation at break, tension set, and tear strength of vulcanized rubber sheets. However, tensile properties alone may not directly relate to the total end-use performance of the product because of the wide range of potential performance requirements in actual use. These properties depend both on the material and the conditions of test and therefore should be compared only when tested under the same conditions. Tensile set represents residual deformation which is partly permanent and partly recoverable after stretching and retraction. For this reason, the periods of extension and recovery must be controlled to obtain comparable results. Since tear strength may be affected to a large degree by stress-induced anisotropy of the rubber as well as by stress distribution, strain rate, size of the specimen, and direction of testing, the results obtained in a tear test can only be regarded as a measure of the strength under particular conditions and not as having any direct relation to service value.
4	Abrasion resistance (ASTM D 5963, ASTM D 1630, ASTM D 2228, SS-ISO 4649) [DIN Abrader]	Abrasion resistance is a performance factor of paramount importance for many rubber products such as tires, conveyor belts, power transmission belts, hoses, footwear, and floor coverings. These tests may be used to estimate the relative abrasion resistance of different vulcanized rubber compounds. Since conditions of abrasive wear in service are complex and vary widely, no direct correlation between these tests and actual performance can be assumed. These tests are suitable for comparative testing, quality control, and research and development work.
5	Compression set (ASTM D 395, SS-ISO 815) [Compression set apparatus at constant stress and at constant strain]	Compression set tests are intended to measure the ability of rubber compounds to retain elastic properties after prolonged action of compressive stresses. These tests are mainly applicable to service conditions involving static stress and are frequently conducted at elevated temperatures. The effects of dynamic stressing are simulated by hysteresis test.

(*Continued*)

TABLE 3.2

Physical Tests of Vulcanized Rubber *(Continued)*

Sl Number	Type of Test (Standards) [Equipment Used]	Significance/Application of the Test
6	Resilience (ASTM D 2632, ASTM D 1054, SS-ISO 4662) [Dunlop Tripsometer, Lupke Pendulum]	Resilience is a function of both dynamic modulus and internal friction of a rubber. It is very sensitive to temperature changes and type of rebound resilience tester. The rebound pendulum (ASTM D 1054) is designed to measure percent resilience of a rubber compound as an indication of hysteretic energy loss that can also be defined by the relationship between storage modulus and loss modulus. The percent rebound measured is inversely proportional to the hysteretic loss. Deflection is determined by measuring the depth of penetration of the rebound ball into the rubber block under test. Percent resilience and deflection are commonly used in quality control testing of polymers and compounding chemicals, especially reinforcing materials. It may not directly relate to end use.
7	Heat buildup (ASTM D 623, SS-ISO 4666) [Goodrich Flexometer, Firestone Flexometer]	This test method describes the heat rise in a vulcanized rubber cylinder after repeated compressive flexing. Because of wide variation in service conditions, no correlation between these accelerated tests and service performance is given or implied. However, the test methods yield data that can be used to estimate the relative service quality of different compounds.
8	Fatigue to failure (ASTM D 4482) [Monsanto Fatigue to Failure Tester]	This test is used for determining fatigue life at various extension ratios. Experience in fatigue testing shows that fatigue life may have a wide, non-normal distribution and therefore a large standard deviation that is compound dependent. Natural rubber, for example, has shown a narrower distribution than many synthetic rubbers. A large number of specimens (usually six or more) may therefore be required to yield the desired precision. This test gives an idea about the crack initiation behavior of a rubber vulcanizate and only a very approximate measure of the crack propagation rate. The information obtained may be useful in predicting the flex-life performance of a compound in active service; however, the user should be aware that in actual use, products are subjected to many other fatigue factors not measured in this test.
9	De Mattia fatigue (ASTM D 813, SS-ISO 132) [De Mattia Fatigue Tester]	This test gives an estimate of the ability of a rubber vulcanizate to resist crack growth of a pierced specimen when subjected to flexing. No exact correlation between these test results and service is implied due to the varied nature of service conditions.
10	Dynamic property (ASTM D 2231, SS-ISO 4664) [Rheovibron Dynamic Viscoelastometer, DMA]	Increased usage of rubber in vibrational and damping applications has resulted in a corresponding increased interest in the characterization of rubbers through dynamic mechanical testing. By this test, one can measure storage modulus, loss modulus, complex modulus, and tan δ which has a relation with tire rolling resistance and in turn fuel economy of the vehicle.

TABLE 3.2
Physical Tests of Vulcanized Rubber *(Continued)*

Sl Number	Type of Test (Standards) [Equipment Used]	Significance/Application of the Test
11	Air aging in test tube enclosure (ASTM D 865, SS-ISO 188) [Multicell Ageing Oven]	Rubber and rubber products must resist the deterioration of physical properties with time caused by oxidative and thermal aging. The isolation of specimens by the use of individual circulating air test tube enclosures prevents any cross contamination from volatile products and permits a more representative assessment of aging performance.
12	Ozone resistance (ASTM D 1149, ASTM D 3395, SS-ISO 1431-1,2) [Ozone Chamber]	The significance of these tests is mainly in their ability to differentiate, in a comparative sense, between different degrees of ozone resistance under the limited and specified conditions of the accelerated tests. But these tests may not give results correlating with outdoor performance because service performance depends on other conditions such as rainfall, ambient temperature, sunlight, etc. These tests basically indicate the protective power of the waxes and anti-ozonants used in rubber vulcanizates.
13	Preparation of test pieces from product (ASTM D 3183, SS-ISO 4661-1) [Splitting machine]	This practice is used when it is necessary to test a product from which specimens cannot be cut directly. The best method is slicing the products of desired dimension or, if required, buffing.

3.4.1.1 Test Procedure

The tensile stress/strain properties of rubber are measured with a tensile testing machine. The early heavy pendulum dynamometers have largely been replaced by inertia-less transducers that convert force into an electrical signal. Measurements of stress and strain are taken continuously from zero strain to breaking point and are recorded graphically.

Dumbbell-shaped test pieces, 10 or 13 cm long, are die-cut from flat sheet and marked in the narrow section with benchmarks 2.5 or 5 cm apart. The ends of the test piece are placed in the grips of the testing machine, and the lower grip is power driven at 50 cm per minute so that the test piece is stretched until it breaks. As the distance between the benchmarks widens, measurement is made between their centers to determine the elongation. The standard temperature for conditioning and testing the samples is 23 ± 2°C.

3.4.1.2 Interpretation

Tensile strength and elongation are useful to the rubber technologist for compound development, for manufacturing control, and for determining a compound's resistance to attack by various chemicals.

Tensile tests are universally used as a means of determining the effect of various compounding ingredients and are particularly useful when such ingredients affect the rate and state of vulcanization of the rubber. Similarly, tensile tests are excellent for controlling product quality once the compound has been selected.

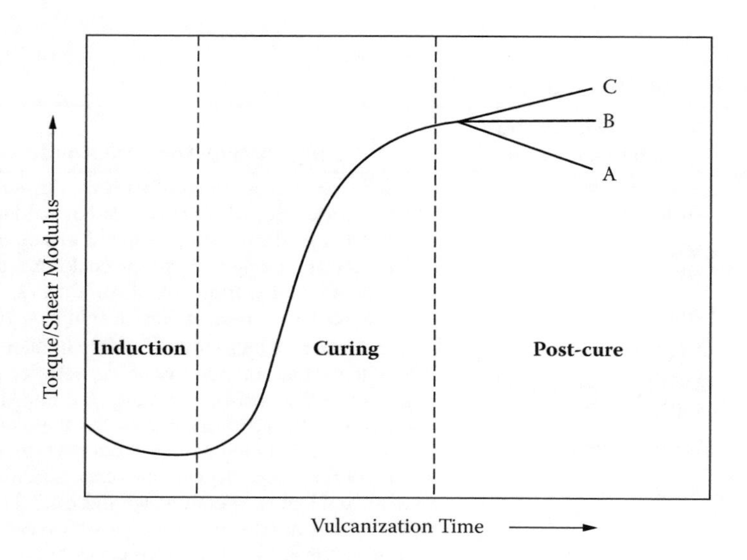

FIGURE 3.5
Typical rheometer curve (sulfur cured): curve A, reversion; curve B, equilibrium cure; curve C, marching cure.

Tensile tests can also be performed before and after an exposure test to determine the relative resistance of a group of compounds to deterioration by heat, oil, ozone, weathering, chemicals, etc. However, it should be noted that the retention of tensile properties is much more significant than the absolute values before and after the exposure test.

3.4.1.3 Design Calculations

Despite their usefulness to a rubber technologist, tensile properties are of limited use to the design or applications engineer. These cannot be used in design calculations, and they

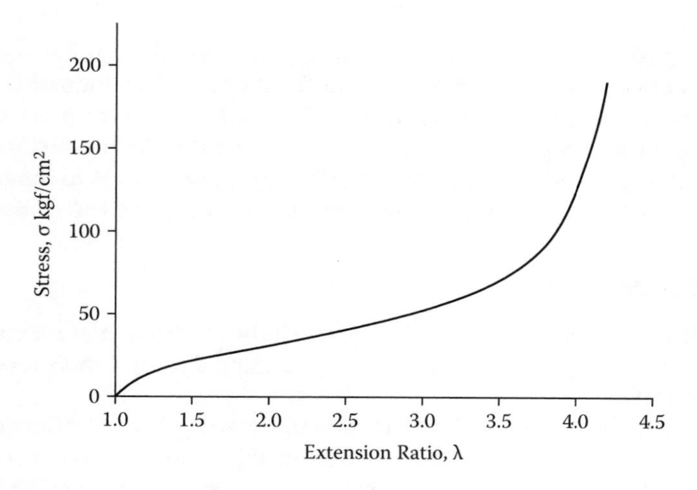

FIGURE 3.6
Typical stress/strain curve of a cured rubber compound.

bear little relation to performance in practice. Rubber components are seldom loaded in tension and never to a degree approaching their ultimate strength or elongation. Belts, hoses, O-rings, mountings, packings, etc., are rarely, if ever, subjected to tension stresses above 1 MPa, so whether their tensile strength is 10 or 20 MPa can hardly affect their ability to perform their functions.

Tensile properties are also not a reliable indication of the quality of the rubber. Compounds with tensile strengths below 7 MPa are usually rather poor in most mechanical properties. Where the tensile strength is greater than 21 MPa, which covers the great majority of rubber products, the correlation between tensile strength and properties such as resilience, abrasion resistance, compression set, and flex life is, at best, haphazard. In the middle range, it is possible to prepare two stocks of compounds with identical tensile strength and elongation and yet find no similarity in service life.

Purchasers of rubber products who over-specify tensile properties may find that the price they pay is unnecessarily high, or that they are sacrificing other properties that may be more important. With established products, buyers often will not consider a new compound, which is actually better for their applications, if it happens to have lower tensile strength and elongation than the original specification. Once again, it is emphasized that rubber specifications should be based on performance rather than on arbitrary physical tests. However, these tests are routinely used to develop rubber compounds as these are quick and easy.

3.4.2 Heat Aging

The properties of an elastomer will generally change after prolonged exposure to high temperatures. Natural rubber, for example, will become soft and gummy, while Neoprene will harden slowly under the same conditions. The extent to which either softening or hardening is taking place will depend upon the particular service required. The rate at which the properties of an elastomer change increases logarithmically with the temperature. Relatively small changes in temperature may, therefore, cause large differences in the degree of deterioration.

Tests for heat aging are carried out for two purposes. First, there are tests to establish the changes in physical properties at elevated service temperatures. Second, there are accelerated aging tests at high temperatures which attempt to predict the long-term life at lower temperatures. The international standard for both is ISO 188 that specifies methods using an air oven or an oxygen pressure chamber. The tests consist of aging test pieces for a given period at a given temperature and then measuring the physical properties that are considered important. In the absence of any specific requirements, the standard suggests measuring tensile strength, stress at an intermediate elongation, elongation at break, and hardness. Comparison is then made with similar test pieces that have not been aged.

With the oven method, the test pieces are exposed to air at atmospheric pressure in either the usual single chamber oven or a multi-cell oven. The latter has the advantage that dissimilar materials can be aged simultaneously without the danger of exchange contamination by plasticizers and other chemicals. The total volume of the test pieces in a single chamber oven should not exceed 10% of the free air space of the oven. The air flow in both types of oven must be steady and at a rate that provides between three and ten complete changes of air per hour. It is also important that no copper or copper alloys be used in the construction of the oven because they could accelerate the aging process. The temperature of the test is as required, but 70°C and 100°C are the most commonly used for general purposes. However, many special-purpose elastomers are tested at higher temperatures. The length of the test is recommended as 1, 3, 7, 10 or a multiple of 7 days.

In the oxygen pressure chamber method, it is recommended that the test pieces be exposed to oxygen at 70°C and be at a pressure of 2.1 MPa. Otherwise the procedure is similar to that of the oven method.

3.4.2.1 Interpretation

The main use of ISO 188 is for quality control, but the detailed requirements of the procedures must be followed closely to obtain good reproducibility. No universal correlation has been found between accelerated tests and natural aging, and the greater the disparity between aging and service conditions, the less reliable any correlation becomes. Experimental difficulties, of course, arise because of the time required to obtain natural aging data and the extent to which extrapolation may be required. Rubber products can and do last 50 years or more.

Aging is also affected by the thickness of the rubber. As the deterioration is mainly due to oxidation, the rate of oxygen diffusion into the bulk of the rubber can have a serious effect on the results. If the rate is too low, during a test or in use, the bulk of the material may experience little or no aging, even though the surface of the same rubber is very badly aged. The rate of oxygen diffusion will become even more significant when the temperature is increased. Comparisons should therefore only be made with results obtained from test pieces of similar shape and size.

It should also be noted that heat aging tests on thin test pieces in the presence of air can give a gross underestimate of the useful life of a product. For example, bridge bearings, which have a small area exposed to the air compared with their total volume, will last many times longer than would be indicated by short-term tests on thin sections.

3.4.3 Set

3.4.3.1 Permanent Set

Rubbers deform under load and rarely return completely to their original dimensions when the load is removed. The difference between the original and final dimensions, expressed in various ways, is known as *permanent set* and can be measured in tension, compression, or shear. However, permanent set in shear is not often required, and there are no recommended standards for its measurement.

In practice, the measurement of permanent set depends upon carefully defining the conditions of the test, the time for which the test is conducted, and the time that is allowed for the test piece to recover. Although standardized, these conditions are arbitrary, and it is sometimes difficult to obtain reproducible results and to relate them to how the rubber will perform under service conditions.

3.4.3.2 Tension Set

The method for measuring tension set consists of stretching a standard strip or dumbbell test piece to a constant strain, holding the elongation for a standard time, removing the load, and allowing the test piece to recover for 30 minutes. The increase in length between the reference marks or narrow section of the test piece, expressed as a percentage of the original length of the narrow section, is the permanent set in tension.

$$\text{Tension Set} = \frac{L_1 - L_0}{L_s - L_0} \times 100 \quad (3.3)$$

where:

L_0 = the original unstrained reference length

L_s = the strained reference length

L_1 = the reference length after recovery

The percentage strain value should be selected in accordance with the final application of the vulcanizate and with reference to its breaking elongation and the test temperature. A value of 100% strain is preferred, but 25, 50, 200, and 300% are acceptable alternatives. The test can be made at a temperature of 23, 70, 85, 100, 125, or 150°C; and the duration of the test can be 24, 72, or 168 h, the test period commencing 30 minutes after specified strain has been reached.

The measurement of tension set under constant stress and the use of ring-type test pieces are no longer recommended in some countries in Europe, although two sizes of rings are standardized in ISO 2285.

3.4.3.3 Compression Set

The practice in Europe is to measure compression set after constant strain at ambient or high temperatures following ISO 815. The measurement of compression set after constant stress is no longer recommended.

Small cylindrical disks of known dimensions are compressed to a fixed height in a simple jig. The jig consists of two or more flat, parallel metal plates that are sufficiently rigid to withstand the stress without bending and are of adequate size to hold the test piece or pieces within the area of the plates. The plates are clamped together by nuts and bolts. Steel spacers of the appropriate thickness, in the form of rings around each bolt, are placed between the plates to control the thickness of the test pieces while compressed.

The recommended sizes for the disks are either 13 mm diameter by 6.3 mm thick or 29 mm by 12.5 mm. The two sizes do not necessarily give the same values for compression set, and comparisons should always be made with similar test pieces. The use of lubricant on the contact surfaces of the plates is optional and may, in some cases, give more reproducible results. However, the lubricant may also affect the compression set values obtained, and again, comparison should only be made when the test conditions are similar.

It is recommended that three test pieces be used, either separately or as a set, for each determination and the results averaged. The bolts are tightened so that the plates are drawn together uniformly until they are in contact with the appropriate spacers, generally sized to give a compression of 25%. The apparatus containing the test pieces is introduced without delay into the central part of an oven that is maintained at test temperature. The recommended temperatures are 23°C or one of nine temperatures between 70 and 250°C. The duration of the test is 24 h for tests at elevated temperatures or 72 h at 23°C.

At the end of the specified time, the test pieces are removed from the jig and allowed to recover at 23°C for 30 minutes before the thickness is re-measured. The compression set is the difference between the original thickness of the test piece and that after recovery expressed as a percentage of the initially applied compression.

$$\text{Compression Set at constant strain} = \frac{t_0 - t_r \times 100}{t_0 - t_s} \tag{3.4}$$

where:

t_0 = the original thickness of the test piece
t_r = the thickness of the test piece after recovery
t_s = the thickness of the spacer

Thus if there is no recovery, the compression set is 100%, and if the test piece fully recovers to its original thickness the compression set is 0%.

3.4.3.4 Interpretation

Permanent set measurements can be useful for production control because they provide an indication of the degree of vulcanization that has taken place. They can also be helpful when selecting a compound from a number of alternatives. Unsuitable compounds can often be identified if a permanent set falls in the high range; differences in values will probably not give a true indication of performance in service.

Probably the most amazing property of rubber is not its elasticity but the degree to which it recovers after being subjected to high strains. Both permanent set and compression set are important laboratory tests that are worthy of close attention. Let us consider the general deformation curve. Over the first infinitesimal strains, the deformation is spontaneously elastic. The next part of the deformation curve is composed of a highly elastic and a plastic deformation, and plastic deformation may not be wholly recoverable.

On removal of the stress, immediate recovery of the spontaneous elastic deformation occurs, followed by the slower recovery of the highly elastic components. The plastic component may be irreversible, and a permanent deformation due to this factor remains. If the temperature is now raised, some further recovery usually occurs but in general, it remains incomplete.

To evaluate the state of cure, better discrimination is obtained when the test is carried out at high elongations, say 75% of the elongation at break. On the other hand, for inner tubes, this test is too severe and it is better to carry out the test at an elongation of 100% or 50%.

The normal compression set test consists of compressing samples by 25% of their height and maintaining them under this degree of compression for 22 h at 70°C.

The conditions have been chosen so as to provide a basis for a good routine test, but where the service conditions (e.g., limits of deflection or load, temperature range, etc.) are known, this test can easily be modified. Too often, compounds that have a high compression set in the 70°C test are rejected for an application involving lower temperatures where the compound might operate satisfactorily.

The term *creep* is applied to the slow overall change in strain that occurs after the completion of the initial elastic deformation, when the rubber is subjected to continued stress. Creep continues more or less indefinitely. As the flow does not take place exactly according to a known mathematical law, it is inadvisable to extrapolate the results; therefore, long-term creep tests are essential.

3.4.4 Hardness

Hardness generally denotes the resistance of material to local deformation, being measured as the resistance to penetration either by a loaded indentor (indentation hardness) or by a loaded sharp point moving over the surface (scratch hardness). Both these forms of tests are discussed below.

Since hardness in this sense is related to the elastic modulus, it increases with increasing crosslink density, so that fully crosslinked polymers (ebonite) are the hardest, and it is reduced by plasticizers.

Measurement of hardness by rebound resilience is done with metals and does not apply to high polymers, because their hardness and resilience are not directly related.

3.4.4.1 Hardness Tests

3.4.4.1.1 Indentation Hardness Test

Deformation under a loaded indentor may be elastic (reversible) and/or plastic (irreversible). Soft vulcanized rubber and rubber-like plastic (e.g., plasticized polyvinyl chloride, PVC) are substantially elastic; hence, the deformation (indention) must be measured while the loaded indentor is pressing on the test piece. Hard plastic and ebonite show both types of deformation; hence, ideally measurements should be made of the total (elastic + plastic) indentation under the load and the residual (plastic) indentation after removing the load, though the latter is not always done. The time of loading, test piece dimensions, and temperatures affect the result, as noted below.

The indentor is usually (1) a ball or a hemispherical-ended plunger (e.g., in the instrument measuring in International Rubber Hardness Degrees [IRHD], and also in the Brinell, Rockwell, Pusey and Jones Plastometer and Wallace Pocket Instrument); (2) a truncated cone (e.g., Shore-A durometer); or (3) a sharp pointed cone (e.g., Shore-D durometer).

The indenting load may be provided by either a dead weight (preferably) or a spring; the former is used in the IRHD instruments and Pusey and Jones Plastometer and the latter in the Shore durometers and Wallace Pocket meters. Dead weight instruments sometimes apply a small, minor or contact load, to give a definite zero position from which to measure the indentation produced by increasing the load to a much larger, major (or total) load.

The measurement made is normally the depth of indentation or penetration of the indentor. For essentially plastic materials, the measured quantity may be the diameter (or area) of the impression or permanent indentation left after removing the load or reducing it to the minor value.

The test result is expressed in various ways:

1. Directly as depth of indentation (e.g., Pusey and Jones Plastometer, BS 903: Part D6 test on ebonite)
2. As hardness degrees, such that 100 = infinitely hard and 0 = either extremely soft (Shore A durometer) or infinitely soft (IRHD). The IRHD value is derived from Young's modulus (E, kgf/mm^2) via a probit curve, being defined as the percentage

frequency corresponding to a probit value of (5.897 + 1.428log*E*). The value of *E* is derived from its relation to the depth of indentation *P* (mm/100) by a ball of radius *R* (mm) under a load *F* (kgf), namely:

$$E = 263\ F/(P^{1.35}\ R^{0.65}) \tag{3.5}$$

where:

E = Young's modulus, kgf/mm^2

F = load, kgf

P = depth of indentation, mm/100

R = the radius of the ball indenter, mm

This relation applies for perfectly elastic isotropic materials; usually there is some departure from this ideal behavior.

Except on very soft materials, IRHD and Shore A durometer values are approximately the same.

3.4.4.1.2 Other Features of Indentation Hardness Tests

With relatively soft elastic materials (e.g., vulcanized rubbers), the result is influenced by the dimensions (especially thickness) of the test piece; the indentation usually increased somewhat with the time of loading and may be influenced by temperature. Hence standard procedures define test piece dimensions, loading time, and temperature.

A micro-test, a scaled-down version of the normal IRHD test, has been developed for use on very small test pieces or finished articles.

3.4.4.2 Hardness Test Apparatus

Dead-load hardness gauge: Designed primarily for testing rubber, this is the best-known apparatus working according to ISO recommendation R48 and giving readings in IRHD. For the usual hardness range (30 to 95 IRHD) it uses a 2.5 mm diameter indenting ball, with dead loads of 30.5 gf (contact) and 580 gf (total). For very soft (10 to 35 IRHD) or very hard (85 to 100 IRHD) rubbers, the ball diameters are, respectively, 5 mm or 1 mm, but with the same loads.

Micro-hardness gauge: This scaled-down version of the above gauge has the ball diameter reduced 1/6, i.e., 0.395 mm and loads to $(1/6)^2$, i.e., 0.85 gf (contact) and 15.7 gf (total). It is for the range 30 to 95 IRHD.

Pocket hardness meter: This pocket-size instrument, with the hemispherical-ended plunger and approximately constant spring loading, reads in IRHD and covers the range 30 to 100 IRHD.

Shore durometers: Several models including pocket size are available for different ranges. Those mostly used are Type A (or A2), with small frustoconical indentor and variable spring loading (i.e., decreasing as indention increases), and the reading is in shore-A degrees (approximately equal to IRHD over the range 30 to 100); Type D, with the sharp conical indentor and heavier (variable) spring loading, giving the more open scale (30 to 100 corresponding to 80 to 100 Shore-A) and hence being suitable for semi-rigid materials.

Plastometer S: This uses a 1/8″ (usually) or 1/4″ diameter ball with dead loads that are 85 gf (contact) and 1085 gf (total); it reads directly as depth of indentation (mm/100). Initially designed to test rubber-covered rollers, it has ball-joined feet to rest on curved surfaces.

3.4.4.3 Compression of Hardness Measurements

The relation between rubber hardness scales (Table 3.3) shows, on the same horizontal line, equivalent readings on the scales most used for vulcanized rubbers and rubber-like materials. The equivalence must be regarded as approximate, especially with very hard materials that show more or less plastic deformation.

3.4.4.4 Interpretation

Table 3.3 shows the different hardness values in different hardness scales. Hardness is one of the most useful and often quoted properties of rubber, but in fact, the figures can be quite misleading. First, the measured values, especially by durometers, are often unreliable because of the mechanical limitations of the equipment and because of operator error. Hardness degrees should therefore never be quoted to better than 5°. Second, the characteristic that is measured, surface indentation, rarely bears any relation to the ability of the rubber product to function properly.

The lack of significance can best be understood by considering three products: a hose cover, a gasket to be used between rough flanges, and an automobile mounting. It is fairly obvious that the ease or difficulty of indenting the surface of the hose cover has nothing to do with utility. The important properties in a hose cover are abrasion resistance and resistance to oil, weather, and other conditions relating to its service. The case of the gasket is unusual because surface indentation has some significance. Indentation by the point of the test instrument is similar, to some extent, to the indentation the gasket will receive from protrusions on the sealing surface.

There is much danger in attributing false significance to hardness in the above cases, but the danger is very real in the case of motor mounting. Motor mounts are typical of many rubber products which are required to carry load and in which the relationship between the load and deformation, called stiffness, is a critical design factor. Hardness

TABLE 3.3

Rubber Hardness Scales

Type of Material	IRHD and Shore-A Durometer	Shore-D Durometer	Pusey and Jones Plastometer	
			1/5-Inch Ball	1/4-Inch Ball
Hard	100	100	0	0
Hard	98	60	—	—
Hard	95	50	14	10
Hard	90	40	27	20
Hard	80	30	48	35
Soft	70	22	68	50
Soft	60	16	92	67
Soft	50	12	125	90
Soft	40	9	170	125
Soft	30	7	260	185

cannot be assumed to be a close measure of stiffness. Hardness and stiffness are both stress/strain relationships, but the relationships are established for two entirely different kinds of deformations. Hardness measurements are derived from small deformations at the surface. Stiffness measurements are derived from gross deformations of the entire mass. Because of this difference, hardness is not a reliable measure of stiffness. Even if hardness and stiffness had a better correlation, the irreducible five-point variation in durometer readings would be equivalent to a 15 to 20% variation in stiffness as measured by a compression-deflection test. Hardness measurement would not, therefore, be sufficiently accurate for design purposes.

The misuse of hardness to measure stiffness is common and causes much confusion. Wherever possible, simulated service tests should be used rather than hardness testers.

3.4.5 Abrasion

Abrasion, the wearing away or "wear" of surfaces is relaxed to and is as important industrially as friction. As a complex property, however, it proves difficult to analyze and measure. Although several machines have been devised for the accelerated abrasion testing of textile, plastics, and rubbers, none gives results in agreement with the performance observed in service. Therefore, they can be used only for comparison purposes between similar materials under particular conditions of abrasion, rather than to obtain absolute values.

Abrasion of rubbers, notably of tire treads, necessarily receives considerable attention, yet no theory to account for the marked improvement effected by incorporation of carbon blacks is entirely acceptable. The observation on unvulcanized crepe rubber subjected to a stream of abrasive particles is a good example. This when used as a lining of conveyer shoots resists abrasion better than the best carbon black reinforced vulcanizates and is an example of the fact that under these conditions the highest wear resistance is shown by materials with low elastic modulus and high elasticity (i.e., recovery).

Several investigators have attempted to separate the effects of the various mechanisms of abrasion such as (1) micro-cutting or scratching on sharp surfaces, (2) plastic deformation occurring progressively in the same place on blunt projections (not with rubbers), (3) fatigue failure caused by repeated elastic deformation on blunt projections (rubbers; this mechanism may be severe in an oxidative atmosphere), and (4) strong adhesion between the surfaces, resulting in welding and pieces being torn out (with rubbers this mechanism is catastrophic, resulting in the formation of rolls of debris and an abrasion pattern of parallel ridges).

In laboratory abrasion tests for rubber and plastics, the test material is abraded by a loose (granular) abradant, abrasive paper or cloth, a bonded abrasive wheel, or knives. However, even the most sophisticated of these have so far failed to correlate satisfactorily with service wear, if only because "service" covers a great variety of conditions of wear, and the relative behavior of different materials depends on these conditions. Hence, laboratory abrasion tests are useful mainly for comparing similar materials intended for one type of service (e.g., tire tread rubber made from the same type of raw rubber), as a pointer to materials with improved abrasion resistance.

3.4.5.1 The ISO Abrader

The abrader described in ISO 4649 (also known as the DIN abrader) is rapidly becoming the standard machine. It is convenient and rapid in use and is well suited to quality control. There is, however, no general close relationship between the results and performance in service.

A disk test piece in a suitable holder is traversed across a rotating drum covered with a sheet of an abrasive cloth. Standard test pieces are cylindrical in shape, 16 mm diameter and not less than 6 mm high. If test pieces of the required thickness are not available, a test piece not less than 2 mm thick may be bonded to a base element of hardness not less than 80 IRHD. The holder moves laterally across the drum at the rate of 4.2 mm per revolution of the drum, and suitable attachments may be provided to rotate the test piece at the rate of one revolution per 50 revolutions of the drum. The test piece is pressed against the drum with a vertical force of preferably 10 N, but sometimes 5 N by means of weights added to the top of the holder.

The drum has a diameter of 150 mm and a length of about 500 mm and is rotated at 40 rev/min. A standard abrasive cloth with corundum particles of grain size 60μm or micrometer is specified, and it is attached to the drum with double-sided adhesive tape. The test piece is automatically applied to and removed from the drum after an abrasion run of 40 m which is equivalent to 84 revolutions. In special cases where the mass loss is greater than 600 mg in 40 m, the run may be reduced to 20 m and the results multiplied by two. The test piece is weighed before and after the run. The weight loss is converted to volume loss by dividing it by the density of the material as determined by the method specified in ISO 2781.

Three test runs for each rubber are required. The same test piece may be used if the mass loss is relatively small. The tests are carried out at laboratory temperature, but in some cases there may be a considerable increase in temperature at the abrading interface. Such a temperature increase is disregarded, but the test piece should be allowed to cool to laboratory temperature between test runs.

To overcome the difficulty of maintaining consistent test conditions, the results are related to those obtained with a standard rubber. Two standard rubbers are specified, one for use when the results are expressed as relative volume loss and the other when they are expressed as an abrasion resistance index.

$$\text{Relative volume loss} = 200\text{ mg} \times \frac{\text{Volume loss of test piece in mm}^3}{\text{Mass loss of standard rubber in mg}} \tag{3.6}$$

Note that this formula requires a load of 10 N and a non-rotating test piece; abrasion resistance is the reciprocal of relative volume loss; and 200 mg is the nominal abrasion.

$$\text{Abrasion resistance index \%} = 100 \times \frac{\text{Volume loss of standard rubber}}{\text{Volume loss of test rubber}} \tag{3.7}$$

Note that the rotating test piece method is preferred, but both tests must use the same procedure.

The difference between the two methods is not immediately obvious. Relative volume loss is used when a consistent abradant is available that produces an abrasion loss of 210 to 220 mg with the standard rubber. The test method can then be reduced to the measurement of the volume loss of the test rubber. The abrasion index is not dependent upon having such a consistent abradant, although in practice, an abrasive cloth meeting the specification in ISO 4649 is usually used.

In Table 3.4, different physical properties along with their test methods are listed as a ready reference.

TABLE 3.4

Physical Properties of Vulcanizates: Standard Test Methods

Sl Number	Testing Method	BIS	DIN	ASTM	BS	ISO
1	Tensile stress-strain	3400 (Pt.-I)	53504	D 412	903 (A2)	37
2	Compression stress-strain			D 575	903 (A4)	
3	Hardness, IRHD	3400 (Pt.-II)	53519 (Tl.1)	D 1415	903 (A26)	48, 1400, 1818
	Microhardness	3400 (Pt.-II)	53519 (Tl.2)	D 1415	903 (A20)	48
	Shore		53505	D 2240	—	—
	of ebonite	—	—	—	—	2783
	on curved surfaces	—	—	—	903 (A22)	—
4	Density	3400 (Pt.-IX)	53479	D 297	903 (A1)	2781
5	Abrasion, DuPont	3400 (Pt.-III)	—	—	903 (A9)	—
	Cylindrical drum	—	53516	D 5963	—	4649
	NBS	—	—	D 1630	—	—
	Akron, Dunlop	—	—	—	903 (A9)	—
	Pico	—	—	D 2228	—	—
	Frack-Hauser	—	53528	—	—	—
6	Resilience, Schob	—	53512	—	—	4662
	Goodyear, Healey	—	—	D 1054	—	—
	Lupke, Dunlop	3400 (Pt.-XI)	—	—	903 (A8)	1767
	Bashore	—	—	D 2632	—	—
	Yerzley	—	—	D 945	—	—
7	Tear strength, Crescent	3400 (Pt.-XII)	—	D 624	903 (A3)	34
	Angle	3400 (Pt.-XVII)	53515	D 624	903 (A3)	34
	Trouser	—	53507	—	—	34
	Needle	—	53506	—	—	—
	Delft	—	—	—	—	816
8	Compression set					
	Constant load	—	—	D 395 (A)	903 (A6)	—
	Constant deflection	3400 (Pt.-X)	53517 (Tl.7)	D 395 (B)	903 (A6)	815
	At low temperature	—	53517 (Tl.2)	D 1229		1653

TABLE 3.4

Physical Properties of Vulcanizates: Standard Test Methods *(Continued)*

Sl Number	Testing Method	BIS	DIN	ASTM	BS	ISO
9	Tension set					
	At constant strain	3400 (Pt.-XII)	53518	D 412	903 (A5)	2285
10	Flex fatigue, De Mattia	3400 (Pt.-VII)	53522	D 430	903 (A10)	132
11	Cut growth, De Mattia	3400 (Pt.-VIII)	53522	D 813	903 (A11)	133
	Ross	3400 (Pt.-XVI)	—	D 1052	—	—
	Texus	—	—	D 3629	—	—
12	Fatigue in strain, De Mattia	—	—	D 430 (B)	—	—
	Monsanto F-to-F	—	—	—	—	6943
	DuPont	—	—	D 430 (C)	—	—
13	Fatigue bend, shoe soling	—	53542	—	—	—
14	Compression Flexometer (Goodrich)	—	53533 (Tl.3)	D 623	—	4666
15	Rotations-Flexometer	—	53533 (Tl.2)	—	—	4666
16	Stress relaxation	—	53537	D 1390	903 (A15)	3384
17	Shear modulus					
	Quadruple shear	—	—	—	903 (A14)	1827
18	Dynamic behavior					
	Torsion pendulum	—	53520	—	903 (A31)	4663
19	Viscoelastic properties					
	Forced vibration	—	53513	D 2231	903 (A24)	4664
20	Dynamic testing					
	General requirements	—	53535	—	—	2856
21	Low temperature tests, Gehman	3400 (Pt.-XVIII)	53545	D 1053	903 (A13)	1432
	Brittle point	—	53546	D 746	903 (A25)	812
	Temperature retraction test	—	—	D 1329	903 (A29)	2921
22	Accelerated aging					
	In air	3400 (Pt.-IV)	53508	D 573	903 (A19)	188
	Air pressure	—	53508	D 454	—	—
	Oxygen pressure	3400 (Pt.-IV)	53508	D 572	903 (A19)	188

(Continued)

TABLE 3.4

Physical Properties of Vulcanizates: Standard Test Methods *(Continued)*

Sl Number	Testing Method	BIS	DIN	ASTM	BS	ISO
23	Ozone resistance, static					
	Flat specimen	3400 (Pt.-XX)	53509	D 1149	903 (A23)	1431
	Triangular specimen	—	—	D 1171	—	—
	Dynamic conditions	—	—	D 3395	—	(1431 P2)
24	Permeability to gas					
	Constant volume	3400 (Pt.-XIX)	53536	—	903 (A17)	1399
	Constant pressure	—	53536	—	903 (A30)	2782
25	Permeability to volatile liquids	—	53532	—	—	—
	Vapor transmission of volatile liquids	—	—	D 814	—	6179
26	Staining tests	—	53540	D 925	903 (A33)	3865
27	Corrosion to metals	—	—	D 1414	—	6505
28	Adhesion props, to textiles	3400 (Pt.-V)	53530	D 413	903 (A47)	36
	To metals, one-plate method	3400 (Pt.-XIV)	53531 (Pt.1)	D 429 (B)	903 (A21)	813
	To metals, two-plate method	—	—	D 429 (A)	—	814
	To metals, conical parts	—	53531 (Pt.2)	D 429 (C)	—	5600
	To rigid plates in shear	—	—	—	903 (A28)	1747
	To textile cord	—	—	D 2138	903 (A27)	—
	To single-strand wire	—	—	D 1871	—	—
	To steel cord	—	—	D 2229	—	—

4

Reverse Engineering Concepts

4.1 General Concepts and Examples

Rubber compound is a complex mixture of different chemical ingredients. Each ingredient has its own function. The ingredients play different roles during compound mixing, processing, and final use of the rubber product. The majority of ingredients used in rubber products are petroleum based. Some of these chemicals change their composition during one of the processing stages, called vulcanization. These chemical changes cause real challenges to a rubber chemist during identification of different chemical ingredients from a finished rubber product. The process of identification of chemical ingredients from the finished product is known as reverse engineering.

Reverse engineering is a well-established technique used to acquire detailed micro information about any product. Different reverse engineering processes are available, viz., material reverse engineering, geometrical reverse engineering, design reverse engineering, etc. Reverse engineering helps design engineers obtain the required information about the benchmark product in a short time span. This process not only helps the design engineers to reduce the design cycle time of any new product in line with the best product available in the market, but also gives them a cutting edge to stay ahead of competitors.

Chemical reverse engineering is a technique used to identify the chemical or material composition of a product. This process is widely practiced in the case of rubber product reverse engineering. However, with any rubber product, being a complex mixture of different materials, the commonly used test methods to identify the chemicals may not be applicable. Worldwide, scientists are working to develop a suitable test method that can be used to identify the trace material present in rubber compounds. Repeatability and reproducibility of such test methods are very important. Although there are several in-house developed and validated test methods, very few of such test methods are available in the public domain.

In this chapter, we discuss the general concept of reverse engineering of a rubber product and the typical test methods used to get information about the material composition of the rubber compound.

Any test requires a representative sample. Test results depend on the way the sample is being prepared. In the case of rubber product reverse engineering, the representative sample plays a very important role. As we have discussed, the rubber product is a mixture of different chemical ingredients, so getting a true sample from a single point of a rubber product is difficult. It is always recommended that more than one sample be collected from different parts of a rubber product in order to obtain acceptable results. The sample numbers will depend on the complexity of the rubber product. In addition to the representative sample, the homogenization of the sample is an important step in the reverse engineering process. The homogenization step in the case of the rubber product reverse engineering process helps to distribute the chemical ingredients evenly in the sample. Besides the representative sample, the condition of the rubber product on which the reverse engineering has to be conducted is also important. Rubber product, whatever way it is designed, shows aging with time. This aging takes place due to various phenomena, some of which are:

- Thermal degradation
- Oxygen/ozone degradation
- Solvent leaching
- Radioactive degradation
- Aerosol degradation
- Environmental stress degradation
- Constant flexing degradation

All such phenomena cause aging of the rubber product, which results in changes in the physical properties as well as loss of different chemical ingredients from the product. Traits noticed in general observation of the aging of rubber products include:

- Increase in hardness
- Reduction in elongation at break
- Appearance of crack on the surface or edge of the rubber product
- Brownish appearance on the surface of the rubber product
- Softening of the rubber product
- Blooming or bleeding on the surface of the rubber product

Therefore, while selecting the rubber product for reverse engineering, it is always recommended that the fresh sample available in the market be used. This can easily be verified by the date of manufacturing, which is available on the surface of the rubber product. A typical process flowchart of reverse engineering is shown in Scheme 4.1.

Various chemical techniques used in reverse engineering are:

1. Solvent extraction
2. Ash content determination
3. Chromatographic separation
4. Chemical digestion

Analytical techniques used for reverse engineering include:

1. Thermogravimetric analysis
2. Differential scanning calorimetry
3. Infrared spectrophotometry
4. Atomic absorption spectrophotometry or inductively coupled plasma spectrophotometry

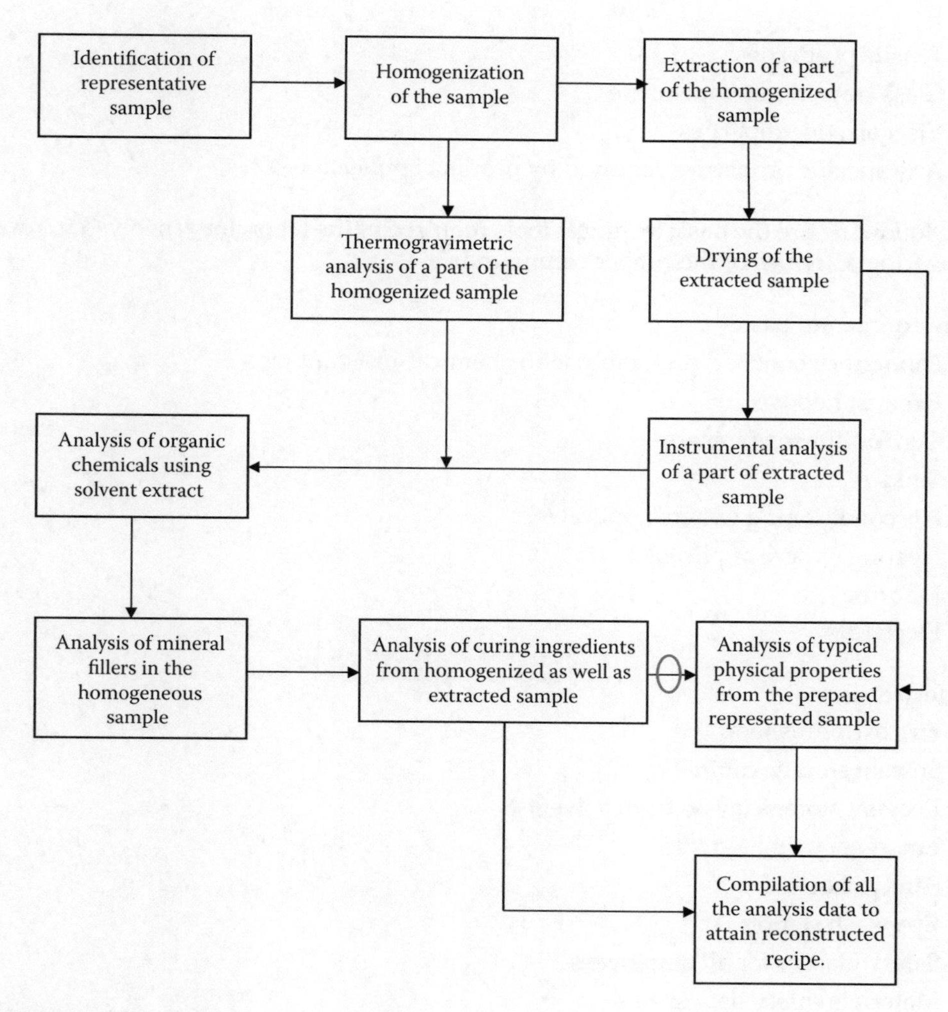

SCHEME 4.1
Typical flow diagram.

5. Gas chromatography
6. Gas chromatography with mass detector
7. High-performance liquid chromatography
8. Elemental analyzer
9. Nuclear magnetic resonance spectroscopy
10. Microscopy

Physical properties that need to be analyzed to support the reverse engineering process are as follows:

1. Hardness
2. Specific gravity

3. Tensile properties
4. Glass transition temperature
5. Viscoelastic properties
6. Any specific properties required by product application

The following are the basic technical tools required at the laboratory involved in reverse engineering activities of the rubber compound:

Environmental Basics
Laboratory benches, preferably with chemical-resistant tops
Exhaust hoods
Gas for Bunsen burners
Sinks
Air conditioning (where applicable)
Heating (where applicable)
Lighting
Electricity

Safety Basics
Fire extinguishers
Solvent storage cabinets
Reagent storage (away from solvents)
Emergency first aid kits
Safety showers
Eye-wash station
Safety glasses for all employees
Materials safety data sheets

Basic Equipment
pH meter
Centrifuge
Muffle furnace
Bunsen burners
Hotplates
Stirrers
Glassware
Ovens
Thermometers
Microscope
Mechanical balance

Strongly Recommended Basics
- Computer
- Word-processing software
- Glassware drying oven
- Distilled/deionized water maker
- Electronic balance
- Laboratory mill
- Vise
- Dissecting tools

Following are brief descriptions of different chemical, instrumental, and physical test techniques used in the reverse engineering process.

4.1.1 Solvent Extraction

Solvent extraction is a process used to separate the highly volatile organic materials present in the rubber products. There are different highly volatile chemicals used in rubber compounds, viz., rubber process oils, antioxidants, anti-ozonants, waxes, organic activators, accelerators, etc. Some of these organic chemicals are present in the rubber product as it is, and some change its chemical structure during the final curing of the rubber product. All of these rubber chemicals, whether they changed their chemical structure or not, are extracted by a suitable solvent during the extraction process. The extraction process generally used is Soxhlet extraction. While conducting the extraction, solvent selection plays an important role. Acetone is the commonly used solvent for extraction. There are several other solvents and mixtures of solvents available for specific extraction of chemical ingredients from the rubber product. Extraction of phthalate-type plasticizer is facilitated by ether-type solvent. However, while doing the extraction with ether solvent, utmost care needs to be taken to avoid flammability. Chilled water flow is required to condense the ether vapor.

Extraction gives rise to two different fractions:

- Extracts—solvent with high volatile organic materials
- Extracted—rubber sample without the extractable organic material

With the extracts, detailed analysis of the organic materials can be performed subsequently; with the extracted sample, analyses of polymer, ash, fillers, etc., can be conducted.

4.1.2 Ash Content Determination

Ash in rubber product is generally inorganic material with melting points that are much higher. Rubber compounds contain different inorganic materials such as zinc oxide (ZnO), silica (SiO_2), clays (different silicates), titanium dioxide (TiO_2), inorganic pigments, etc. Some of these inorganic materials are essential for specific rubbers and some fillers are generally used to extend the rubber product to reduce the cost of the compound. Reverse engineering of a rubber compound requires identification of different types of inorganic materials present in the rubber products. The best way to identify these inorganic materials is by using the ashing technique. Ashing is typically done in the muffle furnace. Different rubbers behave differently while heating at high temperatures. Silicone rubber generates frothing during a high-temperature ashing process. This phenomenon results in loss of material, and quantitative ash content analysis is not possible. So while doing ashing of silicone rubber product, it is always recommended that it should be heated slowly on an oxidizing flame, outside the muffle furnace, followed by high temperature burning inside the muffle furnace. Whatever burning process is being applied will be conducted using the lidded apparatus. Other polymers containing heteroatoms in the polymer chain, like nitrogen (N), halogen (Cl, Br), oxygen (O), etc., generate a lot of charred residue. In such types of rubbers, the ashing is done at a higher temperature.

4.1.3 Chemical Digestion

Chemical digestion is used to separate the polymers from fillers and quantitatively analyze the filler amount. The thermal decomposition technique is generally used to estimate the filler quantity for general-purpose rubbers. However, for some rubbers, the thermal decomposition technique does not give the correct results. Such polymers are nitrile rubbers, polychloroprene rubbers, polyurethane rubbers, etc. These rubbers contain hetero atoms in their chemical backbone. Such hetero atoms cause generation of the charred material during thermal decomposition under nitrogen atmosphere. Due to generation of this charring material, the thermal analysis technique fails to give the filler quantity in the rubber products. Chemical digestion helps the chemists in such cases. Rubber product with hetero atom containing rubbers can be chemically digested using tertiary butyl catechol. During chemical digestion, the rubbers come to the solution and the filler materials including ZnO, silica, and other inorganic pigments remain as is. The digested material is then centrifuged to create two separate layers: one of the solution and the other of the solid material. The solution is then decanted, and the residue is washed several times with solvent and then dried under flow of nitrogen air inside a drafting oven at a temperature of ~120°C. The dried residue is then burnt in a high-temperature muffle furnace. The weight loss during burning gives an idea of the carbon black filler content. The inorganic residue left over after carbon black burning is then chemically analyzed for composition analysis. This method finds very good repeatability in the analysis of carbon black fillers in the rubber compounds that contain hetero atom in the rubber backbone.

Chemical digestion also provides information about the chopped fabric present in rubber products. There are several rubber products available in the market which contain different types of chopped fabric, such as nylon, polyester, rayon, aramid, etc. During thermal decomposition of rubber product containing such chopped fabric, the fabric will also become decomposed. Chemical digestion is one of the best analytical techniques with which to characterize the chopped fabric content in rubber products.

4.1.4 Different Instrumentation Techniques

Different instrumentation techniques are being used in the reverse engineering process. Major instrumental techniques are discussed in previous chapters. Some of the typical thermal degradation curves as well as pyrolysis spectra are included here for reference. These graphs are commonly used in reverse engineering analysis and have been

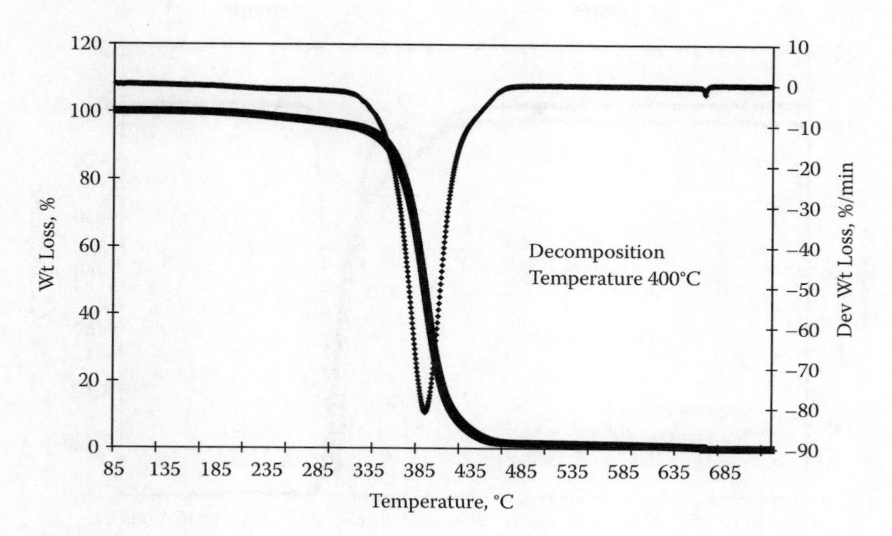

FIGURE 4.1
Decomposition of natural rubber (NR).

explained in previous chapters. Thermograms of various rubbers are given in Figures 4.1 through 4.13. DSC trace of these representative rubbers is shown in Figures 4.14 through 4.17. Decomposition temperature, rate of degradation, initial decomposition temperature, and temperature at which the maximum rate of degradation takes place vary from sample to sample. There may be a slight change in these temperatures when samples are tested in different conditions or machines. Similarly, FTIR spectra of a few rubbers are displayed in Figures 4.18 through 4.27.

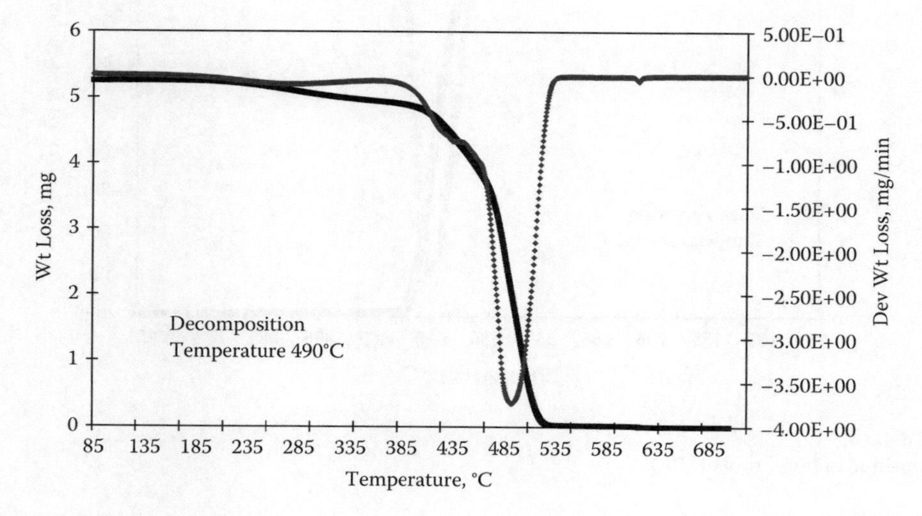

FIGURE 4.2
Decomposition of styrene-butadiene rubber (SBR).

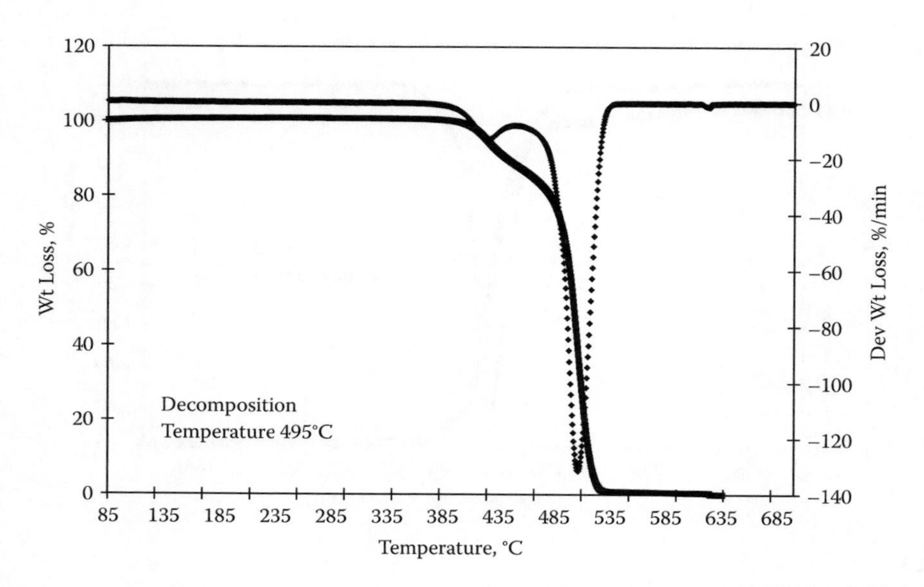

FIGURE 4.3
Decomposition of polybutadiene rubber (BR).

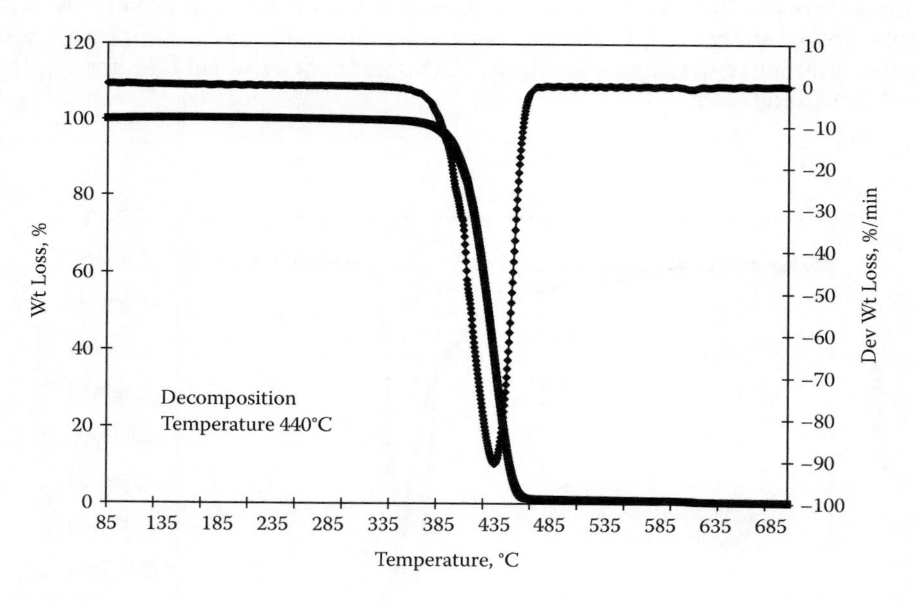

FIGURE 4.4
Decomposition of butyl rubber (IIR).

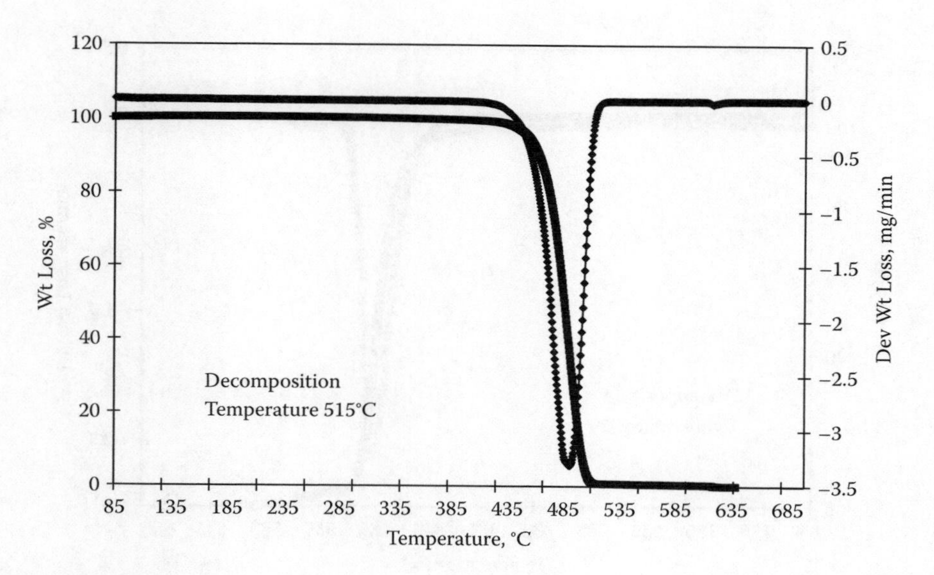

FIGURE 4.5
Decomposition of ethylene propylene diene monomer (EPDM).

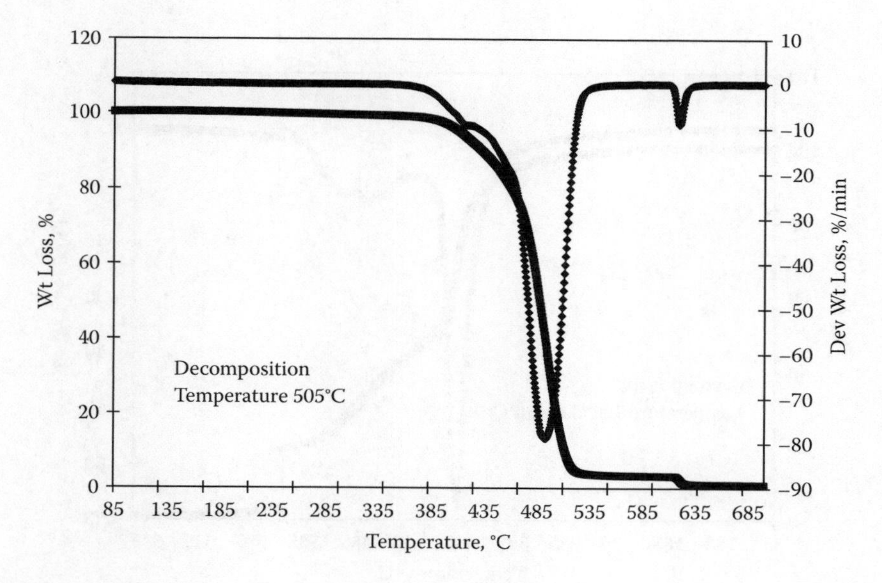

FIGURE 4.6
Decomposition of nitrile butadiene rubber (NBR).

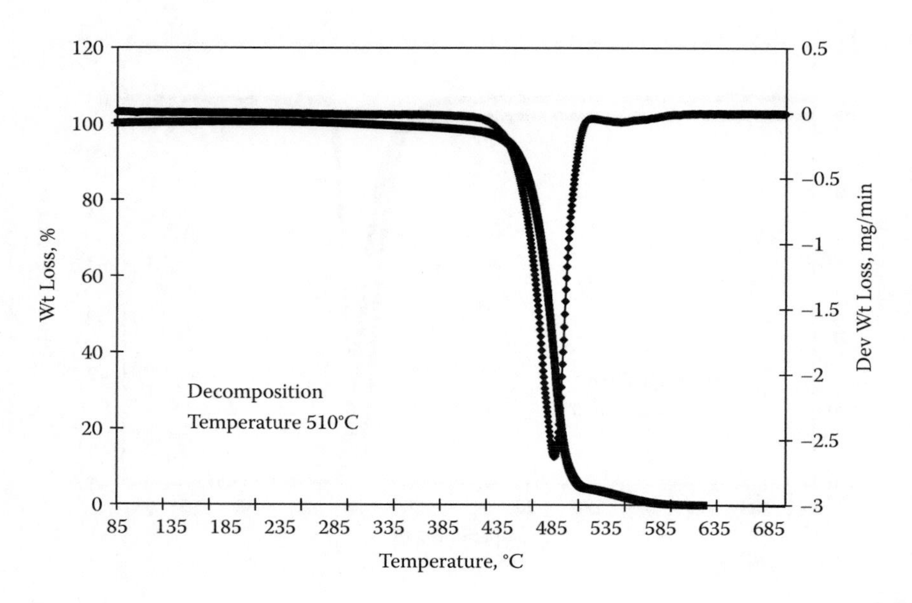

FIGURE 4.7
Decomposition of hydrogenated nitrile butadiene rubber (HNBR).

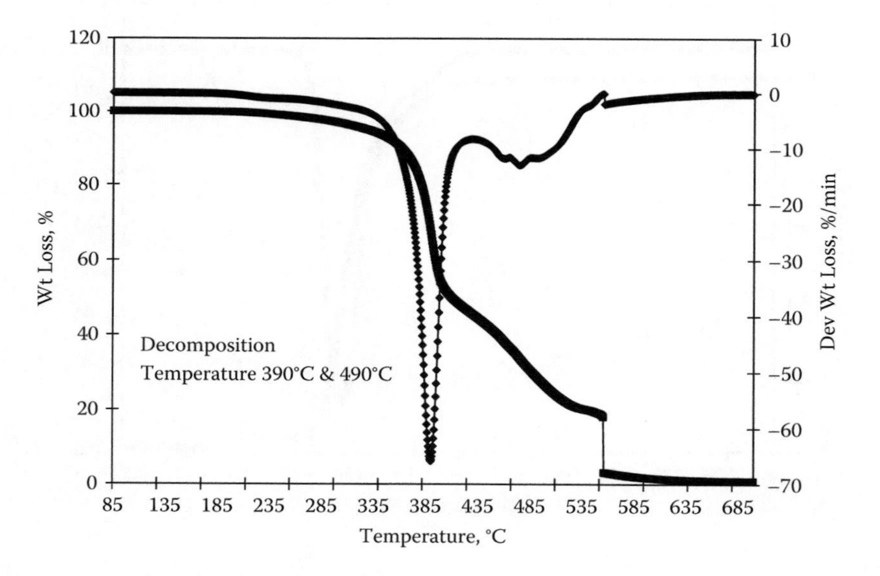

FIGURE 4.8
Decomposition of polychloroprene rubber (CR).

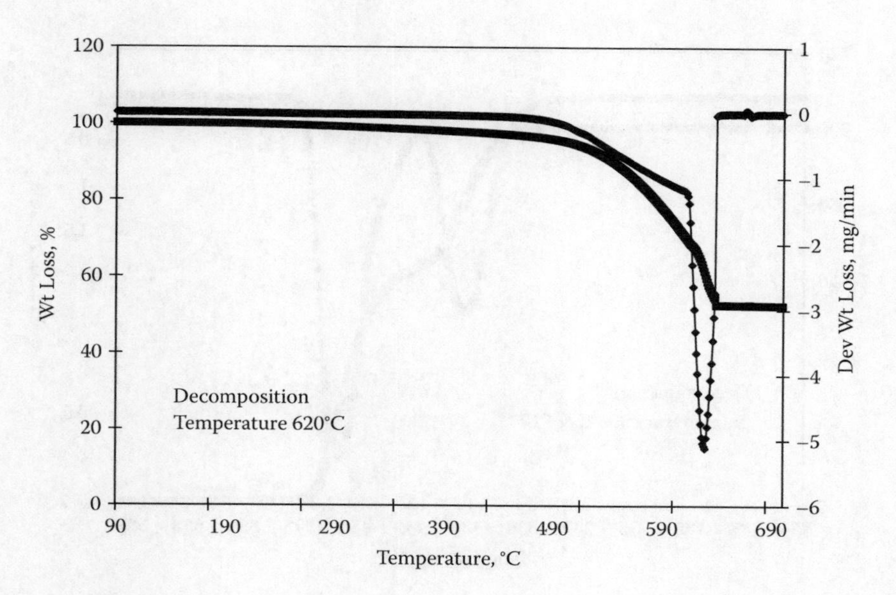

FIGURE 4.9
Decomposition of silicone rubber (MQ).

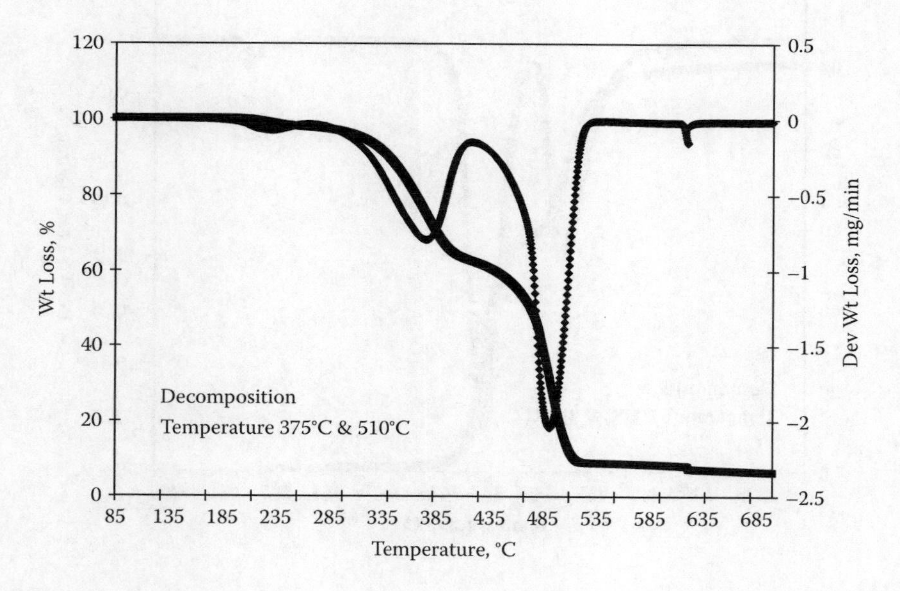

FIGURE 4.10
Decomposition of chlorosulfonated polyethylene rubber (CSM).

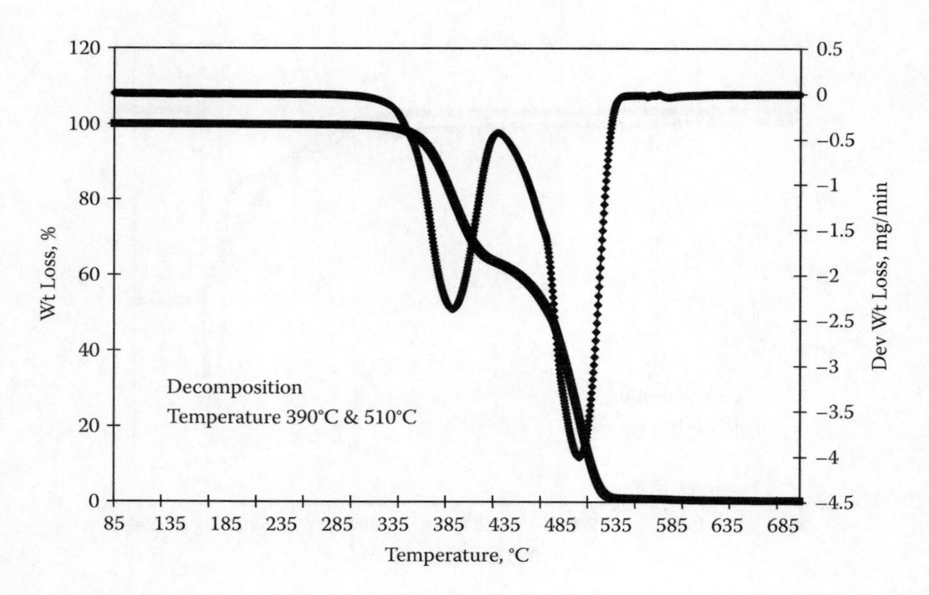

FIGURE 4.11
Decomposition of ethylene vinyl acetate (EVA) copolymer.

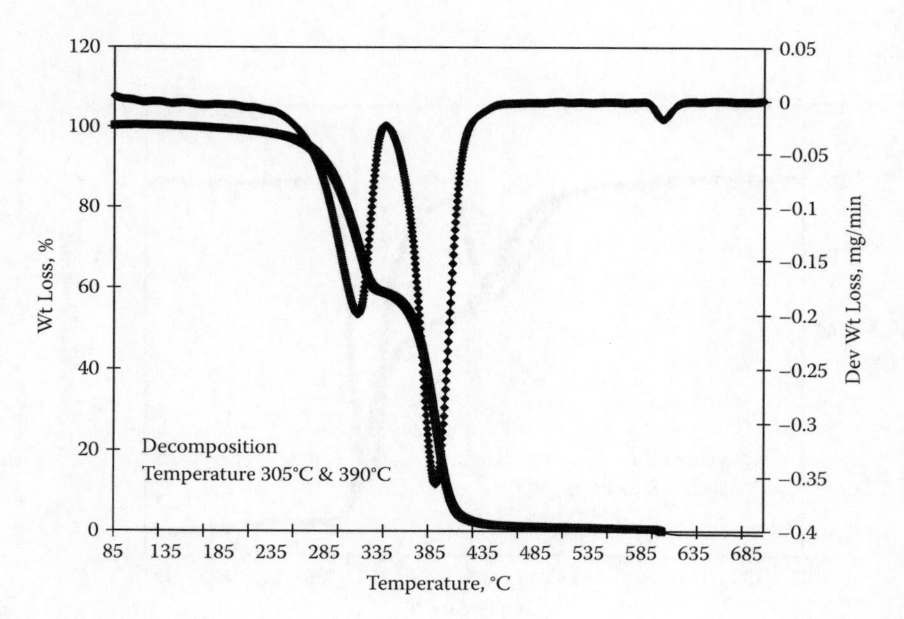

FIGURE 4.12
Decomposition of polyurethane (PU) elastomer.

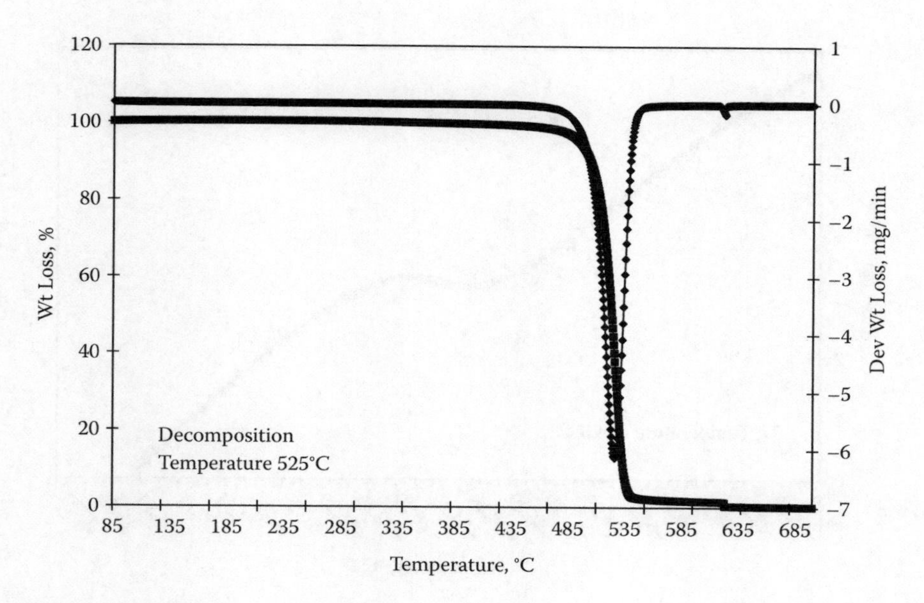

FIGURE 4.13
Decomposition of fluoro elastomer (FKM).

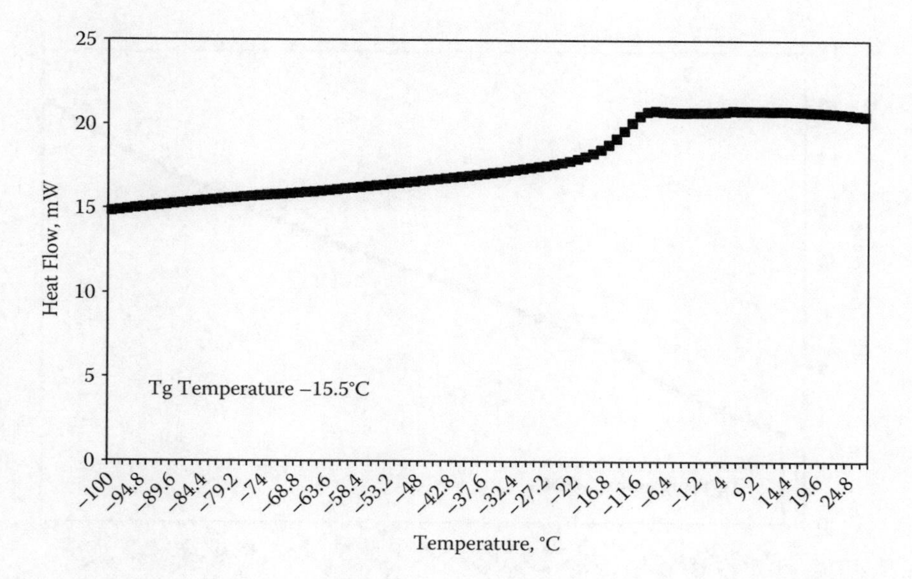

FIGURE 4.14
DSC thermogram of polyacrylic rubber (ACM).

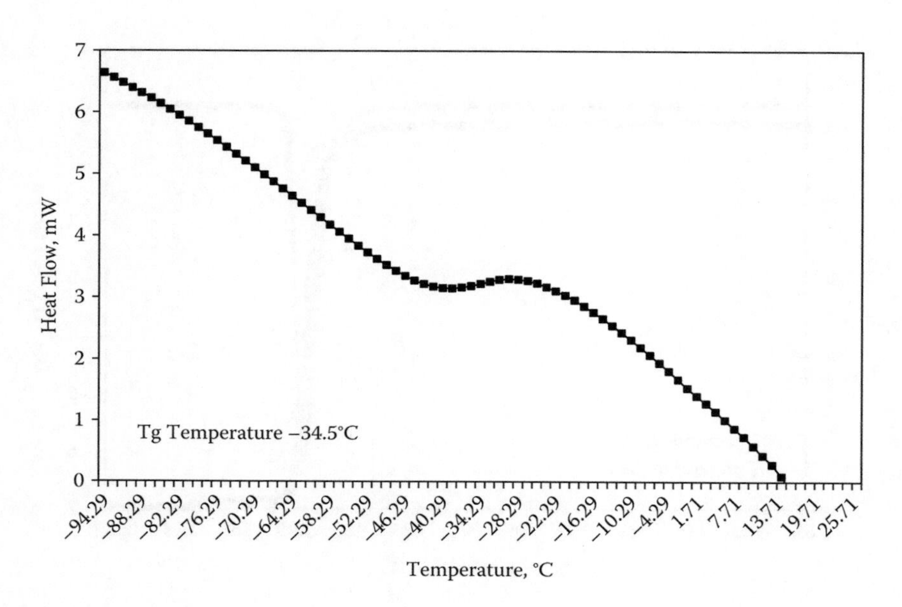

FIGURE 4.15
DSC thermogram of ethylene propylene diene monomer (EPDM).

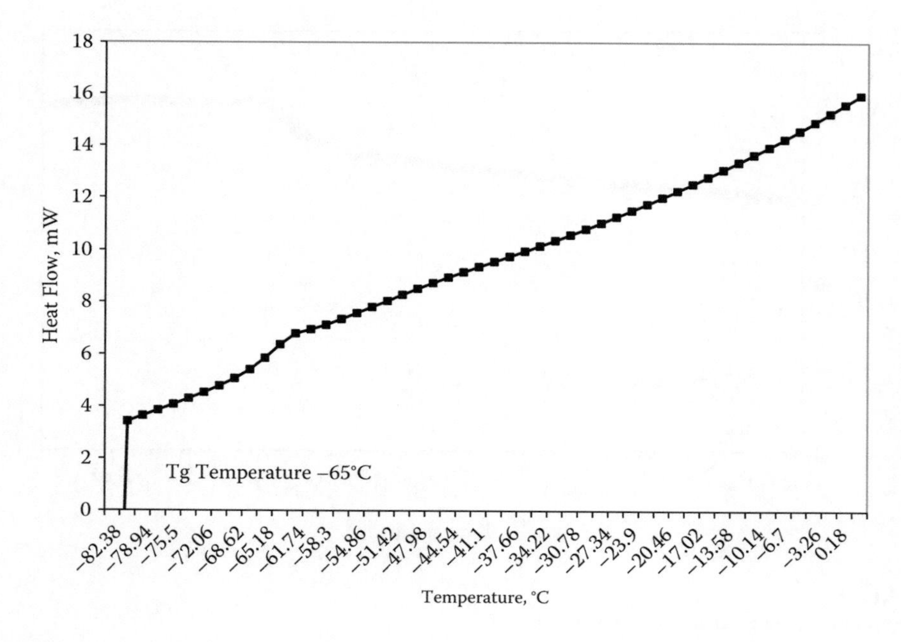

FIGURE 4.16
DSC thermogram of natural rubber (NR).

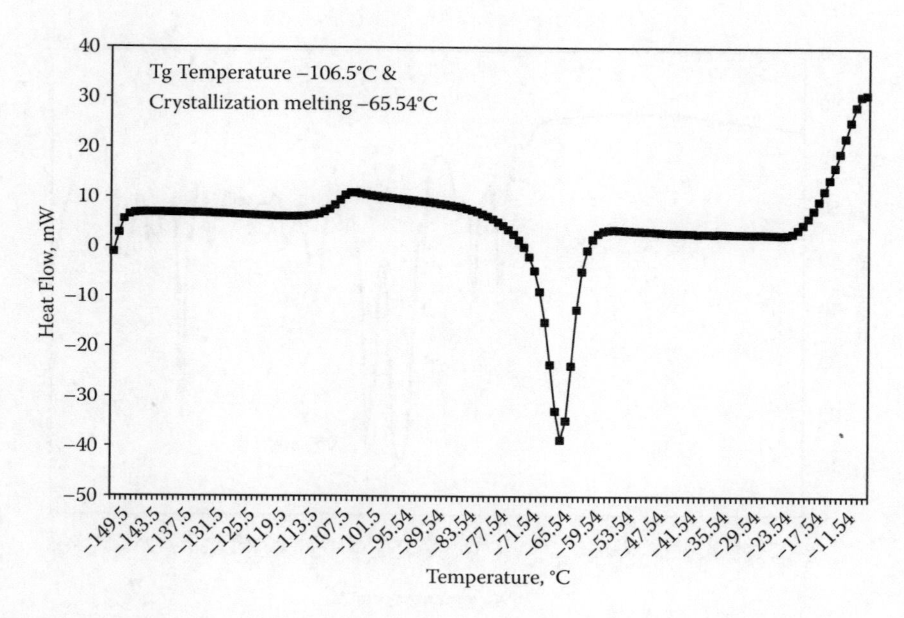

FIGURE 4.17
DSC thermogram of polybutadiene rubber (BR).

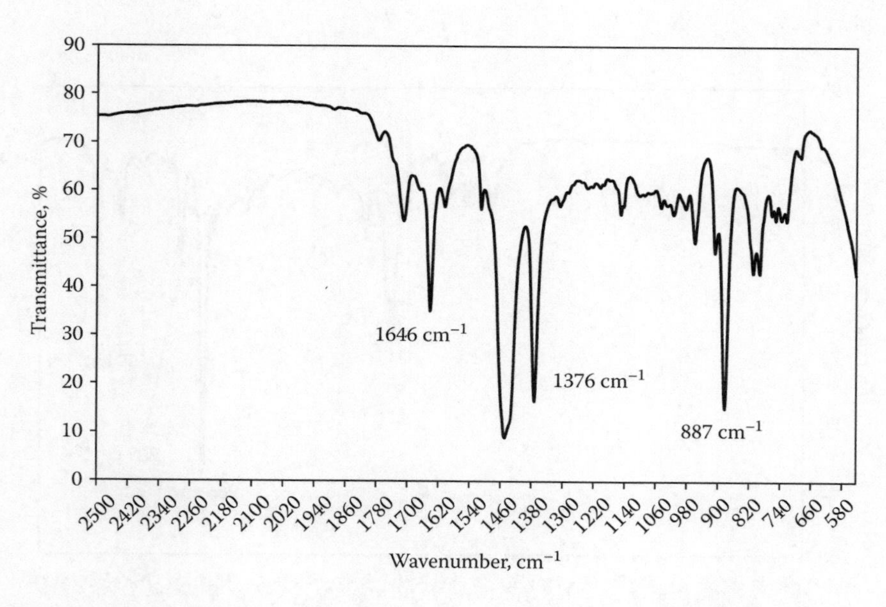

FIGURE 4.18
FTIR spectrum of natural rubber (NR).

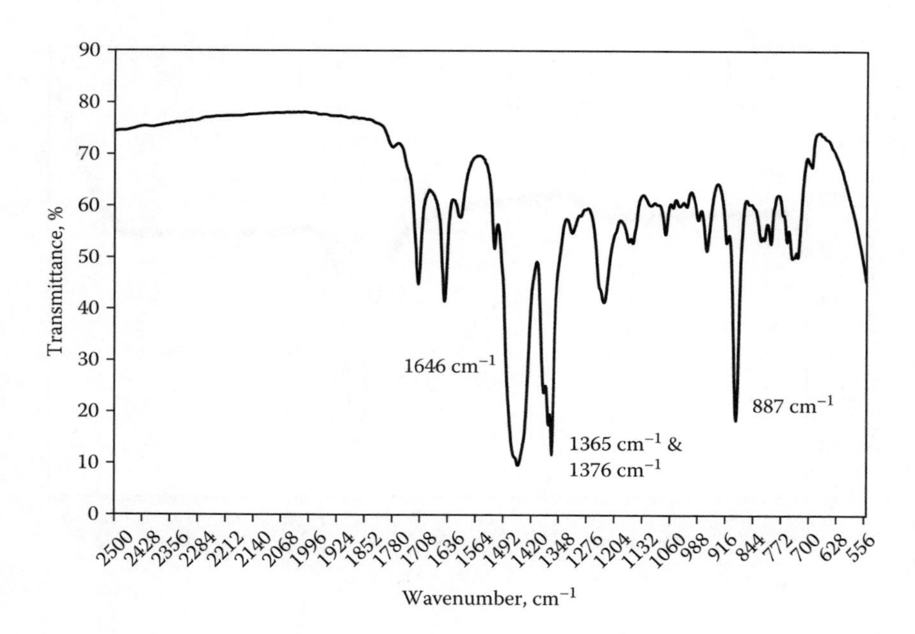

FIGURE 4.19
FTIR spectrum of chlorobutyl rubber (CIIR).

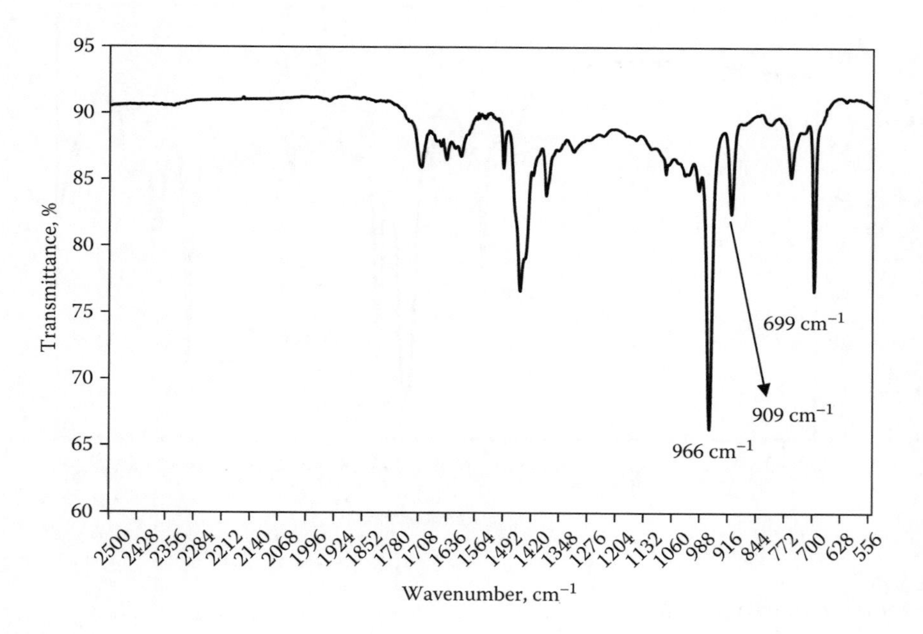

FIGURE 4.20
FTIR spectrum of styrene butadiene rubber (SBR).

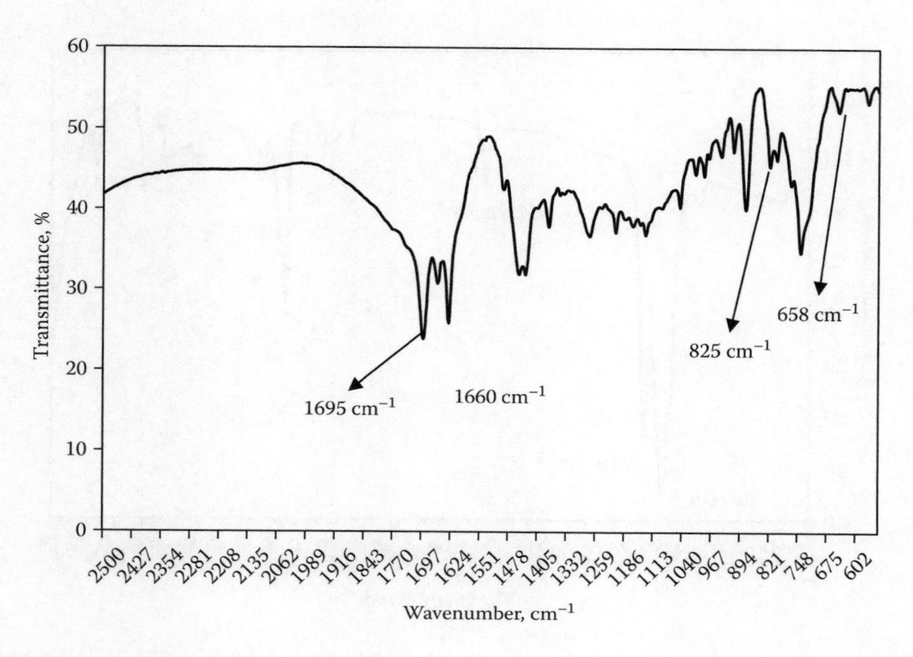

FIGURE 4.21
FTIR spectrum of polychloroprene rubber (CR).

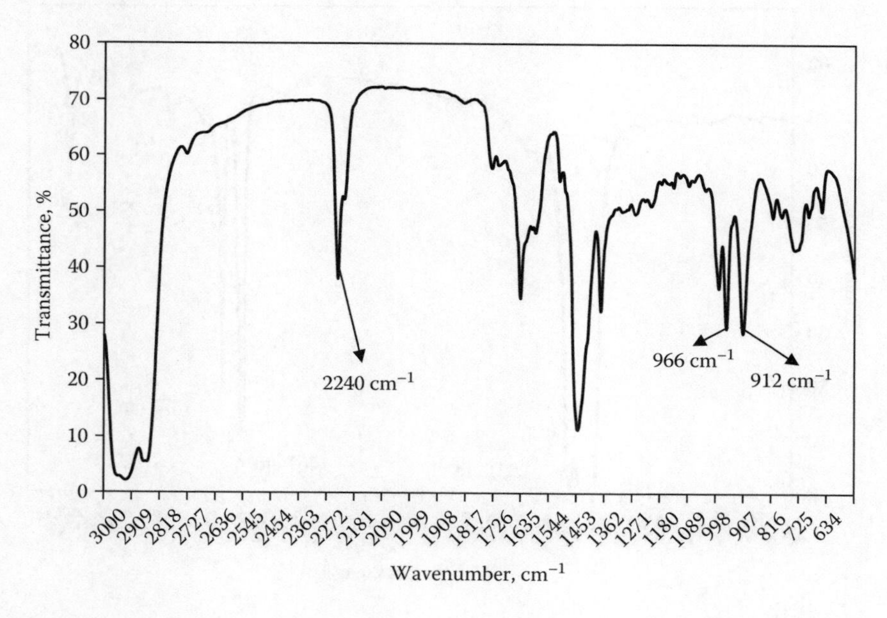

FIGURE 4.22
FTIR spectrum of acrylonitrile butadiene rubber (NBR).

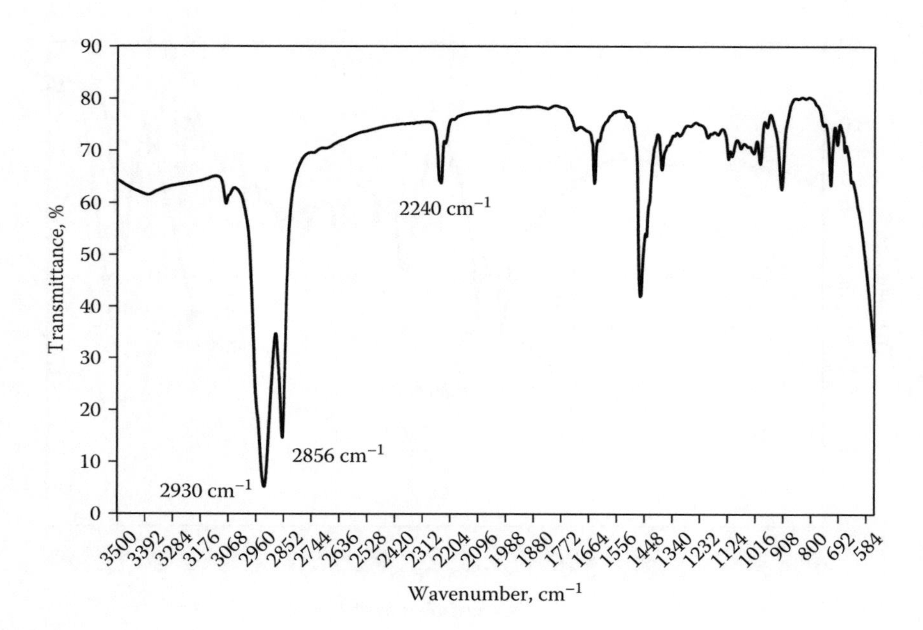

FIGURE 4.23
FTIR spectrum of hydrogenated acrylonitrile butadiene rubber (HNBR).

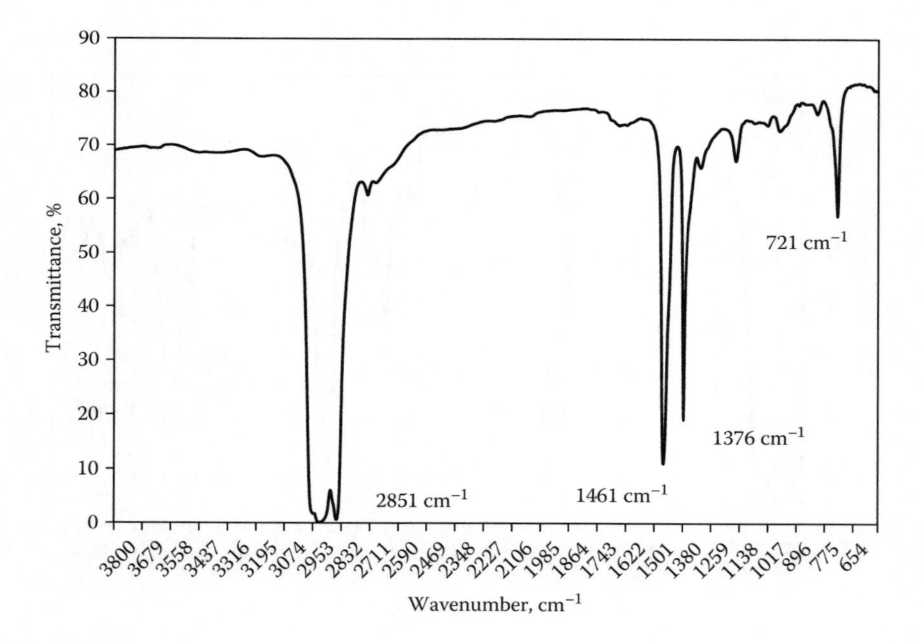

FIGURE 4.24
FTIR spectrum of ethylene propylene diene monomer (EPDM) rubber.

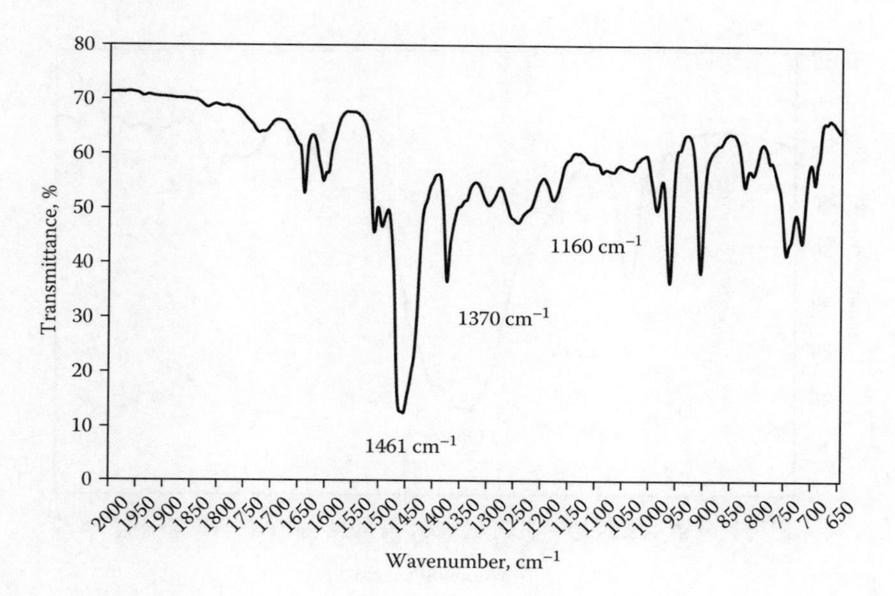

FIGURE 4.25
FTIR spectrum of chlorosulfonated polyethylene rubber (CSM).

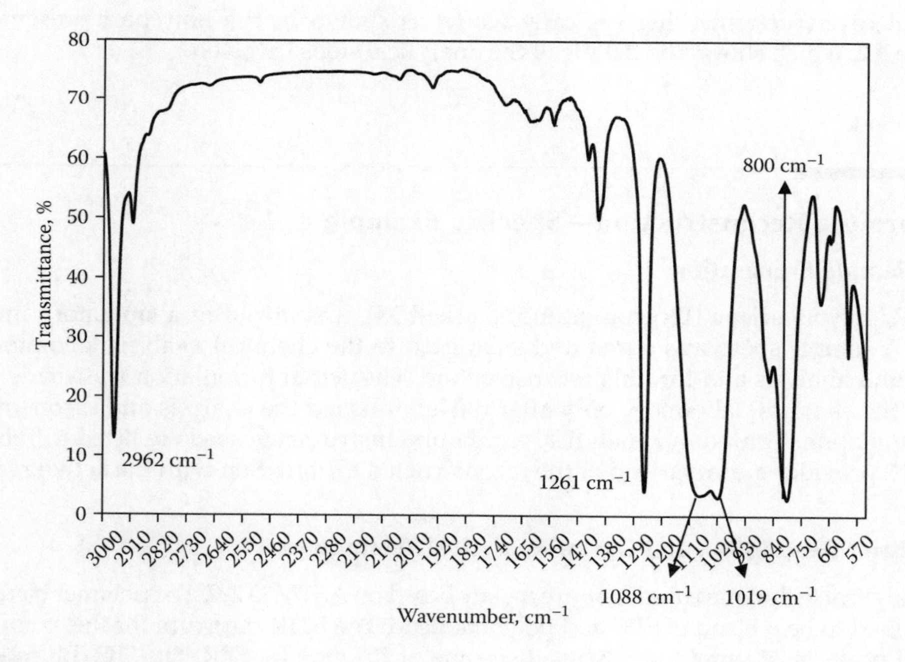

FIGURE 4.26
FTIR spectrum of silicone rubber (MQ) in the region of 570 to 3000 cm^{-1}.

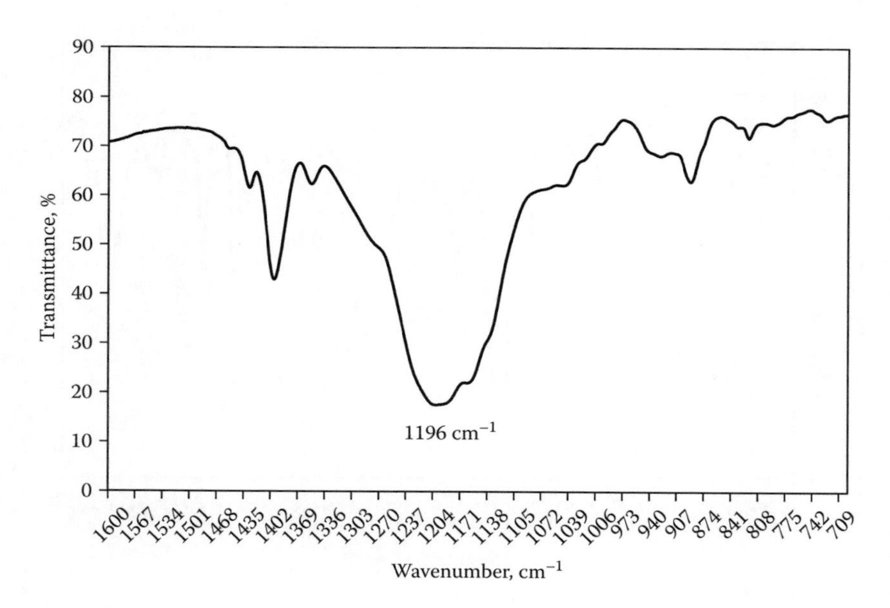

FIGURE 4.27
FTIR spectrum of silicone rubber (MQ) in the region of 700 to 1600 cm^{-1}.

Material reverse engineering typically works as shown in the flowchart presented in Scheme 4.2, which shows the details of the analytical steps involved.

4.2 Formula Reconstruction—Specific Example

4.2.1 Sample Preparation

An SBR/polybutadiene (BR) compound, marked XB, was mixed in a laboratory internal mixer. A sample sheet was cured and submitted to the chemical analysis laboratory for compound analysis and formula reconstruction. The actual formulation was made available to the chemical laboratory only after the lab finished the analysis and reconstructed the formulation. Compound analytical results and instruments used are listed in Table 4.1. Table 4.2 provides a comparison of the reconstructed formulation with the actual recipe.

4.2.2 Brief Description and Use of Analytical Techniques

The analytical techniques used here are mostly based on ASTM D 297. The polymer blend was determined to be a blend of SBR and polybutadiene. The FTIR spectrum for this compound showed peaks at 720 cm^{-1} for polybutadiene and at 707 cm^{-1} for SBR. The SBR/BR ratio was determined by comparing with a previously prepared standard SBR/BR calibration curve.

From TEM particle size image analysis, the average particle size of carbon black was found to be 23.59 microns, which falls within the ASTM standard particle size range of

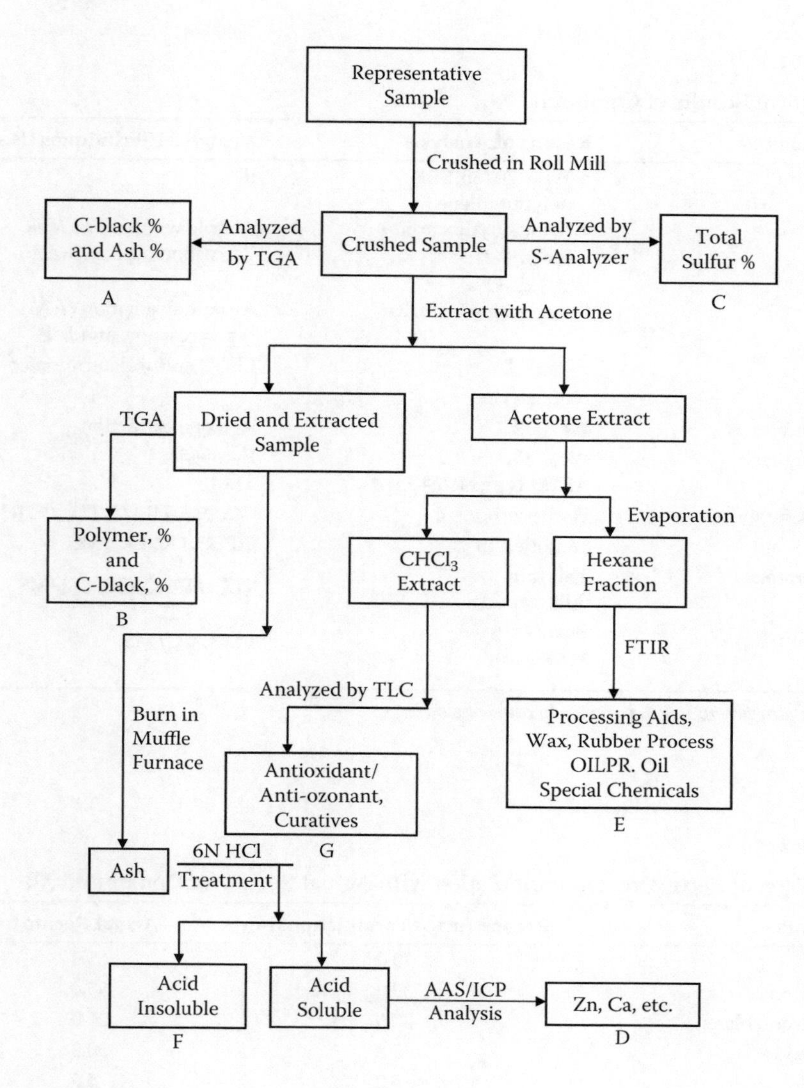

SCHEME 4.2
Typical flowchart of material reverse engineering.

20–25 microns for the N-200 family of carbon black. The carbon black used in compound XB was determined to be N-234.

Acetone extract was analyzed for identification and quantification of hydrocarbon oil, antioxidant, anti-ozonant, and accelerators using gas chromatography-mass spectrometry (GC-MS), high-performance liquid chromatography (HPLC), thin-layer chromatography (TLC), and gas chromatography (GC). These techniques give semi-quantitative analytical results. Hydrocarbon oil was identified to be primarily naphthenic oil. The antioxidant was identified to be ZMTI (zinc *z*-mercaptotolylimidazole), and the anti-ozonant was identified to be 6PPD (*N*-1,3-dimethylbutyl-*N*-phenyl-*p*-phenylenediamine). Based on the detection of methyl groups, the primary accelerator was identified as sulfenamide type, whereas

TABLE 4.1

Analytical Results of Compound XB

Components	Results of Analysis	Analytical Techniques Used
Polymers	• Type, 70/30, SBR/ polybutadiene	IR
	• % total hydrocarbon (HC), 48.4	Sample wt (% extractable + % carbon black + % ash)
Ash	Wt%, 23.8	
	Composition → Si, Zn	Atomic absorption (AA) spectroscopy, and ICP
Sulfur	Wt%, 1.47	LECO sulfur determinator
Wax	Not detected	
Extractables	Wt%, 12.2	Acetone extraction
Carbon Black	Wt%, 13.4	Pyrolysis
	ASTM type N-234	TEM
Extract Analysis	Hydrocarbon oil	GC/MS, HPLC, TLC, FTIR
Antioxidant	Santoflex 13	HPLC, GC, GC/MS
Accelerators	Thiuram Sulfenamide	TLC, HPLC, GC, GC/MS
Other	Phenolic resin Stearic acid	FTIR, GC/MS

HC = hydrocarbon; ICP = inductively coupled plasma.

TABLE 4.2

Comparison of Reconstructed Formulation with Actual Recipe for Compound XB

Ingredients	Reconstructed Formulation (phr)	Actual Recipe (phr)
SBR	70.0	65.0
Polybutadiene	30.0	35.0
N-234 Carbon black	27.6	25.0
Silica dioxide	45.8	50.0
Zinc oxide	5.0	5.0
Hydrocarbon oil	10.8	10.0[a]
Phenolic resins	7.0	6.0
Stearic acid	1.0	1.0
6PPD	2.0	2.0
ZMTI	2.0	1.8
Miscellaneous extractables[b]	1.0	—
TMTM, TMTD, methyl zimate	0.5	0.3 (TMTD)
Sulfenamide accelerator	1.5	1.8
Sulfur	1.7	1.8
Total	205.9	204.7

[a] Naphthenic oil.

[b] Extractables may contain processing aids, etc.

secondary accelerators were identified to be TMTD (tetramethylthiuram disulfide), TMTM (tetramethylthiuram monosulfide), and methyl zimate (zinc dimethyldithiocarbamate). Techniques used here could not distinguish between these accelerators.

The ash content was determined by pyrolyzing a known quantity of sample in air at 800–900°C, and it was found to be 23.8% by weight. The residual ash consisted of mostly silica and zinc as determined by atomic absorption (AA) and inductively coupled plasma (ICP) (see Table 4.1).

4.2.3 Formula Reconstruction

After all analytical results, shown in Table 4.1, were obtained, proprietary software was used to reconstruct the formulation. However, one can also do the formula reconstruction manually as described here.

All rubber hydrocarbon is taken as 100 phr. Other ingredients are calculated on phr basis. For example, in Table 4.1, % carbon black was found to be 13.4 for 48.4 wt% of rubber hydrocarbon. Hence, carbon black content was 27.6 phr. Silica and zinc oxide were calculated (based on AA and ICP data on ash analysis) to be 44.2 phr and 5.0 phr, respectively. Similarly, other ingredients were calculated.

To determine the actual phr of sulfur, some correction needs to be made for other sources of sulfur in the rubber compound. The measured amount of sulfur comes from added sulfur, sulfur from sulfur-donor accelerators, and some sulfur from carbon black. In Table 4.1 sulfur was determined to be 1.47 wt% based on 48.4 wt% RHC. So, on a phr basis, the sulfur content should be 3.03. But correction was made for other sources of sulfur, and reformulated sulfur was 1.7 phr.

4.2.4 Comparison of Reconstructed Formulation with Actual Recipe

Table 4.2 shows the comparison. In general, there is good agreement. The reconstructed ratio of SBR/polybutadiene is within the acceptable range calculated from FTIR data. Carbon black, silica, zinc oxide, hydrocarbon oil, phenolic resin, and stearic acid compare favorably with respective amounts in the actual recipe. After correcting for other sources of sulfur, reconstructed sulfur (phr) with sulfur content in the actual recipe is comparable. We could not, however, determine the exact accelerator based on methyl group detection.

We have found that a good understanding of rubber compounding and compound recipes is helpful in fine-tuning the formulation reconstructed from the analytical results.

4.3 Numerical Problem on Reverse Engineering

A natural rubber (NR) vulcanizate on physical testing gave the following information:

1.	Specific gravity	1.10
2.	International Rubber Hardness Degrees (IRHD) hardness	63
3.	Tensile strength	24.2 MPa
4.	Modulus at 300% elongation	8.1 MPa
5.	Ultimate elongation	500%

Further analysis of the vulcanizate showed:

Ash (all soluble in HCl)	2.5%
Acetone extract	17.5%
Carbon black content	35%
Total sulfur	1.2%

From the above information, deduce the composition of the compound:

Composition of the Mix		%
Ash (all soluble in HCl)		2.5
Acetone extract		17.5
Carbon black content		35.0
Sulfur		1.2
Total		56.2
Rubber hydrocarbon:		100 – 56.2 = 43.8
Constituents (phr)		
Rubber (NR)	—	100
Ash	$\frac{2.5 \times 100}{43.8}$	5.7
Carbon black	$\frac{35 \times 100}{43.8}$	79.9
Acetone extract	$\frac{17.5 \times 100}{43.8}$	39.9
Sulfur	$\frac{1.2 \times 100}{43.8}$	2.7

The ash is acid soluble and hence it may be expected to consist entirely of zinc oxide.

Acetone extract is composed of oil, antioxidants, accelerators, tackifiers, etc.

5

Formulation Reconstruction: Case Studies

5.1 Tire Tread Cap

5.1.1 Objective

The objective of this study is to present a detailed overview of the chemical and analytical techniques and physical testing that are used for analysis of a passenger car tire tread compound. On the basis of the analysis, the tread compound formula is reconstructed.

5.1.2 Experimental Methods

Reverse engineering analysis of a tread rubber sample follows these steps:

- Extraction of rubber compound
- Polymer characterization and quantification
- Determination of carbon black content
- Identification as well as quantification of inorganic materials in the compound
- Characterization of extract for identification of volatile components of the rubber compound
- Determination of sulfur level

Methods such as TGA, FTIR, pyrolysis FTIR, GC-MS, sulfur and ash analysis, discussed in earlier chapters, have been used in analyzing the materials present in the tread rubber sample.

5.1.2.1 Sample Preparation

The tire tread sample with unknown composition was collected from the tire. The tread cap was separated from the tread base using the splitting machine. The tread cap sample was passed through the closed-nip two-roll mill to reduce the sample size. Mechanical sample crushing was important to prepare the homogenized sample.

FIGURE 5.1
Soxhlet extraction setup.

5.1.2.2 Extraction

Extraction of the crushed tread sample was carried out using Soxhlet-type extraction. Figure 5.1 shows a typical Soxhlet extraction setup. The solvent used for tread sample extraction was acetone. Extraction was generally carried out for 16 h at a temperature of 70°C. The extracted rubber sample was heated under vacuum for complete dryness. The solvent along with the extract was stored for further analysis.

5.1.2.3 Composition Analysis

A thermogravimetric analyzer (TGA) was used to determine the composition with respect to polymer, carbon black, and ash content in the tread rubber sample. The crushed sample was heated in TGA from 80 to 850°C at a 40°C/min heating rate. Heating of the sample was performed under nitrogen and oxygen atmosphere. Initially the sample was heated under nitrogen from 80 to 600°C and then the sample was heated under oxygen to 800°C.

5.1.2.3.1 Analysis of Rubber

Rubber of the tread sample was analyzed by pyrolysis FTIR. The extracted tread sample was pyrolyzed in inert atmosphere at 500°C. The condensate of the pyrolysis product was collected on the salt plate and analyzed by FTIR for the wavenumber range of 4000 to 400 cm^{-1}. However, for analysis of the FTIR spectrum, a 2000 to 600 cm^{-1} range was considered.

5.1.2.3.2 Analysis of Volatile Components

Volatile components were analyzed with the extract. After proper drying of the extract, the same was analyzed using GC with mass detector.

5.1.2.3.3 *Analysis of Sulfur*

One of the important curing materials was sulfur. Analysis of sulfur was conducted with the unextracted sample using an elemental analyzer.

5.1.2.3.4 *Analysis of Ash*

For ash analysis, a large quantity of tread rubber sample was burnt inside the muffle furnace and the total ash was measured. This ash was treated with dilute HCl and found totally dissolved in HCl. The acid-dissolved ash was characterized with inductively coupled plasma (ICP) spectroscopy to determine the type of metal present in the ash.

5.1.3 Results and Discussion

From the TGA thermogram (Figure 5.2), the carbon black and ash content were calculated. In Figure 5.2, weight loss represented by "A" is total content of polymer and volatile components, and weight loss represented by "B" is total carbon black content. "C" represents total residue that is also termed *ash content*. Carbon black and ash content of the tread sample from the thermogram were calculated as 30.5% and 2.5%, respectively. A second TGA analysis run of the tread rubber after extraction was also performed to calculate the polymer content and carbon.

From Figure 5.3, the polymer content represented by the weight loss "D" and carbon black content represented by "E" were calculated. The values obtained were 63.2% and 33.9%, respectively. With the help of the data, the polymer, carbon black, volatile matter, and ash content are estimated and represented in Table 5.1.

The derivative of TGA (DTGA) thermogram of the "after extraction" rubber sample was also considered to characterize the presence of polymer blends. The presence of polymer blends was characterized by more than one decomposition peak in the polymer

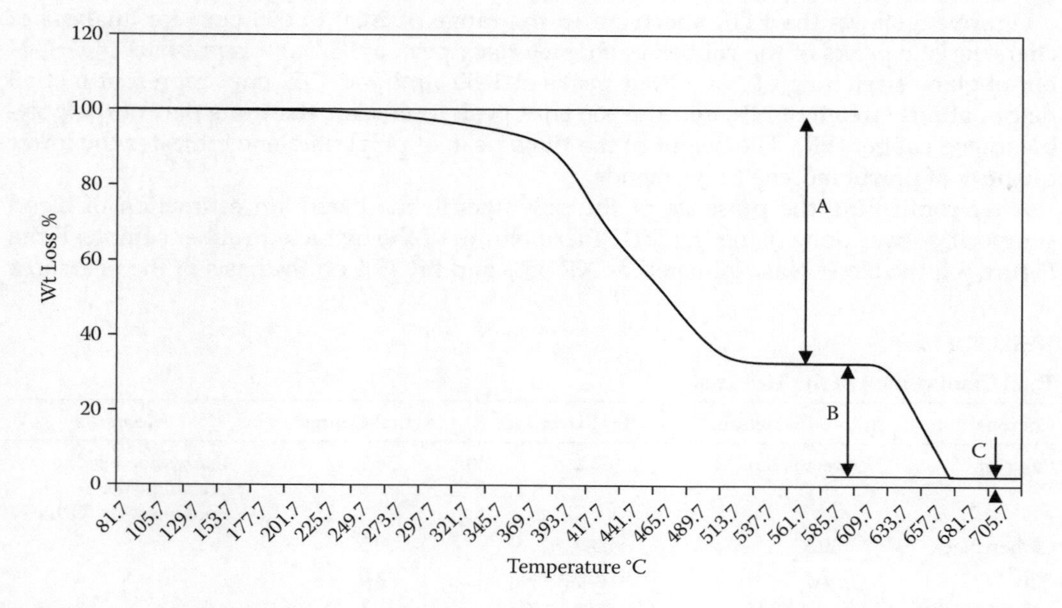

FIGURE 5.2
TGA of tread cap.

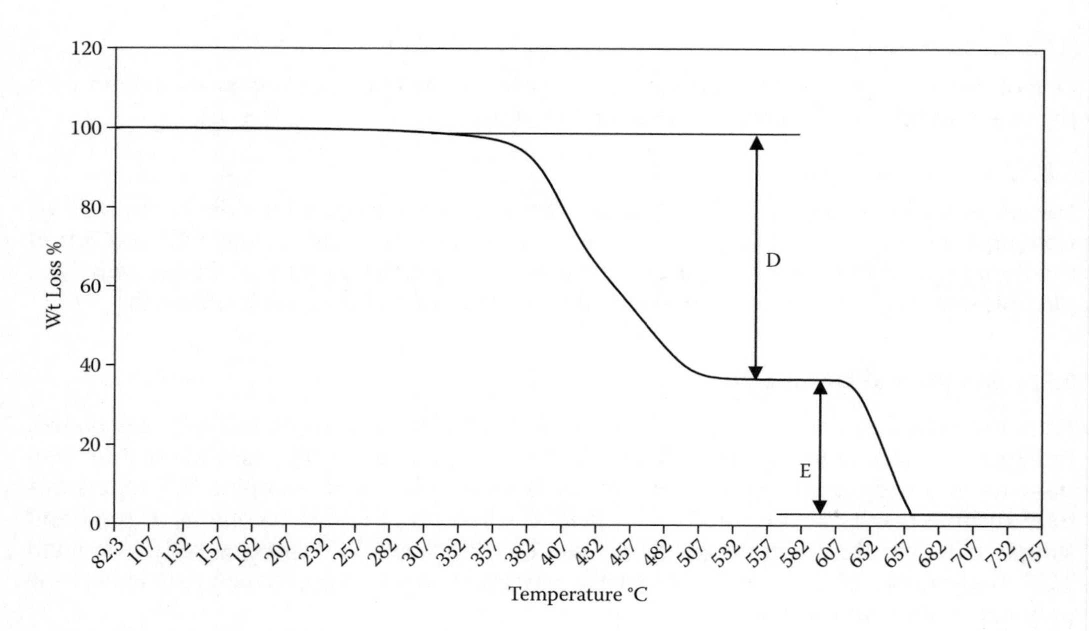

FIGURE 5.3
TGA of tread cap after extraction.

decomposition area (i.e., 80 to 600°C). Figure 5.4 shows the DTGA thermogram of the extracted tread sample.

DTGA was also used to estimate the blend ratio. The areas under the corresponding decomposition peaks were estimated for calculation of blend percent. However, before calculation of the blend ratio, the polymer type was identified by pyrolysis FTIR analysis.

Figure 5.5 shows the FTIR spectrum in the range of 2000 to 600 cm^{-1} for analysis of characteristic peaks of the rubber. A characteristic peak at 887 cm^{-1} represents the =C-H out-of-plane stretching of NR. Other peaks at 1450 cm^{-1} and 1377 cm^{-1} represent methyl deformations stretch of NR, and the 966 cm^{-1} peak represents the trans part of the polybutadiene rubber (BR). The height of the trans peak of polybutadiene indicates the lower quantity of polybutadiene in the blends.

After confirming the presence of the polymers in the blend, an estimation of blend composition was done using the DTG thermogram of the extracted rubber sample. From Figure 5.4, the blend was estimated as NR 65% and BR 35% on the basis of the peak area

TABLE 5.1

Total Composition of the Tire Tread

Composition	Before Extraction, %	After Extraction, %	Actual Content, %	Remarks
Polymer	Not calculated	63.2	56.9	Expression used to calculate the polymer% is (D*B)/E[a]
Carbon black	30.5	33.9	30.5	
Ash	2.5	Not calculated	2.5	
Volatile matter	Not calculated	Not calculated	10.1	

[a] Refer to Figures 5.2 and 5.3

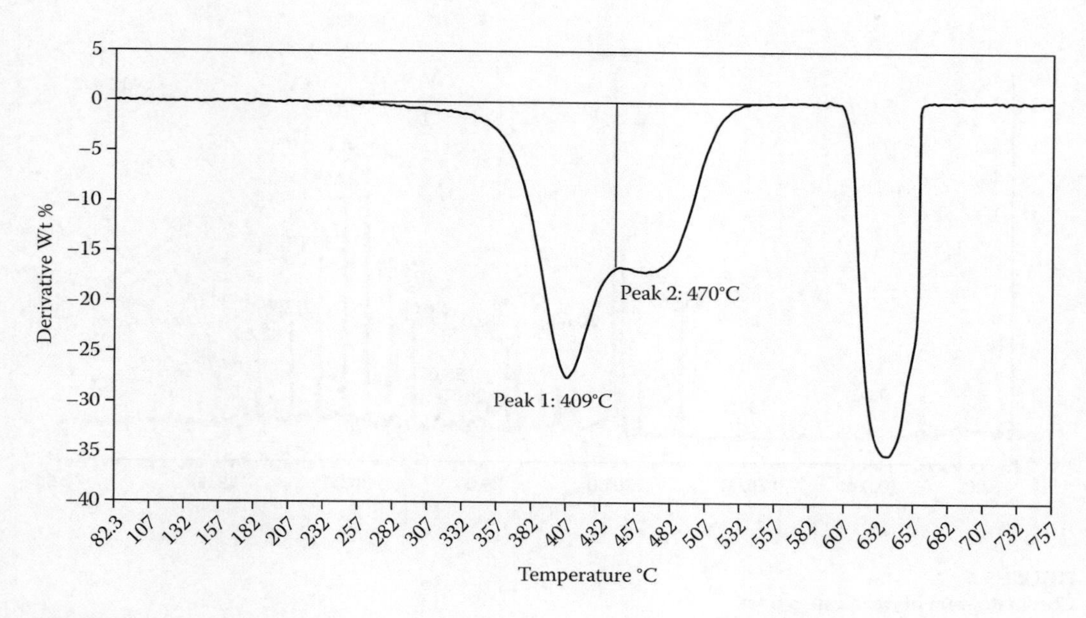

FIGURE 5.4
DTGA of tread cap after extraction.

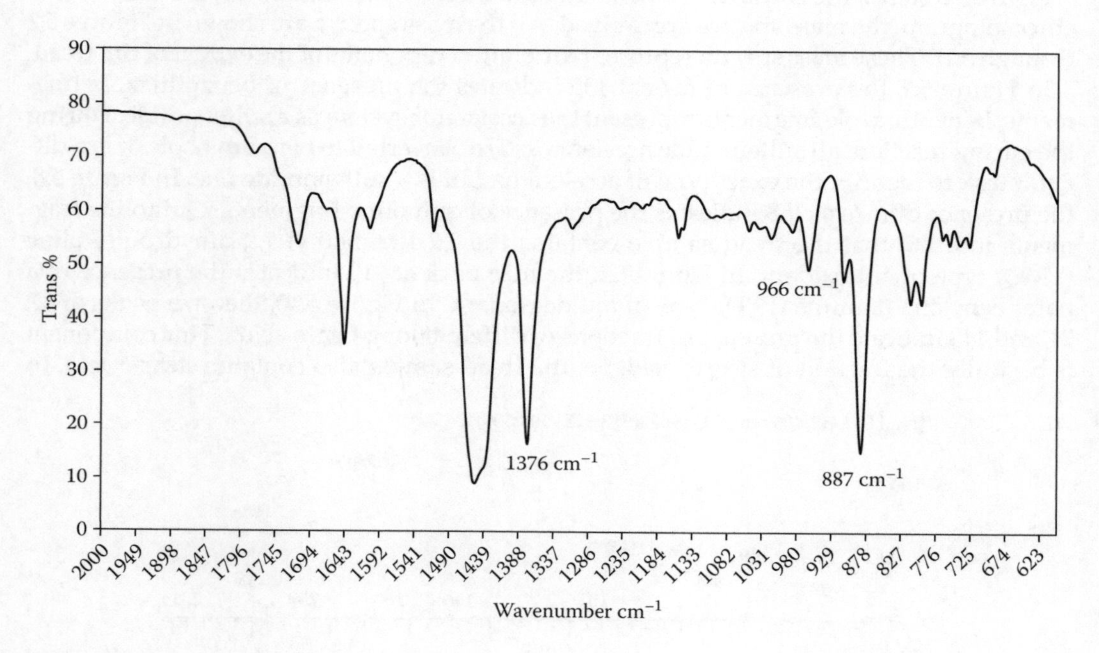

FIGURE 5.5
FTIR spectrum of tread cap after extraction.

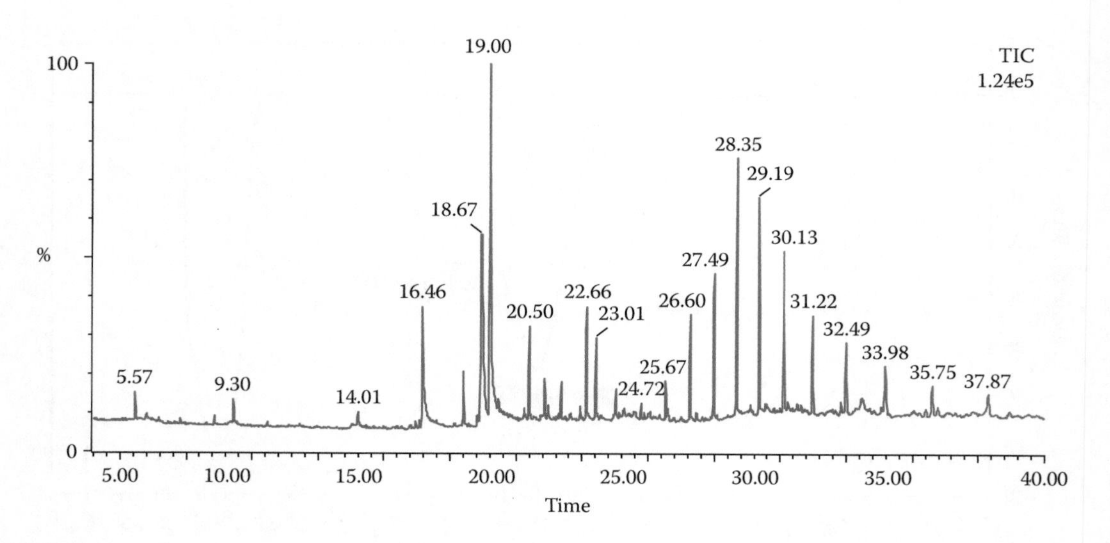

FIGURE 5.6
Chromatogram of tread cap extract.

calculation. For this calculation, thermal software support is necessary. It was observed in different cases that the blend ratio calculated on the basis of this principle is accurate up to a 98% confidence level. However, in the case of tri blends of polymer, this peak area calculation for the determination of polymer blends is not at all accurate. Inaccuracy is also observed for SBR-BR blends in a sample.

Figure 5.6 shows the chromatogram obtained with the tread extract. On the basis of this chromatogram, the mass spectra are derived. All the mass spectra are shown in Figures 5.7 through 5.11. These mass spectra represent different components of the extract of tire tread.

In Figure 5.7, the presence of m/e at 135 indicates the presence of benzothiazole fragments. Benzothiazole fragments represent the accelerator system as a sulfenamide. During the curing reaction, all sulfenamide accelerators are converted to benzothiozole. It is a difficult task to identify the exact type of accelerator, if it is a sulfenamide one. In Figure 5.8, the presence of m/e at 158 indicates the presence of quinoline fragments. Quinoline fragments indicate that the tread sample contains the 2,2,4 trimethyl 1,2 dihydroquinoline (TMQ) type of antioxidant. In Figure 5.9, the m/e peak at 212 indicates the presence of a paraphenylene diamine (PPD)-type of antidegradant. In Figure 5.10, the m/e peaks at 87, 74, and 143 indicate the presence of fractions of different long-chain acids. This component is basically the fraction of stearic acid. So, the tread sample also contains stearic acid. In

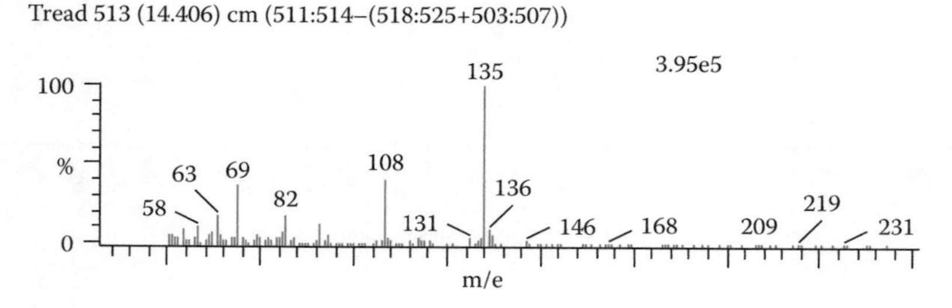

FIGURE 5.7
Mass spectrum of tread cap extract.

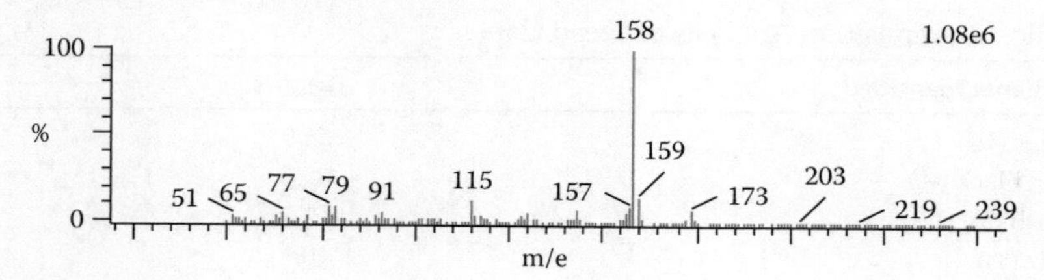

FIGURE 5.8
Mass spectrum of tread cap extract.

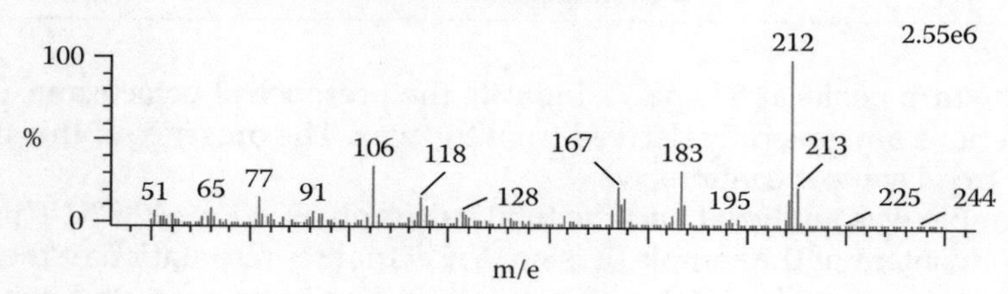

FIGURE 5.9
Mass spectrum of tread cap extract.

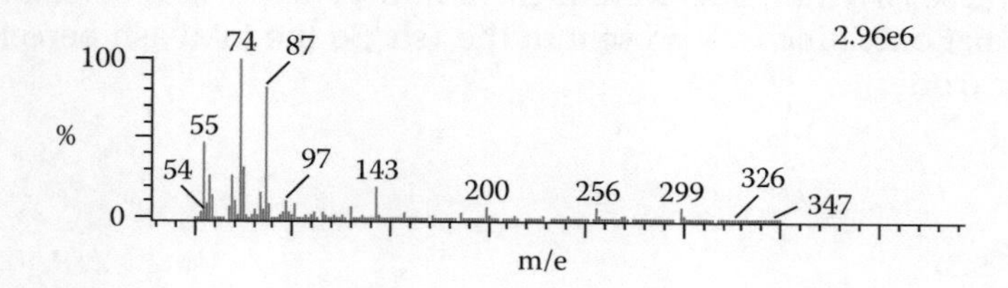

FIGURE 5.10
Mass spectrum of tread cap extract.

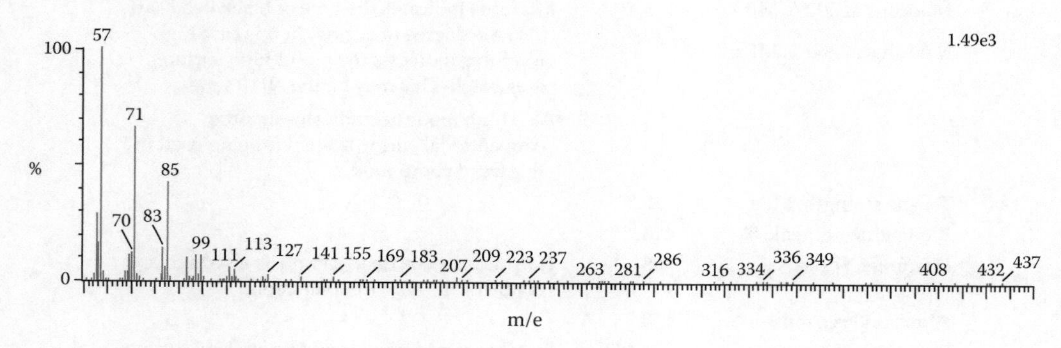

FIGURE 5.11
Mass spectrum of tread cap extract.

TABLE 5.2

Chemical Composition Analysis of Tread Cap

Ingredients Identified	Results
Polymer (%)	56.9
Carbon black (%)	30.5
Total ash (%)	2.5
Volatile (%)	10.1
ZnO (%)	2.5
Polymer type	Blends of NR and BR. The blend ratio is 65:35.
Total sulfur (%) after correction	2.19 (considering 1% correction factor of total carbon black)
Accelerator	Sulfenamide
Antidegradants	TMQ, PPD, wax
Activator organic	Stearic acid
Plasticizer type	Aromatic oil

Figure 5.11, the m/e peaks at 57 and 71 indicate the presence of octacosane, pentacosane, etc. These fractions are generally derived from the wax. The presence of this fraction indicates that the tread sample contains wax.

Crushed sample was analyzed and the total sulfur obtained was 2.5%. This sulfur was the total sulfur content of the sample. It is not the actual representation of the sulfur present in the tread sample. The total sulfur obtained has to be corrected considering the sulfur contributions from carbon black, process oils, and accelerators. On average, a correction factor of 1% of the total carbon black is subtracted from the sulfur content to get the actual sulfur content of the sample. A correction factor for oil and accelerators was not considered, as these two materials were present in the sample in only small quantities. It was identified that only zinc was present in the ash. So the total ash percent in the tread rubber was zinc oxide.

TABLE 5.3

Properties of Tread Cap

Mechanical Properties	Results	Implication in Formula Reconstruction
Modulus at 100%, MPa	4.4	Modulus indicates the type of black used and also the degree of curing. In this case high modulus indicates the use of high surface area black. This may be the N110 series.
Modulus at 300%, MPa	19.9	Also high modulus indicates that the conventional curing system is being used in this tread compound.
Tensile strength, MPa	25.8	—
Elongation at break, %	418	—
Hardness, Shore A	68	Hardness in this tread sample is also high and confirms the above explanation.
Abrasion loss, mm^3	108	—
Specific gravity, gm/cc	1.145	Suitable ingredients should be selected during the formula reconstruction process.

TABLE 5.4
Final Formulation

Ingredients	phr
NR	65
BR	35
N134	54
ZnO	4.5
Stearic acid	2.5
6PPD	2
TMQ	1
Wax	1.2
Process oil	8
Sulfur	2.25
TBBS	1.2
Retarder	0.2

The observations from the above chemical as well as instrumental analyses are tabulated in Table 5.2. Information about chemical composition is sometimes not enough to do the formula reconstruction. In that case, information about mechanical properties is equally important. Important mechanical properties are modulus at 100% and 300%, tensile strength, hardness, and specific gravity. In this case, as it is a tread rubber sample, in addition to the mentioned mechanical properties, other important properties like abrasion resistance, heat buildup, and dynamic mechanical properties are important. In this case the properties measured are shown in Table 5.3. Hence, the final formula may be as shown in Table 5.4.

5.2 Tire Sidewall

5.2.1 Objective

The objective of this analysis is to estimate quantitatively different materials present in the tire sidewall through different macro, micro, and semi-micro analyses.

5.2.2 Experimental Methods

Tire sidewall is a component that protects the carcass from external impact. It also acts as a suspension system of the tire through its flexibility.

5.2.2.1 Sample Preparation

Generally, the sidewall is a very thin, membrane-like rubber part of a tire, and hence the collection of the sidewall sample is critical. One should take care to cut the rubber sample from the tire to avoid the possibility of contamination of sidewall with carcass compound. The proper sidewall component is then crushed for homogenization and extracted with a suitable solvent, acetone. Solvent extraction should be carried out until all of the extractables are extracted by the solvent from the rubber matrix. It was observed that 16 h extraction at 70°C is effective to extract the highly volatile components. The quantitative extraction gives the amount to total extractable present in the sidewall compound. This also represents the volatile components present in the sidewall. After extraction, the rubber is dried under vacuum for further analysis. The extract is also kept for further analysis of the volatile components.

5.2.2.2 Reverse Engineering Analysis

In reverse engineering analysis, several instrumental techniques as described earlier are generally used for the analysis of different chemicals. The dried and extracted rubber part is then analyzed by TGA for macro composition study. Macro composition analysis generally gives quantitative amounts of polymer, carbon black, and ash content. The same rubber is also used to characterize the polymer type by pyrolysis FTIR technique. Pyrolysis FTIR is an absolute technique to identify a polymer from the vulcanized rubber compound. On the basis of the characteristics of absorption or transmittance, the materials are characterized by FTIR. Pyrolysis of the vulcanized rubber is carried out at 500°C under an inert atmosphere so as to protect the oxidation possibilities of the vulcanizate.

5.2.2.3 Micro Analysis

Micro analysis of the volatile components from the extract of the sidewall compound is carried out with GC-MS technique. In this analysis the extract is first dried and the extract is purged into the gas chromatography column with suitable solvent. Column selection is also an important part of the analysis. After chromatographic separation, the separated components are characterized by a mass detector. Mass detection of the separated components is carried out on the basis of the atomic mass unit of fragmented items from different volatile components.

5.2.2.4 Ash Analysis

Ash analysis is carried out by semi micro analysis techniques. The major equipment used for ash analysis is the atomic absorption spectrophotometer (AAS) or the inductively coupled plasma (ICP) spectrometer. However, before instrumental analysis, the ash is treated with different acids to find the solubility. Different acid-soluble fractions are indicators of the presence of different inorganic components in the ash. The HCl-soluble portion of the ash tells about the presence of metal oxide, viz., ZnO, MgO, CaO, etc. H_2SO_4-soluble fraction represents TiO_2. hydrofluonic acid (HF)-soluble fraction of the ash tells about the presence of silica.

5.2.2.5 Elemental Analysis

Sulfur content analysis of the sidewall compound is conducted by elemental analyzer.

5.2.3 Results

The analysis report obtained from the above analysis for a typical tire sidewall is presented in Table 5.5. With the above characterization methods, the results of the analysis are tabulated in Table 5.6.

TABLE 5.5

Reverse Engineering Results of Tire Sidewall

Test Parameters	Results	Explanation
Polymer	57.1%	Polymer content in sidewall compound was identified using TGA. The method of estimation of polymer content was elaborated in the earlier case studies. From the derivative of the TGA thermogram, the numbers of polymers present in the sidewall could be characterized. Please refer to the thermograms in Figures 5.12 through 5.14. Figure 5.12 is the thermogram of the sidewall compound without extraction. Figure 5.13 is the TGA thermogram of the extracted sidewall sample. Figure 5.14 is the first-order derivative of the extracted sidewall sample.
Carbon black	30.4%	Carbon black content was analyzed from the weight loss due to oxidation of the carbon black. This weight loss was obtained from the "before extraction" sidewall sample analysis by TGA. This carbon black content is reported in the composition analysis report. However, carbon black content was also measured from "after extraction" TGA thermogram. This is required to estimate the total polymer content using the empirical equation used in earlier case studies. Figures 5.12 and 5.13 are used to calculate the carbon black content.
Ash	2.5%	Ash content was measured from the TGA thermogram of the "before extracted" sidewall sample. Refer to the thermogram presented in Figure 5.12.
Volatile components	10.0%	Volatile components are basically high-boiling low molecular weight organic materials present in the tire sidewall. The best way to measure the volatile content is through quantitative extraction. Acetone was used in quantitative extraction of the sidewall. Total extraction time was 16 h for effective extraction.
Polymer type	Blend of NR and BR. Blend ratio is NR:BR, 45:55	Polymer type was characterized by pyrolysis FTIR. The pyrolysis FTIR spectrum is shown in Figure 5.15. Methyl deformations at 1450 cm^{-1} and at 1377 cm^{-1} are due to the polyisoprene part of NR/synthetic polyisoprene rubber. Another strong peak at 887 cm^{-1} is also an indication of the presence of polyisoprene unit. This 887 cm^{-1} peak is due to =C-H out-of-plane stretching. Trans and vinyl double-bond out-of-plane C-H bending at 964 cm^{-1} and 994 cm^{-1} and 908 cm^{-1} is due to the butadiene part of BR.
Ash analysis	HCl-soluble ash, 2.5%	Semi micro analysis of ash indicated that the ash was basically a metal oxide, preferably ZnO.
Total sulfur	0.85%	Total sulfur was characterized by the elemental sulfur analyzer. The sulfur content analysis was carried out with unextracted sidewall sample. A correction factor was required to be added in this case as the sample contained carbon black.
Analysis of volatile materials in solvent extract	1. Aromatic oil 2. Sulfenamide accelerator	Volatile components of the sidewall were treated as per earlier case studies. Figure 5.16 shows the FTIR spectrum of the aromatic oil. Figures 5.17 through 5.20 show the mass spectra of different components of the volatile fraction of the sidewall.

(*continued*)

TABLE 5.5

Reverse Engineering Results of Tire Sidewall (*continued*)

Test Parameters	Results	Explanation
	3. Stearic acid 4. PPD 5. TMQ	Solvent extract was dried with constant flow of nitrogen so as to avoid any chemical reaction of the extracted chemicals. Dried extract was swirled with hexane and separated through column chromatography. The toluene fraction was collected and prepared for FTIR analysis. The absorbance spectrum of the fraction was obtained. The characteristic aromatic peak was at 1600 cm^{-1} wavenumber. The oil was basically a mixture of all the components of aromatic, naphthanic, as well as paraffinic fraction. Characteristic peaks of paraffinic oil at 720 cm^{-1} were also present. The exact type of oil was estimated after calculation of C_A (aromatic) and C_P (paraffinic) content of the oils. Toluene fraction of the extract was analyzed by GC-MS. Mass spectrum is shown in Figure 5.17. The m/e at 268 was due to 6PPD mass fraction, and the corresponding m/e at 211 was due to TMQ mass fraction. Figure 5.18 is also the mass spectrum of toluene fraction. The m/e at 173 was also the fraction of TMQ antioxidant. The presence of mass fractions at m/e of 268, 211, and 173 indicated the presence of 6PPD and TMQ in the sidewall compound. Methanol fraction of the extract was analyzed by GC-MS. The mass spectrum is shown in Figure 5.19. The intense peak at m/e = 135 at low voltage (14 to 16 eV) indicated the presence of marcaptobenzothiazole. Marcaptobenzothiazole is the ultimate breakdown product of most of the sulfenamide accelerators. The 135 mass fraction was a clear indication of the presence of a sulfenamide accelerator. Figure 5.20 shows the GC-MS spectrum of methanol fraction of the sidewall extract. The m/e values at 172, 270, 284, 256, and 282 were due to different higher fatty acid fractions, viz., decanoic acid, octadecanoic acid, heptadecanoic acid, hexadecanoic acid, tetradecanoic acid, etc. All these acid fractions indicated the presence of stearic acid.

TABLE 5.6

Reverse Engineering Results

Ingredients Identified	Results
Polymer (%)	57.1
Carbon black (%)	30.4
Total ash (%)	2.5
HCl-soluble ash (%)	2.5
ZnO (%)	2.5
Volatile (%)	10.0
Polymer type	NR:BR (45:55)
Total sulfur (%), corrected	0.85
Plasticizer type	Aromatic oil
Antidegradants	6PPD, TMQ
Activator	Stearic acid
Accelerator	Sulfenamide

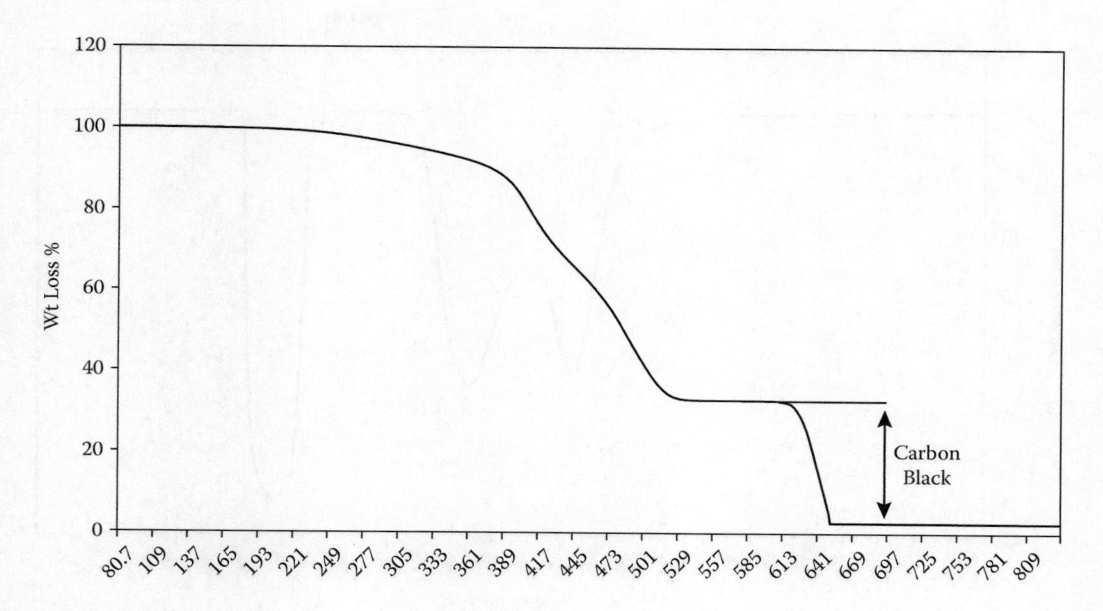

FIGURE 5.12
TGA of tire sidewall before extraction.

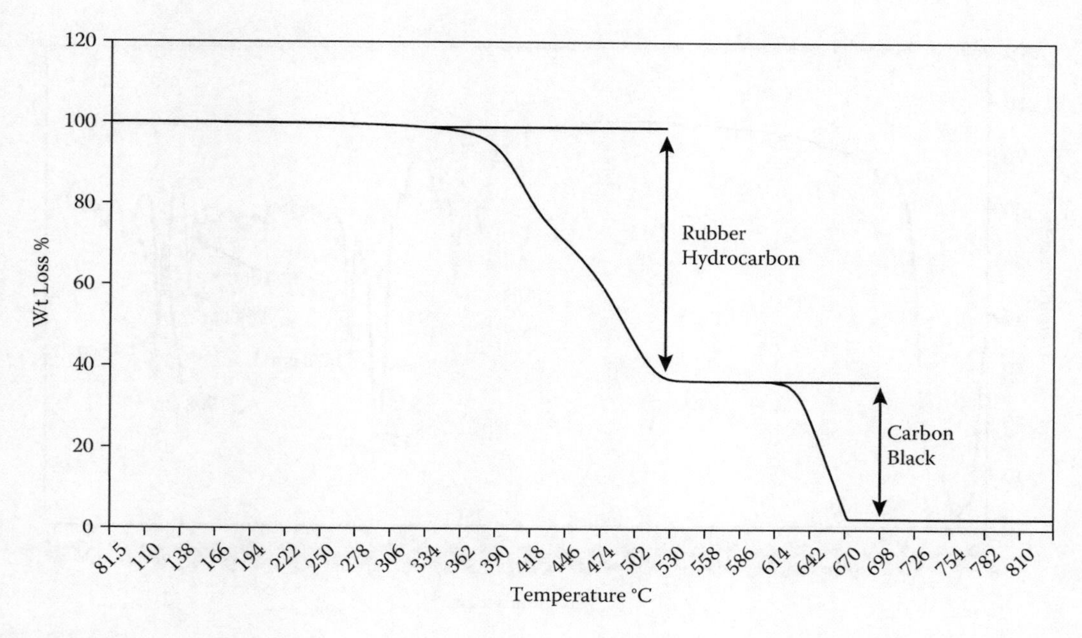

FIGURE 5.13
TGA of tire sidewall after extraction.

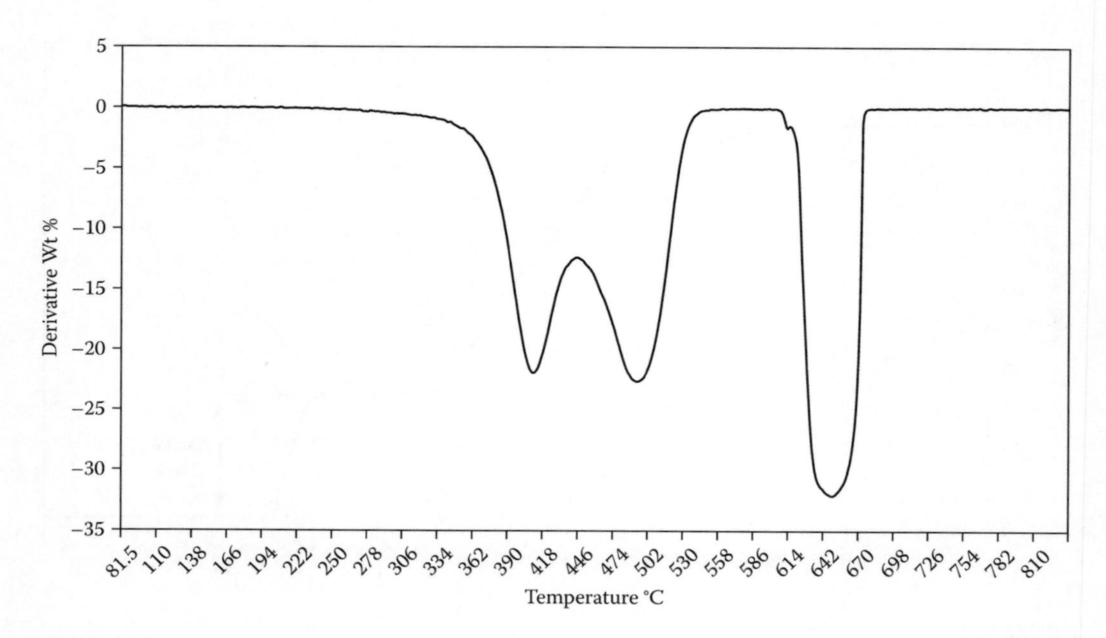

FIGURE 5.14
DTGA of tire sidewall after extraction.

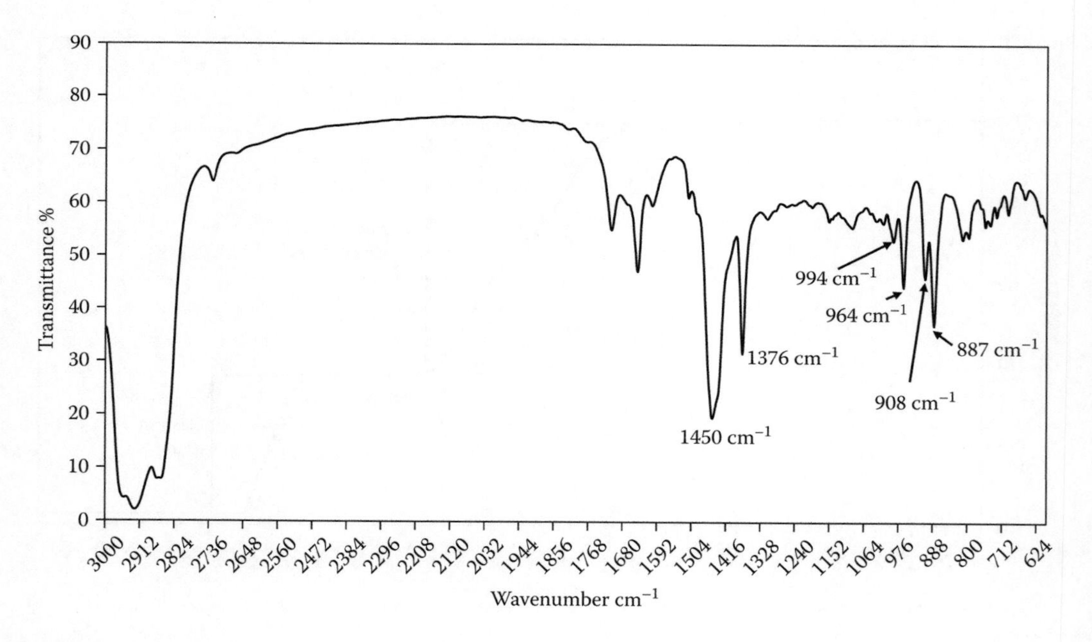

FIGURE 5.15
FTIR spectrum of tire sidewall after extraction.

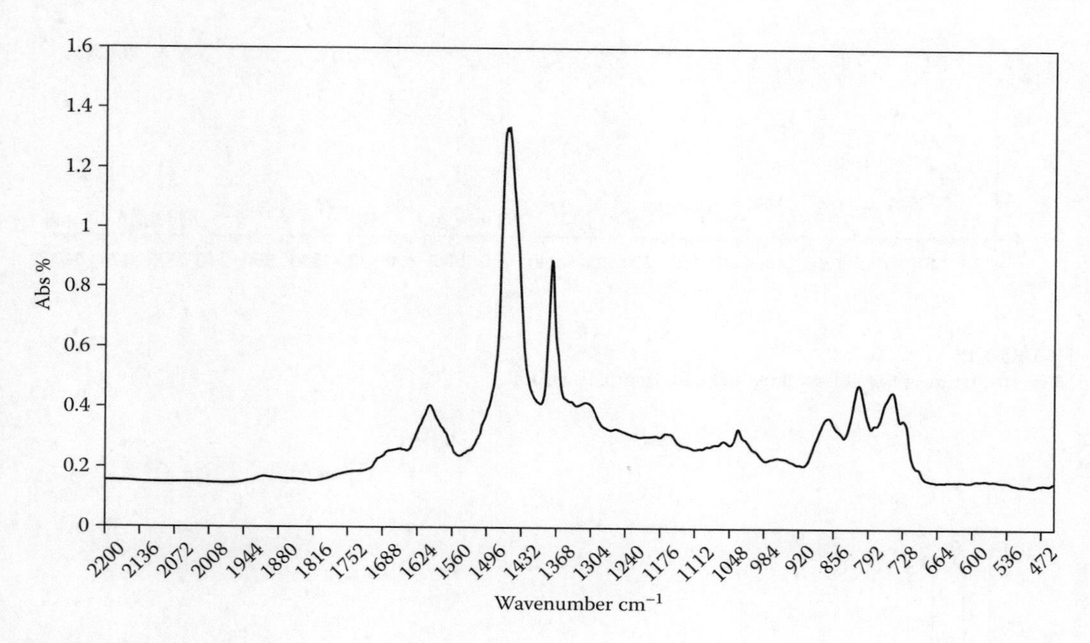

FIGURE 5.16
FTIR spectrum of aromatic oil in tire sidewall after extraction.

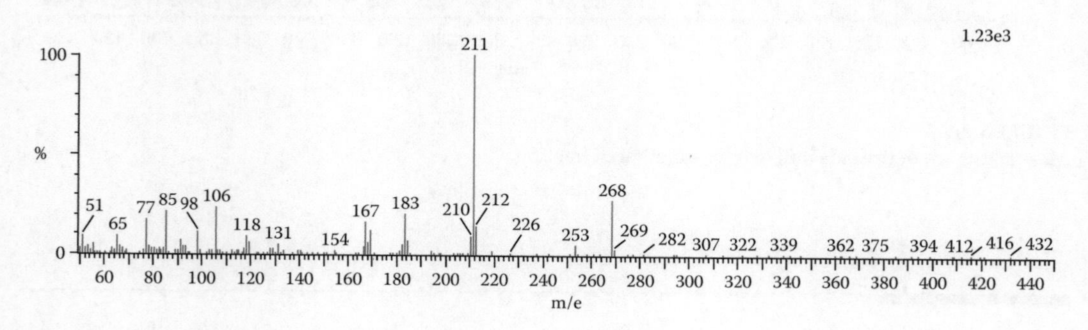

FIGURE 5.17
Mass spectrum of tire sidewall extract—toluene fraction.

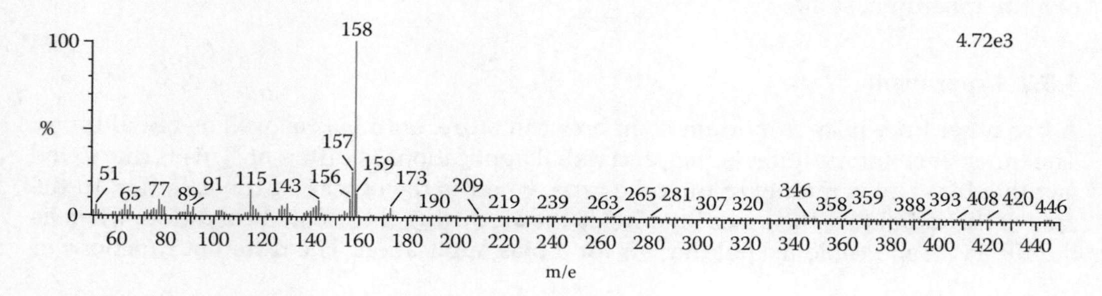

FIGURE 5.18
Mass spectrum of tire sidewall extract—toluene fraction.

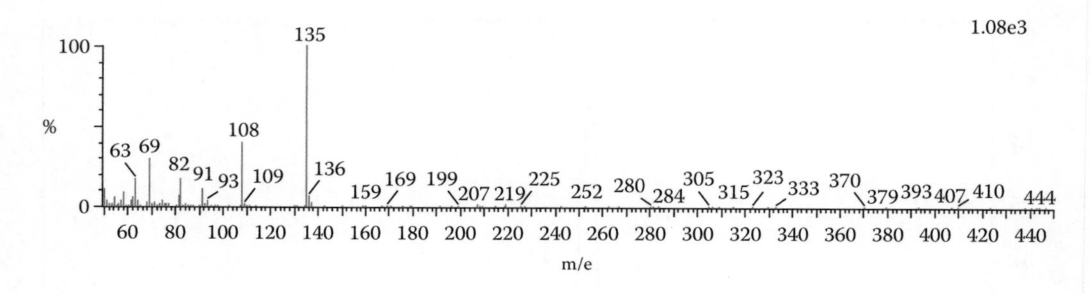

FIGURE 5.19
Mass spectrum of tire sidewall extract—methanol fraction.

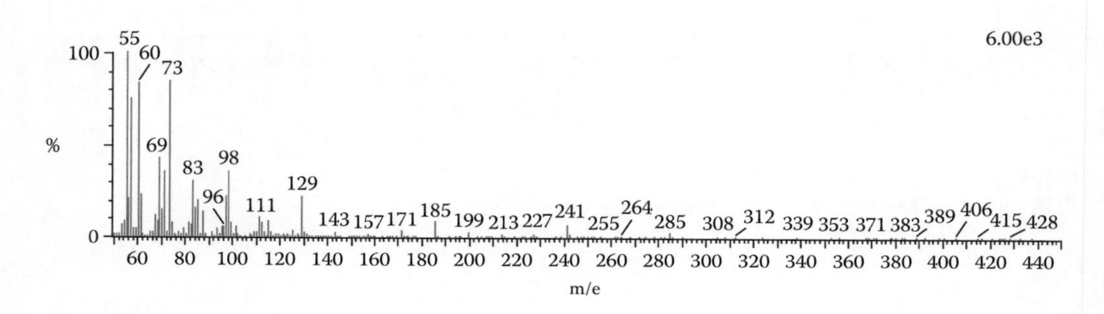

FIGURE 5.20
Mass spectrum of tire sidewall extract—methanol fraction.

5.3 Tire Inner Liner

5.3.1 Objective

The objective of this case study was to find out the composition of the rubber compounds of a tire inner liner sample.

5.3.2 Experiment

A tire inner liner is an important component in a tire, both bias as well as radial types. The inner liner composition in bias and radial applications is different. This is due to the fact that bias tire is mainly of the tube type, whereas the other is a tubeless tire. In the case of the radial tire, the inner liner holds the air during the inflated condition, whereas the tube is responsible for holding air for a bias application. The different functions of the inner liner make it necessary for the material composition to be different in the two applications.

5.3.2.1 Sample Preparation

In this case study, a typical radial inner liner composition was analyzed. The sample preparation for an inner liner of a passenger tire as well as a truck/bus tire is critical. The inner liner gauge is very thin as compared to the other components of the tire. Expertise is needed to collect the inner liner sample from a tire. There is a possibility of contamination of squeeze or ply compound during collection of the inner liner compound. After collecting the sample, crushing in a closed-nip two-roll mill was performed. The homogenized sample was analyzed as follows.

5.3.2.2 Pyrolysis

Polymer type was analyzed by using pyrolysis FTIR. Pyrolysis was done at 500°C under an inert atmosphere. For pyrolysis, the extracted rubber sample was considered to avoid any interference from organic materials present in the rubber sample.

5.3.2.3 GC-MS Analysis

For identification of volatile components, the extract was analyzed by using GC-MS, on the basis of the atomic mass unit of the separated fraction of the materials.

5.3.2.4 Ash Analysis

Ash analysis was done to identify the components of ash. A good amount of rubber sample was taken for combustion in a muffle furnace. The ash was quantitatively measured and treated with different acids, *viz.*, dilute hydrochloric acid followed by dilute sulfuric acid followed by hydrofluoric acid. With the acid dissolved part, the analysis was done either by AAS or by ICP spectrometry.

5.3.2.5 Elemental Analysis

The total sulfur content in the rubber sample was determined with the help of elemental analysis.

5.3.3 Results

The report obtained from the above experiments is presented in Table 5.7. The results from the above experimental analysis are tabulated in Table 5.8.

TABLE 5.7

Reverse Engineering Results of Tire Inner Liner Compound

Test Parameters	Results	Explanation
Polymer	50.4%	Polymer content was estimated from TGA. Thermal decomposition studies were conducted to get the polymer content. One decomposition of the sample was done without the extracted sample, and the second decomposition analysis was performed with the sample after extraction as shown in Figures 5.21 through 5.23.
Carbon black	30.8%	Carbon black content was also determined from the TGA thermogram of the non-extracted inner liner sample. Weight loss B, in Figure 5.21, indicated the quantity of carbon black in weight percent. Refer to the thermogram presented in Figure 5.21.
Ash	13.2%	Ash content was calculated from the TGA thermogram of the non-extracted inner liner sample. Weight loss C, in Figure 5.21, gave the amount of ash content in the inner liner sample. Refer to the thermogram presented in Figure 5.21.
Volatile content	5.6%	Volatile content was measured by quantitative extraction of the inner liner sample. In this case the extraction was conducted using acetone as the extracting solvent.
Polymer type	Halo butyl rubber	For identification of the polymer in the inner liner sample, the extracted inner liner sample was pyrolyzed under an inert atmosphere. The pyrolyzed condensate was collected on the salt plate and was analyzed by FTIR technique, following ASTM D3677 standard. The spectrum of the sample was then characterized on the basis of its characteristic peaks (Figure 5.24). The following peaks in the fingerprint region of the spectrum are of interest to characterize this polymer: 1. =C-H stretching at 887 cm^{-1}, typical for isoprene unit. 2. Splitted peak of -C-H stretch at 1376 cm^{-1} and 1366 cm^{-1}, typical characteristics of the butyl unit. On the basis of the FTIR pattern, the polymer was a butyl type. However, a simple butyl rubber compound was not compatible with other rubber compounds of the tire, *viz.*, squeeze or ply compounds. These compounds are basically NR or general-purpose synthetic rubber based compound. So butyl rubber will show a cure incompatibility with these NR or similar types of synthetic rubber. Because of this fact, the compound was analyzed further to check whether this butyl rubber was halogenated or not. In order to study the same, the widely used "congo red" paper color test was performed. The sample was burnt in the test tube in a fume hood. The gas coming out of the pyrolyzed sample was passed through the congo red paper. The color of the paper was changed from red to blue. That indicated the presence of the acidic group, which in this case might be HCl (hydrochloric acid) or HBr (hydrobromic acid).

TABLE 5.7

Reverse Engineering Results of Tire Inner Liner Compound

Test Parameters	Results	Explanation
		Further, the sample was analyzed by chemical method to identify the presence of halogen. The most preferred chemical test is Lassaigne's test. The following steps were followed: 25 to 30 mg of fresh sodium metal were placed into a dry Pyrex test tube. 10 mg of the extracted sample were added into the tube. The test tube was heated over a small flame, just enough to melt the sodium. Then the test tube was heated strongly so that the test tube became glowing red. In this condition, the test tube was held under a flame for 10 min to complete the fusion reaction. Then, the test tube was cooled down to room temperature. Carefully, 1 mL of methanol was added, and the test tube was heated. When all of the sodium had reacted, 60 mL of water were added. The mixture was heated until it boiled and was then filtered. The filtrate, an alkaline solution, possibly contained sodium cyanide (originated from the elements of the sample), sulfide, and halogens. The filtrate was used for the following tests: 20 mL filtrate was taken into a test tube and acidified by adding 2M sulfuric acid. The solution was boiled in a fume cupboard for 5 min to remove HCN and H_2S. One drop of silver nitrate solution was added into the test tube. The appearance of a white or yellowish precipitate that was insoluble in nitric acid confirmed the presence of halogen. The halogens present were identified according to the following tests: A second sample in sulfuric acid solution was placed into a test tube. To it, chloroform (1 mL) and chlorine water or 1% sodium hypochlorite solution (two drops) were added. The mixture was mixed well and the chloroform phase was allowed to separate. The color of the chloroform solution gave an indication of the halogen present. Brown means that there is bromine, a violet color points to iodine, while chlorine is present if the color of the solution does not change. Following the above process, bromine was identified in the sample. Refer to the pyrolyzed spectrum (Figure 5.24) of the rubber sample.
Ash analysis	HCl-soluble ash, 3.5% HCl-insoluble ash, 9.7% HF-soluble ash, 0.0%	The rubber sample, without extraction, was weighed and placed inside a muffle furnace at 650°C under constant supply of air. The rubber sample was burnt along with the volatile components of the sample. The leftover material was ash. The ash was then treated quantitatively as per the following scheme:

(*continued*)

TABLE 5.7

Reverse Engineering Results of Tire Inner Liner Compound (*Continued*)

Test Parameters	Results	Explanation
		Ash → Treated with HCl → The acid solution was quantitatively filtered → The residue was washed, dried, and weighed → HCl-soluble ash was determined → HCl-insoluble part was treated with concentrated H_2SO_4 → H_2SO_4 acid solution was quantitatively filtered → The residue was washed, dried, and weighed → H_2SO_4-soluble ash was determined → H_2SO_4-insoluble ash was treated with HF acid → Quantitatively the HF solution was filtered → The residue was washed, dried, and weighed.
		From the above semi micro ash analysis, different materials of the ash were analyzed.
Total sulfur	0.75%	Total sulfur of the rubber sample before extraction was analyzed by elemental analyzer. Elemental analyzer burnt the rubber sample at high temperature in the presence of an inert gas, mainly helium. During the burning process the sulfur was converted to sulfur dioxide. The sulfur dioxide was then reduced to sulfur which was determined by a suitable detector, preferably TCD. There are several instruments available based on different detector systems. The most common detectors are the IR detector and the thermocouple detector. One can also determine the sulfur content using ion chromatography.
		In this rubber sample, total sulfur was directly read out from the instrument. However, in the case of analysis of carbon black filled rubber sample, total sulfur obtained from the instrument is not the actual sulfur used in the rubber compound, but also includes sulfur from carbon black, processing oils, etc. A proper precaution is needed in that case while reporting the total sulfur of the sample.
Analysis of volatile materials in solvent extract	1. Stearic acid 2. Thiazole accelerator	Volatile components of the rubber sample were analyzed by using the solvent extract. The extract was dried under controlled condition to avoid any oxidation as well as degradation of the extract material. The residue of the extract was dissolved in methanol and injected to GC with mass detector. Accelerator system analysis is one of the most difficult parts of the reverse engineering. This is due to the fact that the accelerators generally lose their original chemical structure during the vulcanization process. The overall GC chromatogram is shown in Figure 5.25.
		In this case, the presence of the thiazole accelerator was confirmed by GC with mass detector. The mass spectrum is shown in Figures 5.26 through 5.28. The intense peak at m/e = 135 at low voltage (14 to 16 eV) indicated the presence of marcaptobenzothiazole. In addition to the thiazole fraction, another fraction was identified as stearic acid on the basis of the m/e at 157, 171, etc. The peak at m/e = 115 confirmed the presence of indene type of resin.

TABLE 5.8

Reverse Engineering Results of a Tire Inner Liner Compound

Ingredients Identified	Results
Polymer (%)	50.4
Carbon black (%)	30.8
Total ash (%)	13.2
HCl-soluble ash (%)	3.5
HF-soluble ash (%)	9.7
ZnO (%)	2.5
MgO (%)	1.0
Clay-type filler (%)	7.2 (This filler may be used to improve the processability as well as to reduce the air permeability.)
Volatile (%)	5.6
Polymer type	Blends of NR and bromobutyl rubber. Bromobutyl content in the blend is high as compared to NR. NR may be present at the level of 20 phr. The small hump in the DTGA thermogram confirms the presence of NR.
Total sulfur (%), after correction with respect to carbon black	0.75
Plasticizer, resin type	Naphthenic oil: This oil was selected on the basis of the compatibility with the polymer obtained through analysis. CI resin: This is used to improve the tack of the inner liner compound in the green stage.
Accelerator type	Thiazole accelerator: This accelerator is basically used as a retarder in bromobutyl application.

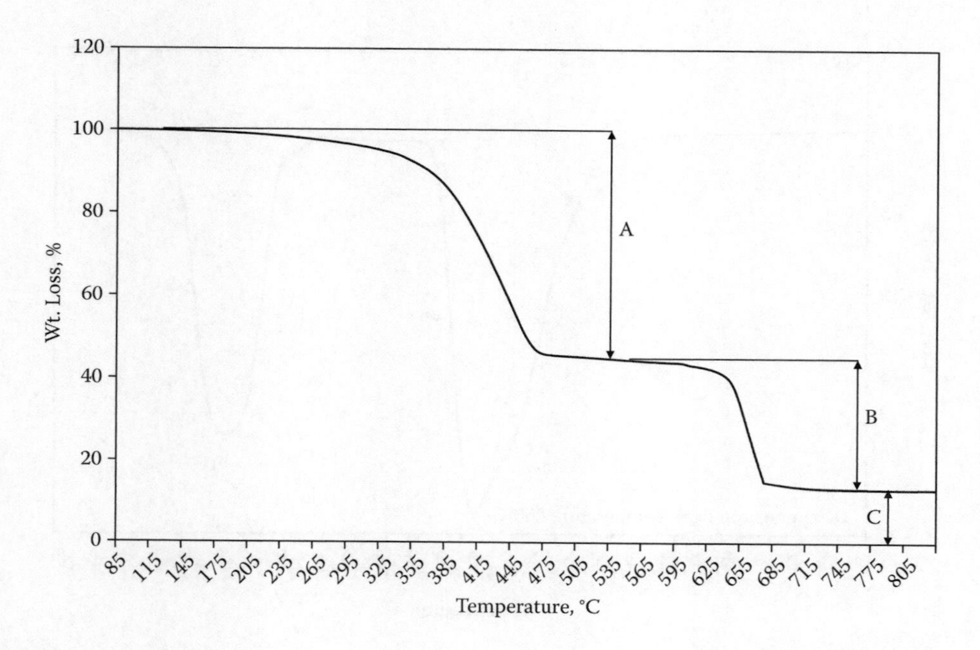

FIGURE 5.21
TGA thermogram of inner liner sample before extraction.

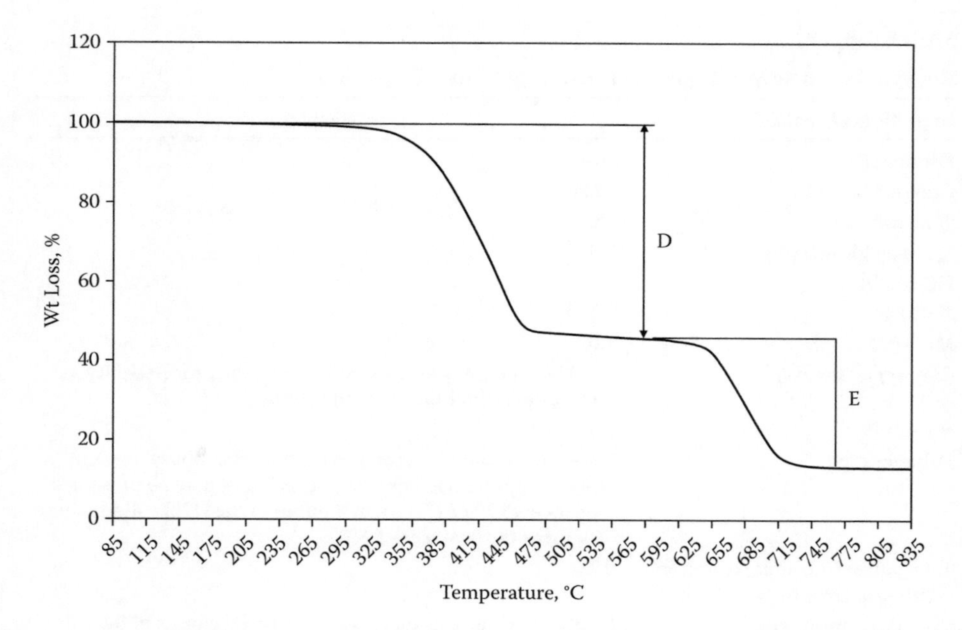

FIGURE 5.22
TGA thermogram of inner liner sample after extraction.

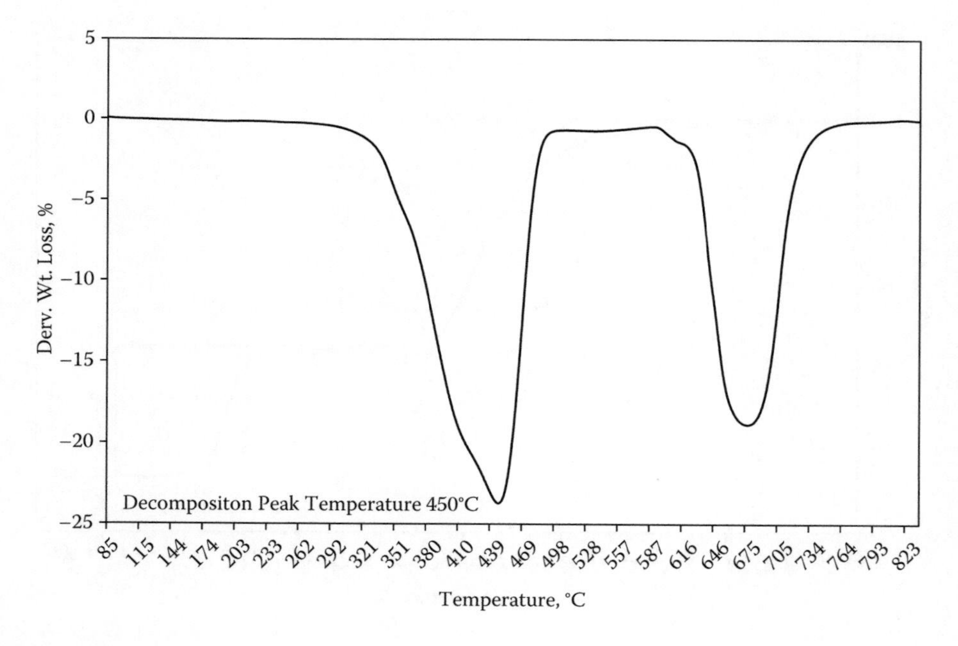

FIGURE 5.23
DTGA thermogram of inner liner sample after extraction.

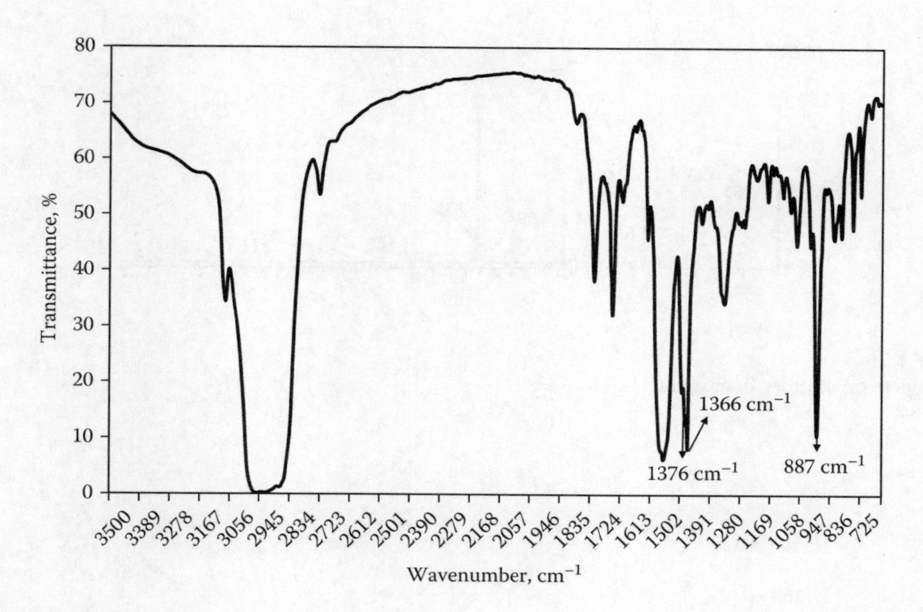

FIGURE 5.24
FTIR spectrum of pyrolyzed inner liner sample.

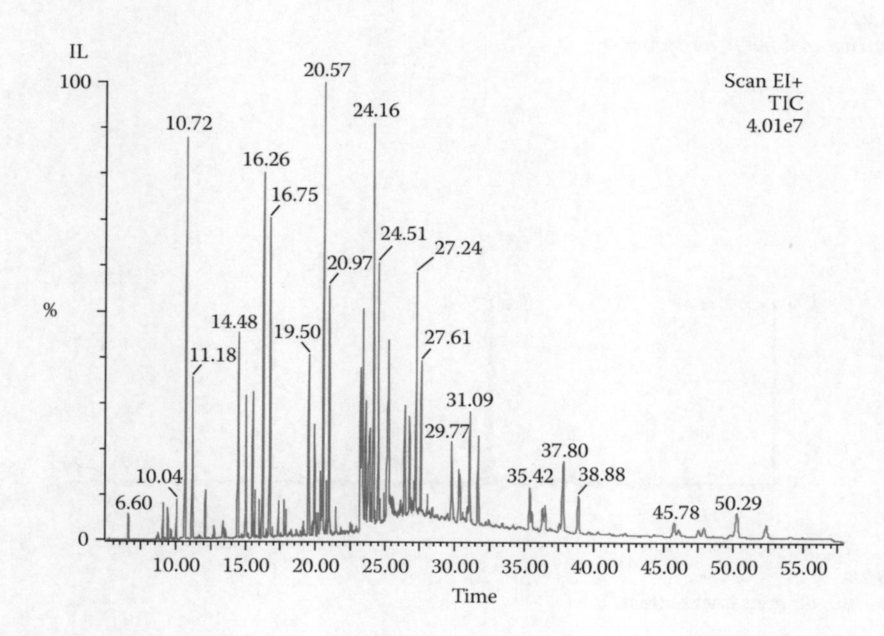

FIGURE 5.25
Gas chromatogram of inner liner extract.

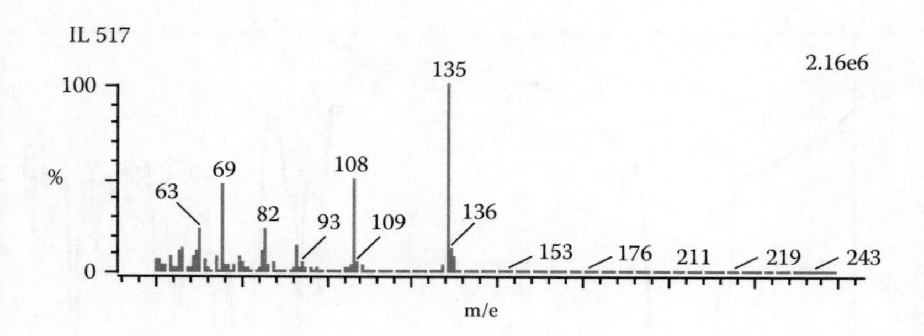

FIGURE 5.26
Mass spectrum of inner liner extract.

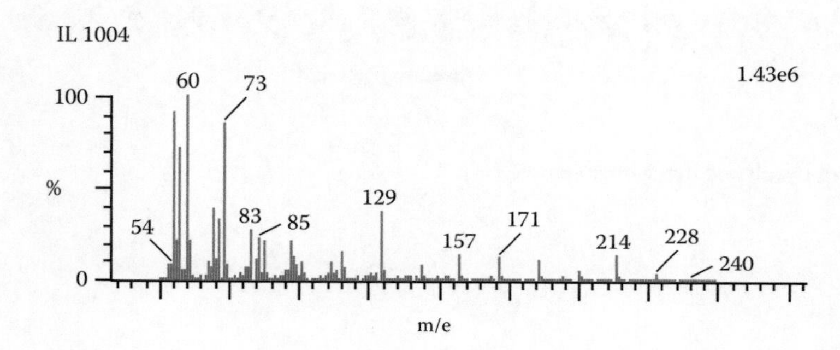

FIGURE 5.27
Mass spectrum of inner liner extract.

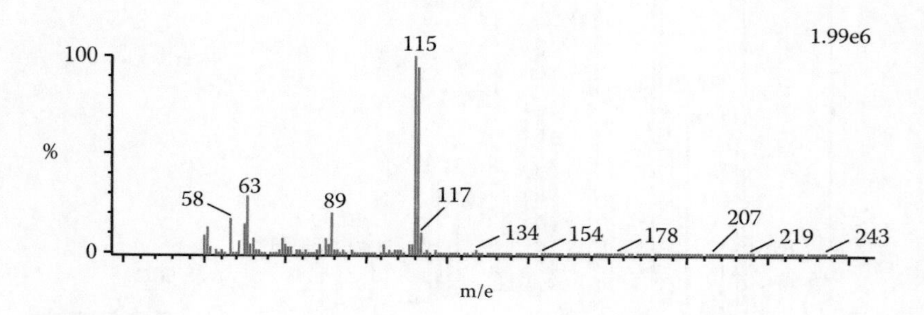

FIGURE 5.28
Mass spectrum of inner liner extract.

5.4 Heat Resistance Conveyor Belt Cover Compound

5.4.1 Objective

The objective of this study is to present a detailed overview of the chemical and analytical techniques that are used for analysis of a heat resistance conveyor belt cover compound sample. On the basis of the results, the cover compound formula is reconstructed.

5.4.2 Experiment

Reverse engineering analysis of the sample was done as follows:

- Extraction of rubber compound
- Polymer characterization and quantification
- Estimation of carbon black
- Identification as well as quantification of inorganic materials in the compound
- Characterization of extract for identification of volatile components of the rubber compound
- Determination of sulfur level

5.4.2.1 Sample Preparation

The conveyor belt cover compound sample was collected and passed through the closed-nip two-roll mill to reduce the sample size. Mechanical crushing was important to prepare the homogenized sample. While collecting the sample it is to be remembered that the cured rubber conveyor belt should not be very old. This will change its mechanical properties as well as composition.

5.4.2.2 Extraction

Extraction of the crushed cover compound sample was carried out using Soxhlet-type extraction and acetone as solvent for 16 h at a temperature of 70°C. An extracted rubber sample was heated under vacuum for complete dryness. The solvent along with the extract were stored for further analysis. All the tests to analyze the composition, determine the polymer type, perform volatile components analysis, perform sulfur content analysis, etc., were carried out in this case.

5.4.2.3 Composition Analysis

A thermogravimetric analyzer (TGA) was used to determine the composition with respect to polymer carbon black and ash content in the rubber sample. The crushed sample was heated in TGA from 80 to 800°C at 40°C/min heating rate. The detailed thermal analysis condition was explained in Section 5.1.

5.4.2.3.1 Analysis of Rubber

Rubber of the extracted cover compound was analyzed by pyrolysis FTIR in an inert atmosphere at 500°C. The condensate of the pyrolysis product was collected on the salt plate and analyzed by FTIR.

5.4.2.3.2 Analysis of Volatile Components

Volatile components were analyzed from the extract. After proper drying of the extract, the same was estimated using GC with mass detector.

5.4.2.3.3 Analysis of Sulfur

Analysis of sulfur was conducted with the unextracted sample using an elemental analyzer.

5.4.2.3.4 Analysis of Ash

For ash analysis, a large quantity of sample was burnt inside the muffle furnace, and the total ash was measured. This ash was treated with dilute HCl and was fully dissolved in it. The HCl dissolved ash was characterized by ICP spectroscopy.

5.4.3 Results and Discussion

Carbon black and ash content were calculated from the TGA thermogram. In Figure 5.29, weight loss represented by "A" is total content of polymer and volatile components and weight loss represented by "B" is carbon black content. "C" represents total residue which is also termed as ash content. Weight loss due to carbon black (B) and ash content (C) of the cover compound from the thermogram was calculated as 20.5% and 12%, respectively.

A second TGA analysis run of the cover compound rubber after extraction with acetone was also performed to calculate the polymer, carbon black, and ash content. The thermogram is represented in Figure 5.30. From Figure 5.30, the polymer content represented by the weight loss "D" and weight loss due to carbon black represented by "E" were calculated. "F" represented the residue. The values obtained for D and E were 66.7% and 22.5%, respectively.

With the help of the data, the polymer, volatile matter, carbon black, and ash content were estimated. These are represented in Table 5.9.

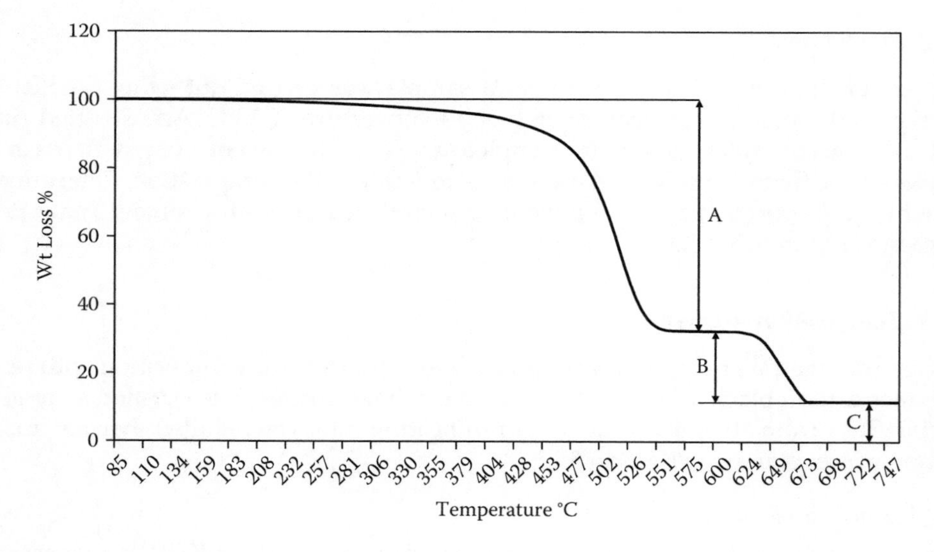

FIGURE 5.29
TGA of conveyor belt cover compound.

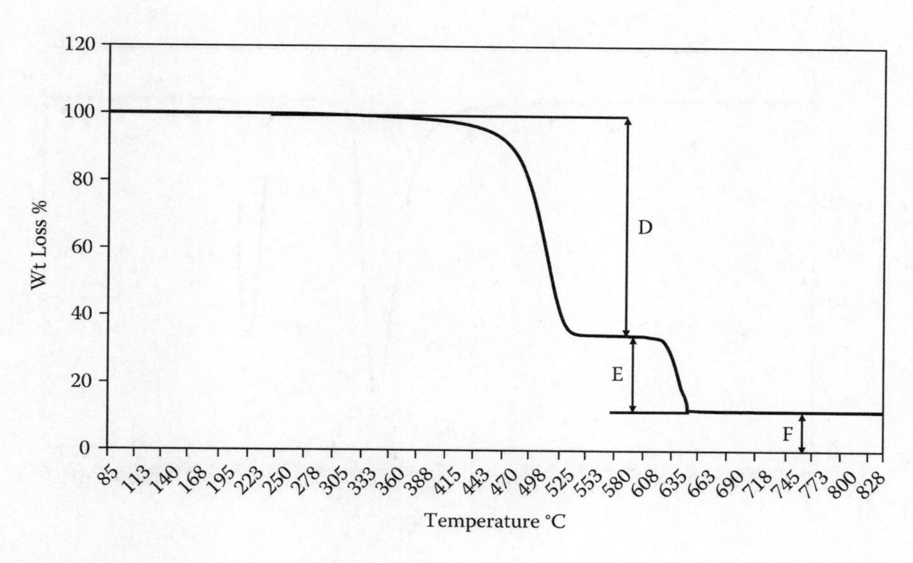

FIGURE 5.30
TGA of conveyor belt cover compound after extraction.

Derivative of TGA (DTGA) thermogram of extracted rubber sample was also considered to characterize the presence of polymer. Presence of polymer was characterized by the characteristic decomposition peak of the polymer which is generally between polymer decomposition area (i.e., 80 to 600°C). Figure 5.31 exhibits the DTGA thermogram of the extracted conveyor belt cover compound sample. This was then compared with the TGA library of standard rubbers available with the manufacturer.

Figure 5.31 shows that the cover compound sample contains single polymer. The polymer type was identified by pyrolysis FTIR analysis. Figure 5.32 shows the FTIR spectrum in the range of 4000 cm^{-1} to 450 cm^{-1} for analysis of characteristic peaks of the rubber.

The peak at 720 cm^{-1} represents the -CH_2 stretching, characteristics of ethylene propylene diene monomer (EPDM). Other peaks at 1380 cm^{-1} for CH_3 stretch of EPDM, and 1460 cm^{-1} due to -CH_2 stretching of EPDM are observed. The 1460 cm^{-1} peak height of –CH_2 stretching is higher than the 1380 cm^{-1} peak height of –CH_3 stretching. This is one indication that the EPDM of cover compound contains high ethylene content.

TABLE 5.9
Total Composition of the Conveyor Belt Cover Compound

Composition	Before Extraction, %	After Extraction, %	Actual Content, %	Remarks
Polymer	Not calculated	66.7	60.8	Expression used to calculate the polymer % is: $(D*B)/E$ (Refer to Figures 5.29 and 5.30.)
Carbon black	20.5	22.5	20.5	
Ash	12.0	Not calculated	12.0	.
Volatile matter	Not calculated	Not calculated	6.7	

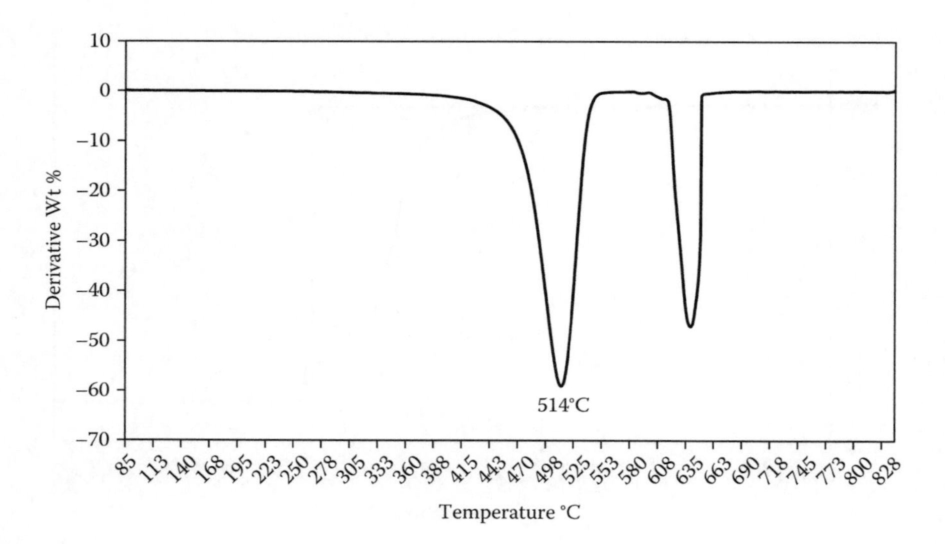

FIGURE 5.31
DTGA of conveyor belt cover compound after extraction.

The chromatogram was obtained with the cover compound extract. On the basis of this chromatogram, the mass spectra are derived. These are shown in Figures 5.33 through 5.36, which represent different components of the extract of the cover compound.

In Figure 5.34, the presence of m/e at 91 and 106 indicates zinc diethyl dithiacarbamate. This fragment is an accelerator fragment for EPDM-based cover compound. In Figure 5.35, the presence of m/e at 108 indicates the presence of phenol fragments, which is different from other phenol fractions at m/e = 165 and m/e = 183. It is inferred that phenolic resins are present in the recipe. In Figure 5.36, the m/e peaks at 87 and 74 are attributed to

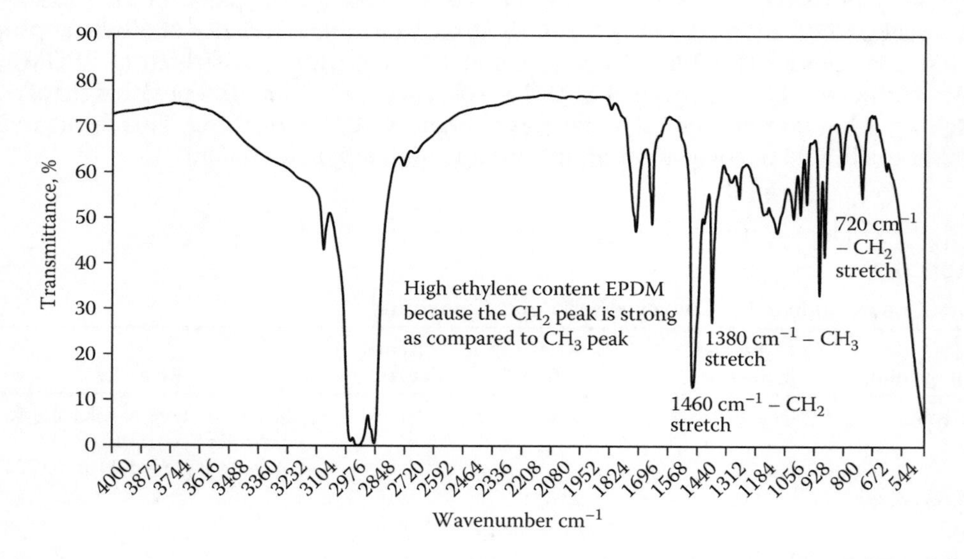

FIGURE 5.32
FTIR spectrum of conveyor belt cover compound after extraction.

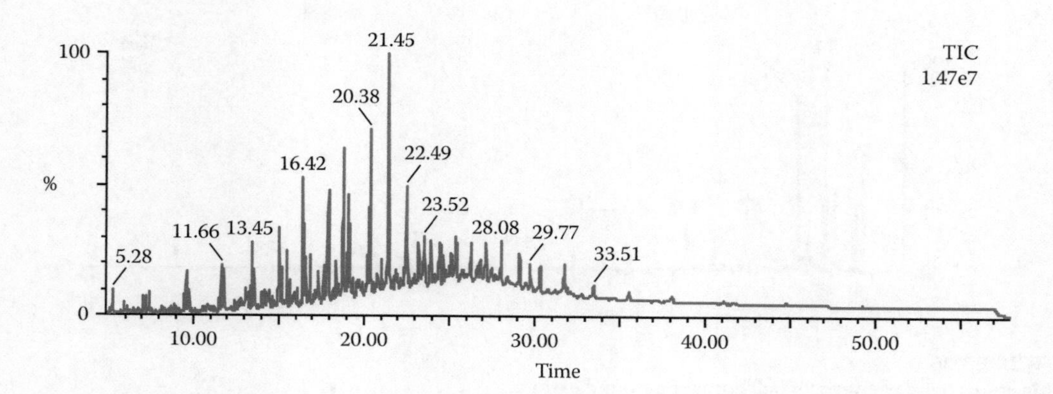

FIGURE 5.33
Chromatogram of conveyor belt cover compound extract.

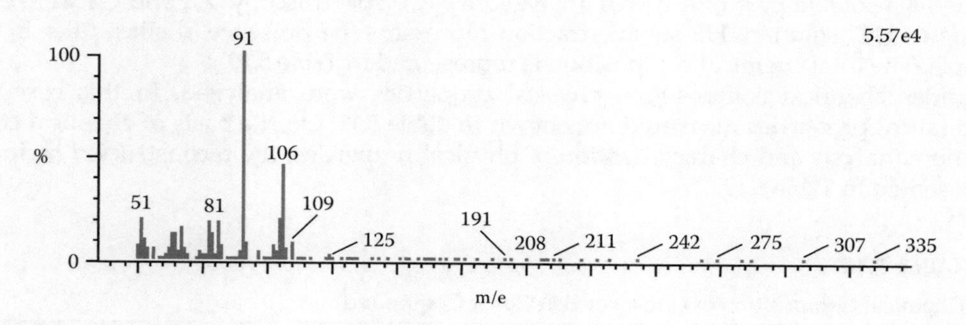

FIGURE 5.34
Mass spectrum of conveyor belt cover compound extract.

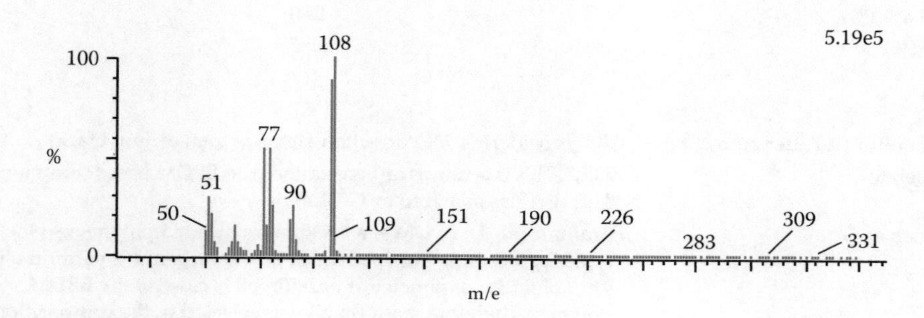

FIGURE 5.35
Mass spectrum of conveyor belt cover compound extract.

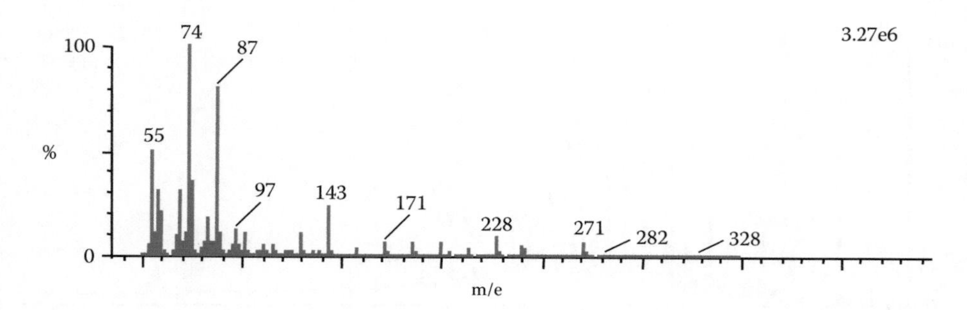

FIGURE 5.36
Mass spectrum of conveyor belt cover compound extract.

octadecanoic acid. This component is basically the fractions of stearic acid. So, inclusion of stearic acid in the cover compound was confirmed. Crushed sample was analyzed, and the total sulfur was obtained as 0.88%. This sulfur is the total sulfur content of the sample. A correction factor was applied in this case, as this sample contained carbon black.

The HCl-soluble part of ash was analyzed by ICP spectroscopy. Zn and Ca were estimated in HCl solution. HF-soluble fraction represents the presence of silica filler in the sample. The total chemical composition is represented in Table 5.10.

Besides chemical composition, physical properties were analyzed. In this case the mechanical properties measured are shown in Table 5.11. On the basis of chemical composition analysis and characterization of physical properties, the reconstructed recipe is represented in Table 5.12.

TABLE 5.10
Chemical Composition of Conveyor Belt Cover Compound

Ingredients Identified	Results
Polymer (%)	60.8
Polymer type	EPDM with high ethylene content
Carbon black (%)	20.5
Volatile (%)	6.7
Total ash (%)	12.0
ZnO (%)	5.5
CaO	0.3
Silica	6.2
Total sulfur (%) after correction	0.68 (considering 1% correction factor of total carbon black)
Accelerator	ZDEC. This is acting as an accelerator in an EPDM-based compound. This was also analyzed by GC-MS.
Plasticizer	Paraffinic oil. As EPDM is a backbone saturated polymer, so the typical processing aid used for EPDM compound is paraffin oil. The solubility parameter of paraffin oil is close to the EPDM polymer. Therefore, paraffin oil was selected in the composition.
Activator organic	Stearic acid
Resin	Phenolic resin. This resin is basically used as tackifier in the EPDM compound. This was also characterized by GC-MS in the extract.

TABLE 5.11

Mechanical Properties of the Conveyor Belt Cover Compound

Mechanical Properties	Results	Implication in Formula Reconstruction
Modulus at 100%, MPa	3.0	Modulus indicates the type of black used and also the degree of curing. In this case high modulus indicates the use of high surface area black. This may be the N110 series.
Modulus at 300%, MPa	7.5	
Tensile strength, MPa	15.0	
Elongation at break, %	650	
Hardness, Shore A	70	

TABLE 5.12

Reconstructed Formula for Conveyor Belt Cover Compound

Ingredients	phr
EPDM	100.00
N330	35.00
Silica	10.00
ZnO	9.00
Stearic acid	2.50
CaO	0.50
Phenolic resin, as tackifier	8.00
Paraffinic oil	2.00
Heat-resistant antioxidant mainly amine type due to high heat resistance application, though this was not being characterized in the sample.	1.50
Sulfur	1.00
ZDEC	1.80

5.5 Fuel Hose Cover

5.5.1 Objective

The objective of this study was to present a detailed overview of the chemical and analytical techniques and physical testing that were used for analysis of the fuel hose cover sample. On the basis of the analysis, the hose cover compound formula was reconstructed.

5.5.2 Experiment

Reverse engineering analysis of a fuel hose cover compound sample was done as follows:

- Extraction of rubber compound
- Polymer characterization and quantification

- Determination of carbon black content
- Identification as well as quantification of inorganic materials in the compound
- Characterization of extract for identification of volatile components of the rubber compound
- Determination of sulfur level

5.5.2.1 Sample Preparation

The fuel hose cover sample with unknown composition was passed through the closed-nip two-roll mill to reduce the sample size. Mechanical sample crushing was important to prepare the homogenized sample.

5.5.2.2 Extraction

Extraction of the crushed fuel hose cover sample was carried out using Soxhlet-type extraction. Quantitative extraction was performed by acetone for 16 h at a temperature of 70°C. Extracted rubber sample was heated under vacuum for complete dryness. The solvent along with the extract were stored for further analysis.

Methods such as TGA, FTIR, pyrolysis FTIR, GC-MS, sulfur analysis, ash analysis, etc., discussed in earlier chapters, were used in analyzing the materials present in the sample.

5.5.2.3 Composition Analysis

A thermogravimetric analyzer (TGA) was used to characterize the polymer present in the sample and its content in the compound. The composition was analyzed mainly by chemical means. The crushed sample was refluxed with *p*-dichlorobenzene for 30 min. After reflux, the solution was cooled and tert-butyl hydroperoxide was added into the mixture. The total mixture was further boiled for 30 min and then cooled and diluted with toluene. The total solution was placed in the centrifuge tube for centrifuging the whole mixture at high speed, around 14,000 rpm. The clear solution from the centrifuge tube was separated, and the solid part was washed with toluene, acetone, and water. The solid part from the centrifuge tube was dried and heated in the muffle furnace. Weight loss due to burning in the muffle furnace was measured.

5.5.2.3.1 *Analysis of Rubber*

Rubber of the fuel hose cover sample was analyzed by pyrolysis FTIR. The extracted insulation sample was pyrolyzed in an inert atmosphere at 500°C. The condensate of the pyrolysis product was collected on the salt plate and analyzed by FTIR. However, for analysis of the FTIR spectrum, a wavenumber range of 2000 to 600 cm^{-1} was considered.

5.5.2.3.2 *Analysis of Volatile Components*

Volatile components were analyzed with the extract. After proper drying of the extract, it was analyzed using GC attached with a mass detector.

5.5.2.3.3 *Analysis of Sulfur*

Sulfur, one of the important curing materials, was analyzed from the unextracted sample using an elemental analyzer.

5.5.2.3.4 *Analysis of Ash*

For ash analysis a large quantity of insulation sample was burnt inside the muffle furnace, and the total ash was measured. This ash was treated with dilute HCl. The residue obtained from the HCl solution was treated with HF, and it was found that the residue was completely dissolved in HF. The ash that was dissolved in HCl was characterized by ICP spectroscopy for elemental analysis.

5.5.3 Results and Discussion

The TGA thermogram presented in Figure 5.37 indicates the initial steep weight loss due to dehydrohalogenation. This kind of degradation is typical for chlorine-containing polymers, mainly polychloroprene rubber (CR) and polyvinyl chloride (PVC). However, the percent dehydrohalogenation is high in the case of PVC material as compared to CR, due to the low content of chlorine in CR as compared to PVC. Another important observation with this thermogram is the continuous degradation during the changeover of analysis atmosphere from inert to air/oxygen. This is due to the presence of a heteroatom. From this thermogram, estimation of polymer and other materials like carbon black and ash would not give accurate results. TGA analysis was performed to characterize the polymer only.

Due to difficulty in estimating polymer and carbon black in this sample, chemical digestion technique was used. Weight loss (A) obtained by heating the digested sample in a muffle furnace was measured as carbon black content of the sample. Quantitative extraction (B) will give the total volatile content. The residue after extraction, (C), will give the ash content of the sample. With the aid of the data, shown in Table 5.13, the polymer, carbon black, volatile matter, and ash content were estimated.

Figure 5.38 shows the FTIR spectrum from 2000 to 600 cm^{-1}, which indicates characteristic peaks of the rubber. The peak at wavenumber 825 cm^{-1} represents the characteristics = C-H bending of CR. Other peaks at 1453 cm^{-1} and 1436 cm^{-1} represent methylene deformation.

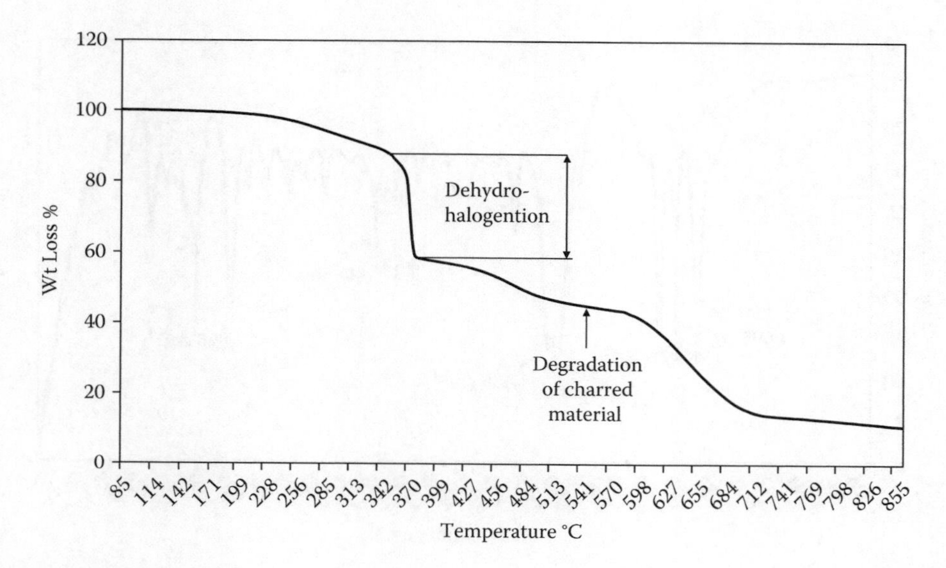

FIGURE 5.37
TGA of fuel hose cover.

TABLE 5.13

Total Compositions of the Hose Cover Insulation Compound

Composition	Before Extraction, %	After Extraction, %	Actual Content %	Remarks
Polymer	Not calculated	Not calculated	45.3	Polymer was estimated as [100– (*A* + *B* + *C*)]
Carbon black	Not calculated	Not calculated	21.6	Carbon black was calculated through chemical digestion followed by ignition in muffle furnace
Ash	Not calculated	Not calculated	17.4	Ash was estimated on the basis of the total residue after ignition
Volatile matter	Not calculated	Not calculated	15.7	Volatile matter was calculated by quantitative extraction

These characteristic peaks are split, indicating the polychloroprene rubber. The 1134 cm^{-1} wavenumber peak reflects methylene waging deformation. Other split peaks at 1660 cm^{-1} and 1695 cm^{-1} are due to the C=C stretch. These are characteristic peaks of polychloroprene rubber.

Figure 5.39 shows strong peak at 1728 cm^{-1} wavenumber which represents -C=O stretching. The peak at 1122 cm^{-1} is due to -C-O-stretching of -C-O-. The 1072 cm^{-1} peak indicates C-O (with =O beneath C) stretching of -O-CH_2. The 742 cm^{-1} peak appears due to out-of-plane bending of aromatic groups. The above characteristic peaks help to identify the type of phthalate. In this case, it is likely to be dioctyl phthalate type plasticizer.

Crushed sample was analyzed, and total sulfur was obtained as 0.43%. The sulfur was the total sulfur content of the sample. It was not the actual representation of the sulfur

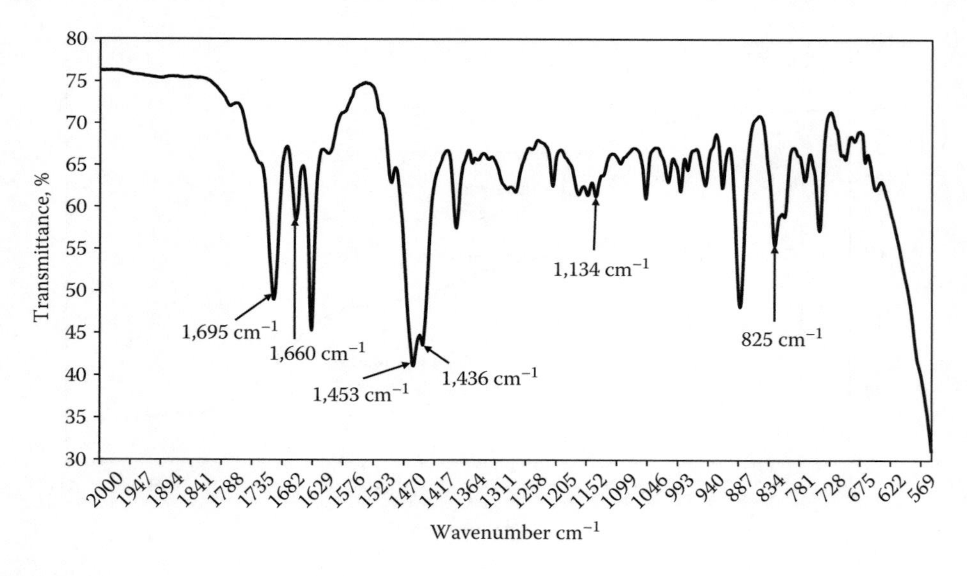

FIGURE 5.38
FTIR spectrum of fuel hose cover after extraction.

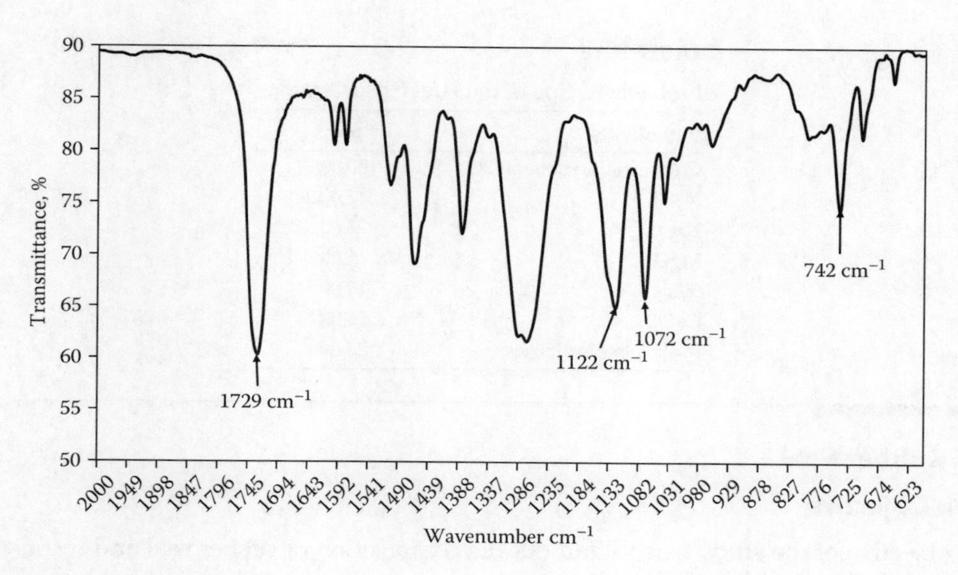

FIGURE 5.39
FTIR spectrum of the extract of the fuel hose cover sample.

present in the tread sample. The total sulfur obtained was corrected considering the sulfur contribution from carbon black, process oils, and also accelerators. On average, a correction factor of 1% of the total carbon black was subtracted from the sulfur content to get the sulfur content of the sample.

The HCl-soluble part of ash was analyzed by ICP spectroscopy, which confirmed the presence of Zn, Ca, and Mg. From the HF-soluble fraction, silica filler was detected in the sample. The observations from chemical as well as instrumental analysis are tabulated in Table 5.14. On the basis of the chemical analysis the probable recipe was formulated in Table 5.15.

TABLE 5.14
Total Composition of the Fuel Hose Cover Compound

Ingredients Identified	Results
Polymer (%)	45.3
Carbon black (%)	21.6
Total ash (%)	17.4
Volatile (%)	15.7
ZnO (%)	2.5. In this recipe, ZnO is mainly used as the vulcanizing agent.
MgO (%)	0.5.
Silica (%)	7.0
$CaCO_3$ (%)	7.5
Polymer type	Polychloroprene rubber
Total sulfur (%) after correction	0.21 (considering 1% correction factor of total carbon black). This sulfur source was mainly the polychloroprene rubber. It was either the sulfur modified or mercaptan modified polychloroprene rubber used in the hose compound.
Plasticizer type	Phthalate type

TABLE 5.15
Probable Recipe of the Fuel Hose Cover

Ingredients	phr
Chloroprene rubber (CR)	100.00
N330	47.00
ZnO	5.00
MgO	1.00
Silica	15.00
$CaCO_3$	16.00
Phthalate plasticizer	25.00

5.6 Rubber Seal

5.6.1 Objective

The objective of the study was to find out the composition of rubber seal and reconstruct the formulation based on the results.

5.6.2 The Experiment

5.6.2.1 Sample Preparation

Unlike the product in earlier chapters, the unknown compound chosen in this example was an unvulcanized one from which the product, a seal in a critical part of an aircraft, was made. The compound was procured from an external agency, stored, and used only when there was a requirement. Development of the compound was necessary, as the product performed poorly in actual operation.

5.6.3 Analysis

The unknown compound supplied was first analyzed using TGA and FTIR. The results are shown in Figures 5.40 and 5.41. It can be concluded using the knowledge gained in the

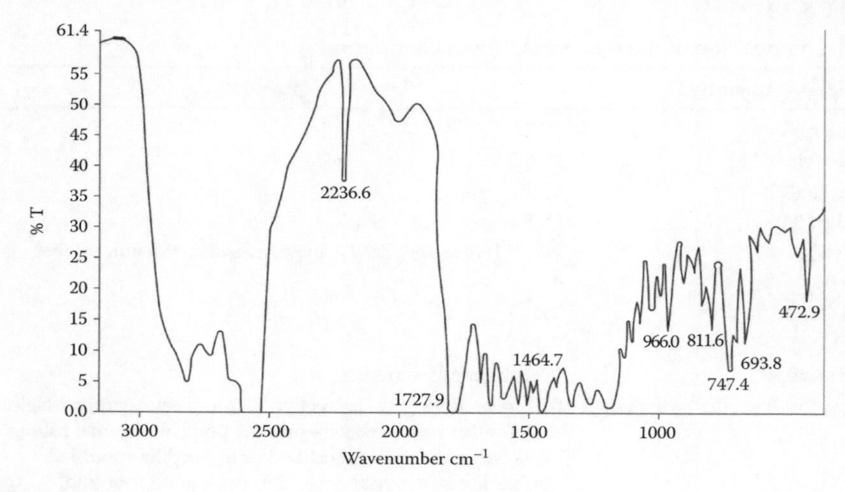

FIGURE 5.40
FTIR spectra of unknown compound.

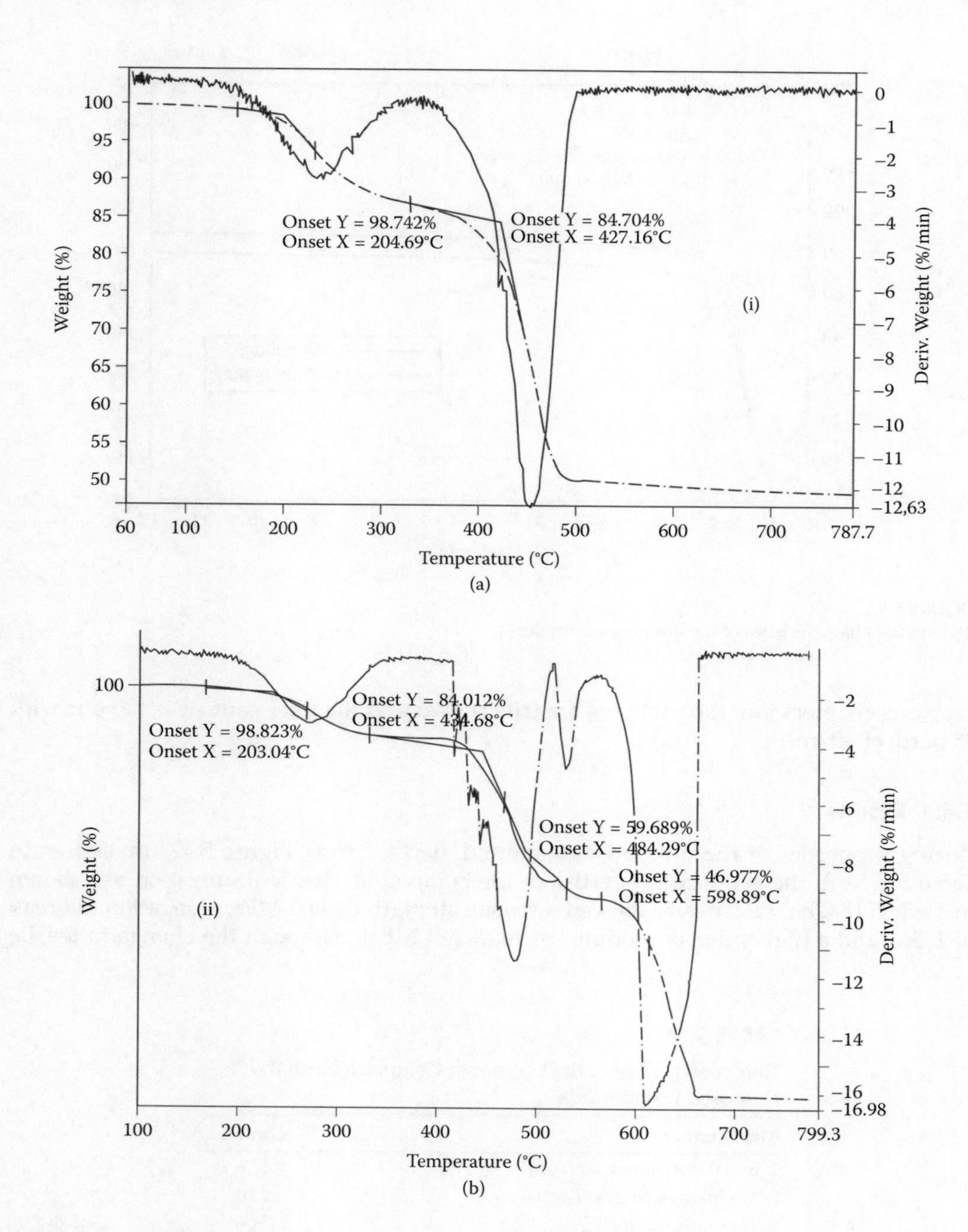

FIGURE 5.41
Thermograms of unknown compound (a) in nitrogen and (b) in air.

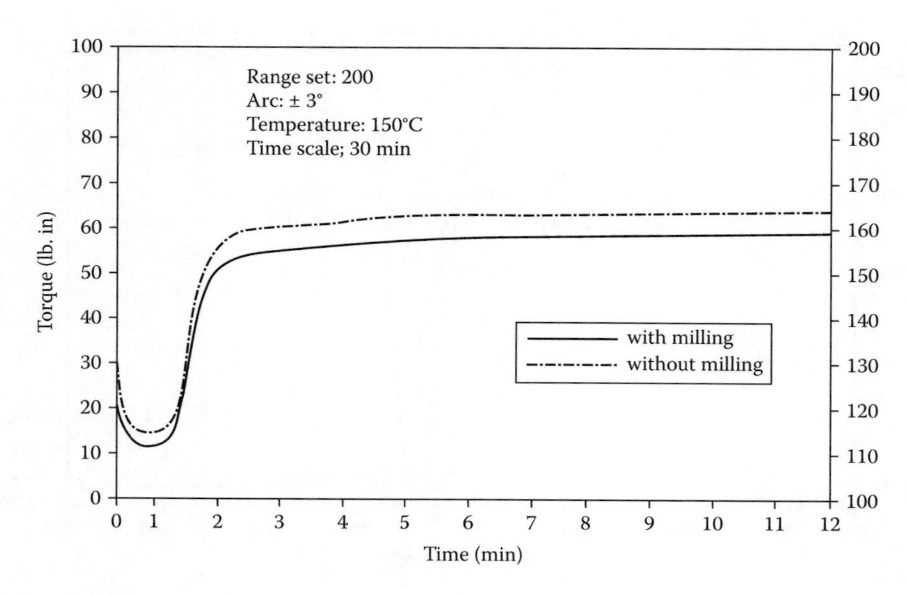

FIGURE 5.42
Rheometric characteristics of the unknown compound.

previous chapters that the rubber is a nitrile rubber and the filler content is 125 phr with 35 parts of oil/resin.

5.6.4 Results

Curing properties of the unknown compound, derived from Figure 5.42, are shown in Table 5.16, and the physical properties of the compound after vulcanization are shown in Table 5.17. The vulcanizate showed a tensile strength of 14.3 MPa, elongation at break of 175%, and a high value of modulus at 100% (9.4 MPa). Although the change in tensile

TABLE 5.16
Rheometric Data of the Unknown Compound and the Proposed Compounds

Mix Number	Control
1 Initial viscosity: $L_o(dN-m)$	47.6
2 Minimum viscosity: $L_i(dN-m)$	23.0
3 Thermoplasticity: $T_p = (L_o - L_i)$	24.6
4 Induction time: t_i (time for one unit rise above L_i)	2′20″
5 Scorch time: t_2 (time for two unit rise above L_i)	2′35″
6 Maximum cure: $L_f\,(dN-m)$	118.4
7 Optimum cure: $0.9(L_f - L_i) + L_i(dN-m)$	54.43
8 Optimum cure: t_{90} (minutes) time	6′00″
9 Time given for molding of 2 mm sheet	6.00′

Notes: Temperature, 150°C; chart motor: 30 min; range scale: 200; arc ±: 3°.

TABLE 5.17
Physical Properties of the Unknown Compound

Before Aging at RT				After Aging at 100°C, 36 h				Tear Strength (N/mm)				
TS (MPa)	EB %	Modulus (MPa) @%		TS (MPa)	EB %	Modulus (MPa) @%		Before Aging	After Aging	Hardness Shore A°	Compression Set (Constant Strain) at 70° for 22 h	Mooney Scorch Time at 120°C t_5 (Min)
		50	100			50	100					
14.3	175	4.3	9.4	14.5	113	8.6	13.6	38.9	33.4	78–79	12.8	8

TABLE 5.18

Representative Formulations of the Mixes

Polymer and Ingredients	A	B	C
NBR	100	100	100
Sulfur	—	0.5	—
ZnO	7.0	5.0	5.0
MgO	4.0	4.0	4.0
Stearic acid	1.0	1.0	1.0
Carbon black	70	70	70
Plasticizer	10	10	10
Antioxidant	2.5	2.5	2.5
Combination of accelerators	4.0	3.5	4.5

strength was negligible on aging, there was considerable deterioration of elongation at break with increased modulus at 100% (45%). A hardness value of 78 to 79 Shore A was recorded with a moderate tear strength of 38.9 N/mm. The Mooney scorch value of 8 min (rheometric scorch time of 2′35″) indicated that the compound was scorchy either while designed or during aging in storage.

As the compound provided might deteriorate/age with time, the primary purpose of the design was to choose a series of compounds with variation in hardness, tensile strength, elongation at break, and swelling resistance. Initially, few compounds were designed (three representative formulations are given in Table 5.18). These compounds had variation in black, sulfur/accelerator, etc.

The curing properties are shown in Figure 5.43. It was noted that these compounds had much higher scorch time with optimum cure slightly higher than the control compound. Physical properties are shown in Table 5.19. The tensile strength was close to the specification (discussed later) with much better elongation at break. The retention of tensile

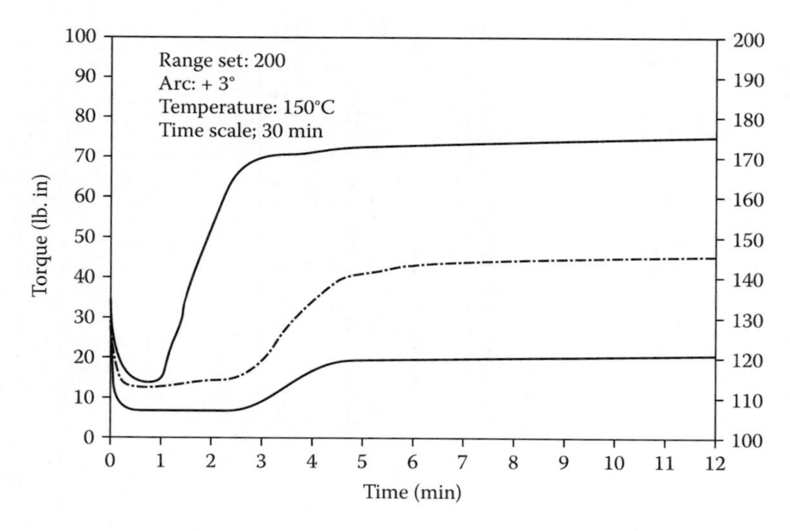

FIGURE 5.43
Rheograph of different representative compounds (formulation given in Table 5.42).

TABLE 5.19
Physical Properties of Various Compounds

Sample ID	Before Aging at RT				After Aging in Air for 72 h at 100°C				After Aging in Oil (FH-51) for 24 h at 100°C			
	TS (MPa)	EB%	Modulus (MPa) at 100% and 300%		TS (MPa)	EB%	Modulus (MPa) at 100% and 300%		TS (MPa)	EB%	Modulus (MPa) at 100% and 300%	
Unknown compound	14.0	172	9.7	—	18.0	93	—	—	13.0	137	9.6	—
A	11.0	780	1.8	5.0	15.0	628	3.0	8.3	11.0	775	1.5	4.5
B	13.0	455	3.2	9.6	16.0	334	5.1	15.0	13.4	458	2.8	9.5
C	11.5	750	1.8	5.4	15.0	587	3.2	9.0	12.5	781	1.6	5.2

TABLE 5.20

Hardness and Density Values

	Before Aging at RT	After Aging in Oil (FH-51) for 24 h at 100°C	
Sample ID	**Hardness Shore A**	**Hardness Shore A**	**Density (g/cc)**
Unknown compound	78–79	76–77	1.32
A	63–64	62–63	1.27
B	68–69	66–67	1.27
C	65	61–62	1.26

strength and elongation at break at 100°C for 72 h aging was also much better than the control compound. However, the modulus of the designed compound was lower and the hardness and density given in Table 5.20 are lower. Low modulus compound might give better sealing characteristics because of the elasticity. All the compounds were further subjected to swelling in FH-51, ATFK, and a mixture of isooctane and toluene. The results are given in Table 5.21. It was found that some of the designed compounds exhibited excellent behavior in ATFK-50, OM-51, and FI-51 and met the specification. Although the results of the designed compounds were slightly higher only in the isooctane and toluene mixture, these were much better than the specification given.

Hence a new series of compound was designed, mostly by variation of black, resin, and nitrile content, to improve hardness and modulus. All the formulations are not shown here. The results of a few compounds are given in Tables 5.22 and 5.23. It was observed that all the properties of the developed compounds were equivalent or much better than the unknown compound. It was interesting to note that swelling resistance of the developed compounds was improved several-fold. Hence the seal was made from the final compounds by manufacturing of which would correctly assess the composition and performance.

However, continuous research on newer compounds having variation of filler and nitrile contents was taken up for understanding. The outline of the formulation of the compound, E_f, is given in Table 5.24. The properties of the final compound are given in Table 5.23.

TABLE 5.21

Swelling Properties of Different Compounds

Sample ID	Volumetric Swelling in Oil OM-15 (FH-51) at 70°C FPR 24 h (Volume Swell%)	Gravimetric Swelling in Oil OM-15 (FH-51) at 70°C for 24 h (Weight Swell%)	Gravimetric Swelling in Oil MS-20 at 130°C for 24 h (Weight Swell%)	Gravimetric Swelling in Fuel Oil T-1 (ATFK-50) at Room Temperature for 24 h (Weight Swell%)	Gravimetric Swelling in a Mixture of Isooctane + Toluene (70:30) at Room Temperature for 24 h (Weight Swell%)
Unknown compound	9.6	7.3	–6.2	3.4	12.5
A	5.5	2.2	–5.9	1.4	16.7
B	3.3	2.2	–5.2	1.3	15.6
C	17.8	1.9	–6.3	1.3	16.2

TABLE 5.22

Comparison of Properties of the Developed Compounds with Specifications

Properties	Compounds						Specification	Remarks
	D	E	F	G	H	Unknown		
Tensile strength before aging, MPa	14.2	14.2	13.3	12.9	15.1	14.0	12	Passed
Elongation at break before aging, %	245	209	352	291	229	172	150	Passed
Hardness before aging, Shore A	76–77	79–80	75–76	78–79	84–85	78–79	78–85	Passed
Density, g/cc	1.32	1.33	1.27	1.31	1.32	1.32	128 ± 0.05	Passed
Volumetric swelling in oil OM-15 (FH-51) at 70°C for 24 h, %	5.4	4.5	3.5	6.3	0.2	9.6	4–14	Passed
Gravimetric Swelling in Media%								
(a) In mixture of isooctane + toluene at RT for 24 h	9.9	9.2	16.2	12.6	8.7	12.5	35	Passed
(b) In fuel T1 (ATFK-50) at RT for 24 h	0.6	0.6	1.1	0.86	0.55	3.4	0–15	Passed
(c) In oil FH-51 at 70°C for 24 h	–0.6	–0.4	1.9	0.8	1.0	7.3	–1 to 10	Passed
(d) In oil MS-20 at 130°C for 24 h	–5.9	–5.1	–2.5	–4.0	–1.5	–6.2	–12	Passed
Change in % elongation during aging at 100°C in air for 72 h	–38	–40	–36	–44	–39	–46	–55 to –5	Passed
After Aging in Oil FH-51 at 100°C for 24 h								
(a) Change in hardness	0	1	1	2	0	1	10	Passed
(b) % change in tensile strength	6.0	5.6	16.5	7.7	11.2	7.1	20	Passed
(c) % change in elongation at break	7.7	3.3	19.3	13.7	16.1	20.3	20	Passed

In order to meet the frost resistance and brittleness temperature, the aim was to decrease T_g by various means (i.e., by blending with low amounts of BR, NBR having low ACN content or completely replacing the NBR-33 with NBR-18). When the BR was introduced in the recipe, the swelling was affected. However, the changes in properties of the compound (not shown here) were well within the specification. Total replacement with NBR-18 offset the swelling properties, and the compound did not meet the specification. Even partial replacement was not favorable. The most sensitive property which most of these compounds did not meet was the volumetric swelling in OM-15 (FH-51) oil at 70°C for 24 h. Even an increase of the filler loading did not bring down the swelling within the specification for such compound. It was interesting to note that tensile properties of these modified compounds were comparable with those of the control compound. Hardness and density also lie in the range of specification.

There were only a few compounds including E_f that met the specification (Table 5.23). Hence, detailed analysis was done using E_f. The tensile properties, hardness, density, and swelling were all within the specification provided (Table 5.23). The formulation of the compound is given in Table 5.24.

TABLE 5.23

Properties of E_f Compound

Sl Number	Test Parameters	Requirement	Result
1	Tensile strength, kg/cm^2 (Min)	120	Avg. 122
2	Elongation at break, % (Min)	150	Avg. 132
3	Residual elongation at break % Max	8	4
4	Hardness, IRHD	78 to 85	85
5	Density, gm/cm^3	1.28 ± 0.05	1.34
6	Volumetric swelling in oil OM-15 (FH-51) at 70°C for 24 h, %	4 to 14	14
7	Relative residual deformation with 30% compression at 100°C for 72 h, in oil OM-15 (FH-51), % (max)	35	16
8	Gravimetric swelling in media, %		
(a)	In a mixture of isooctane + toluene (70:30) PBV at RT for 24 h (Max.)	35	20
(b)	In fuel T1 (ATFK-50) at RT for 24 h	0 to 15	6
(c)	In oil OM-15 (FH-51) at 70°C for 24 h	−1 to 10	8
(d)	In oil MS-20 at 130°C for 24 h not less than	−12	−2
9	Change in percent (%) elongation during aging at 100°C in air for 72 h	−55 to −5	−33
10	Brittleness temperature, °C (not above)	−48	Passes
11	Coefficient of frost resistance at −45°C (Min)	0.2	0.2
12	Ozone resistance test at RT for 25 ppm, 20% stretching for 72 h with relative humidity of 65%	Appearance of cracks is not allowed	—
13	Aging coefficient as per RAPRA at 90°C in air for 21 d (Min)	0.5	—
14	After aging in oil FH-51 at 100°C for 24 h (as compared to original)		
(a)	Change in hardness, IRHD (Max)	10	−14
(b)	Change in tensile strength, % (Max)	20	+3
(c)	Change in elongation at break, % (Max)	20	20
15	Flex cracking test (De Mattia) as per ISO 132 or equivalent (crack initiation)	Minimum 25,000 cycles	—
16	Conformation of base rubber compound	Record the value	Nitrile

TABLE 5.24

Formulation of E_f Compound

Ingredient	phr
Nitrile rubber	100
Sulfur	0.5
ZnO	5.0
MgO	4.0
Stearic acid	1.0
Carbon blacks	155
Plasticizer	20
Antioxidant	2.5
Accelerators	4.0

5.7 V-Belt Compound

5.7.1 Objective

The objective of this investigation is to identify the polymer present and the rough composition of the given rubber product (top and bottom rubbers for a V-belt) using elemental, TG, and DSC analysis, and IR spectroscopy. The intention of this section is to demonstrate that even without the use of very sophisticated equipment, an estimate on the composition could be made.

5.7.2 Experiment

5.7.2.1 Sample Preparation

The rubber, after removal of the fabric coating by buffing, was subjected to acetone extraction for 24 h. After acetone extraction, the sample was used for elemental analysis and IR spectroscopic analysis. The portions before extraction were used for TG and DSC analyses.

5.7.2.2 Elemental Analysis

The given compound was analyzed for the presence of nitrogen, halogen, or sulfur using Lassaigne's test. A small portion of the compound, after acetone extraction, was fused with metallic sodium in a fusion tube until the tube was red hot. It was then plunged into distilled water and extracted. The water extract was tested for nitrogen, halogen, and sulfur.

5.7.2.3 Thermogravimetric Analysis

Approximately 10 mg of the compound was heated from room temperature to 800°C in nitrogen atmosphere in a thermogravimetric analyzer (TGA Q 50, TA Instruments, New Castle, Delaware) at a heating rate of 20°C/min.

5.7.2.4 Differential Scanning Calorimetric Analysis

Approximately 6 mg of the compound was heated in nitrogen atmosphere in a differential scanning calorimeter (DSC Q 100, TA Instruments) at a heating rate of 20°C/min.

5.7.2.5 Infrared Spectroscopy

A small portion of the compound, after acetone extraction, was pyrolyzed in a pyrolyzing tube. The vapors were directly collected in the IR-cell for infrared spectroscopic analyzer (PerkinElmer Infrared Spectrophotometer, model 843, Waltham, Massachusetts).

5.7.2.6 Compositional Analysis

The probable compositions of the rubber compounds were estimated from the TGA curves obtained by carrying out the experiments in nitrogen and oxygen atmospheres. The amounts of polymer and low molecular weight components like oil and plasticizers were estimated from the TGA curve in nitrogen atmosphere, whereas the amounts of organic and inorganic fillers present in the compounds were assessed from the TGA curve in oxygen atmosphere.

5.7.3 Observations

5.7.3.1 Top Rubber

5.7.3.1.1 Elemental Analysis

The water extract after sodium fusion was analyzed for nitrogen, halogen, and sulfur. There was no indication of the presence of any of the above elements.

5.7.3.1.2 Differential Scanning Calorimetric Analysis

The DSC curve of the compound shown in Figure 5.44 shows two baseline shifts: at –63°C and at –50°C. This trace shows that the top rubber compound consists of a blend of two rubbers.

5.7.3.1.3 Infrared Spectroscopic Analysis

The IR spectrum of the compound is given in Figure 5.45, and the results of the IR analysis are summarized in Table 5.25. Based on the IR spectrum, it was assumed that the given compound contains SBR and NR.

5.7.3.1.4 Elemental Analysis (Spot Test)

In a Weber reaction, 0.5 g of dried, acetone-extracted compound was placed in a test tube. Then 5 ml of bromine solution (10% in $CC1_4$) was added to it, and the mixture was heated in a water bath to boiling. Then 0.5 to 0.6 ml of 10% solution of phenol in $CC1_4$ was added,

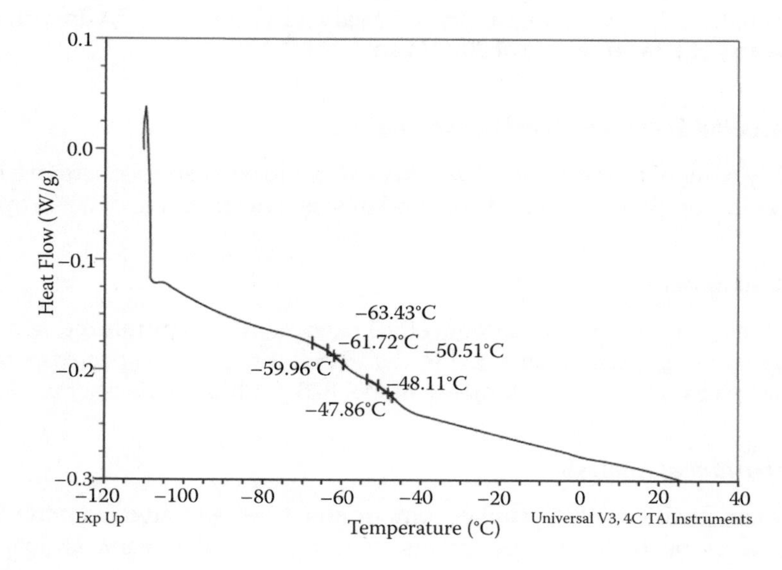

FIGURE 5.44
DSC trace of the unknown top rubber.

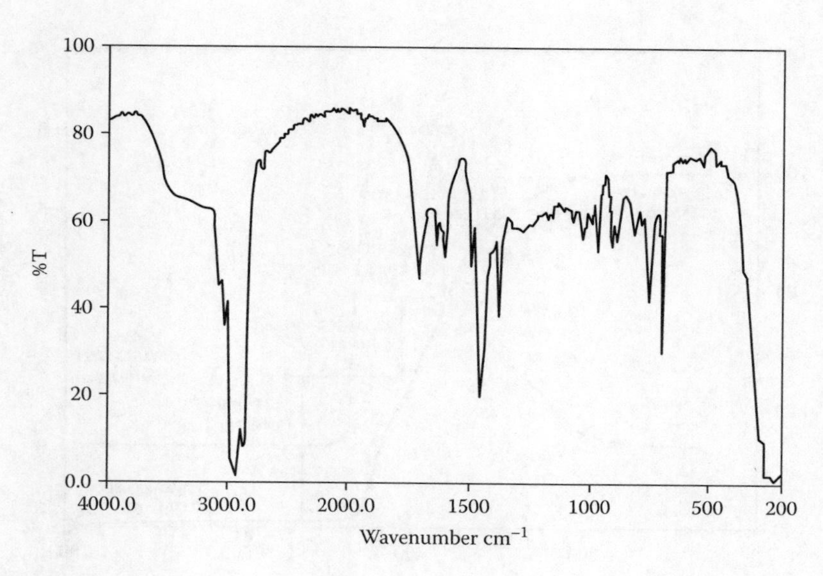

FIGURE 5.45
IR spectra of the top rubber.

and heating was continued for 15 min. A greenish gray color was generated, indicating the presence of SBR in the given compound.

5.7.3.1.5 Estimation of Probable Composition of the Compound

The TGA curves of the compound in nitrogen and oxygen atmospheres are given in Figures 5.46 and 5.47, respectively. The derivative thermogram shows multiple peaks, which indicate that the top rubber compound in the sample is a blend of different rubbers.

TABLE 5.25
Results of IR Spectroscopy Analysis of the Top Rubber

Sample 3—Top Rubber	Assignment
3031	Olefin CH stretching (NR)
2924	CH_2 asymmetric stretching (NR)
2862	CH_2 and CH_3 asymmetric stretching (NR)
2725	Overtone of CH_2 umbrella (NR)
1638	C=C stretching (NR and SBR); benzene ring substitution (SBR)
1491	C-C stretching vibration in benzene ring (SBR)
1450	CH_2 symmetric and CH_3 asymmetric deformation (NR and SBR)
1375	CH_3 symmetric deformation (NR)
992	-C-H out-of-plane bending, trans butadiene unit (SBR)
966	-CH out-of-plane bending, trans butadiene unit (SBR)
908	1,2-vinyl (SBR); butadiene (SBR)
887	-C-H stretching of isoprene unit
755	-CH out-of-plane deformation and ring deformation of monosubstituted benzene (SBR); due to butadiene part
699	-CH out-of-plane deformation and ring deformation of monosubstituted benzene (SBR)

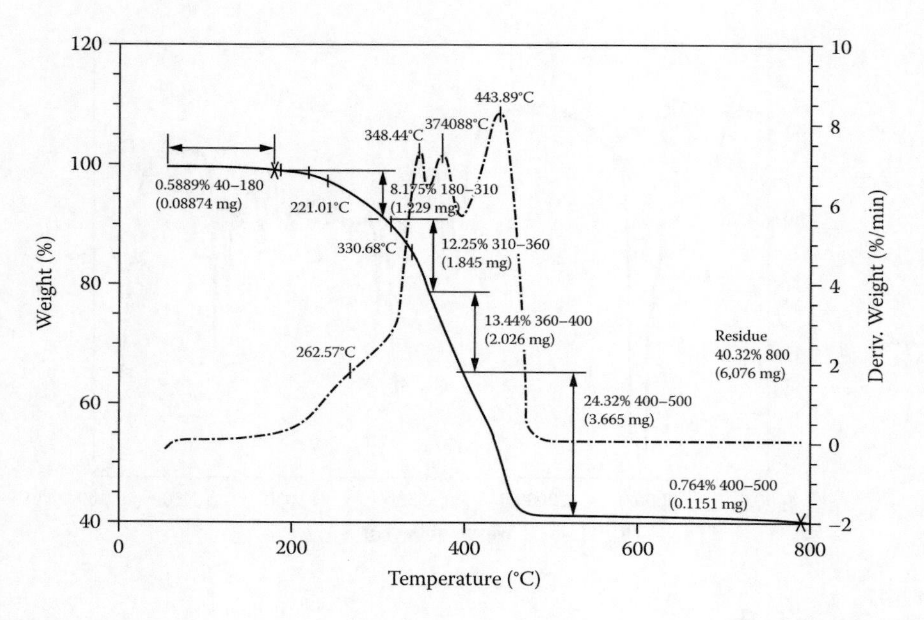

FIGURE 5.46
TGA and DTGA curves of top rubber in nitrogen.

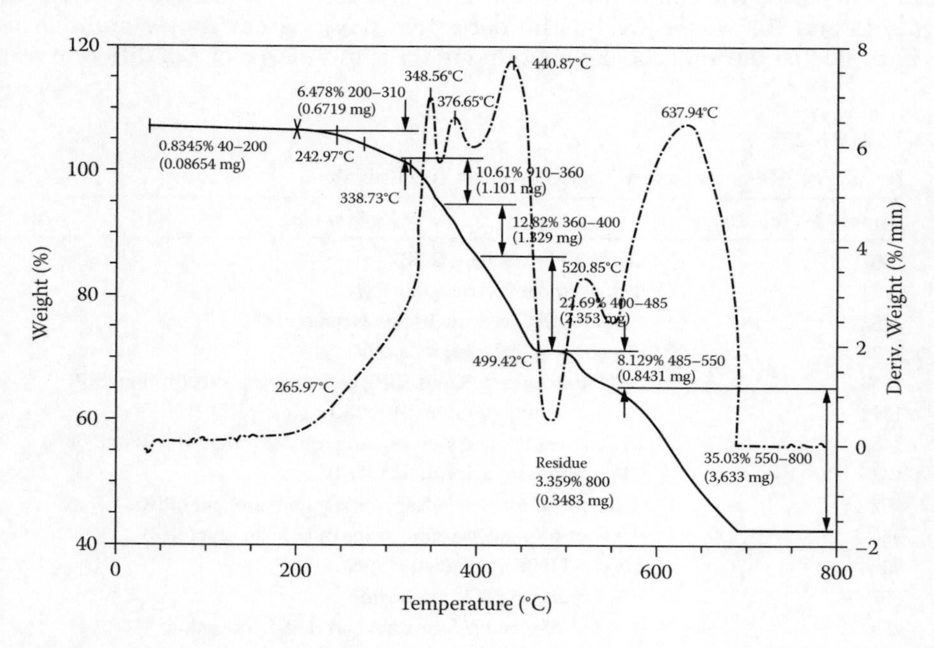

FIGURE 5.47
TGA and DTGA curves of top rubber in oxygen.

The peak maximum temperatures suggest that NR and SBR are present in the compound. Based on the TGA curves in nitrogen and oxygen atmospheres, given in the figures, the probable composition of the given compound was suggested as follows:

Polymer[a]	–	100
Low molecular weight ingredients (oils, plasticizers, etc.)	–	20
Carbon black	–	100
Inorganic residue	–	10

[a] Assumed to be a blend of 65 parts SBR and 35 parts NR

5.7.3.2 Base Rubber

5.7.3.2.1 Elemental Analysis

The water extract after sodium fusion was analyzed for nitrogen, halogen, and sulfur, but there was no indication of the presence of any of these elements.

5.7.3.2.2 Differential Scanning Calorimetric Analysis

The DSC curve of the compound given in Figure 5.48 shows two baseline shifts at –61°C and –47°C, indicating that the base rubber compound consists of a blend of two rubbers.

5.7.3.2.3 Infrared Spectroscopic Analysis

The IR spectrum of the compound as shown in Figure 5.49 was analyzed. The results of the IR spectroscopy analysis are summarized in Table 5.26. The given compound contained SBR and NR.

5.7.3.2.4 Elemental Analysis (Spot Test)

In a Weber reaction, a greenish gray color was generated, indicating the presence of SBR in the given compound.

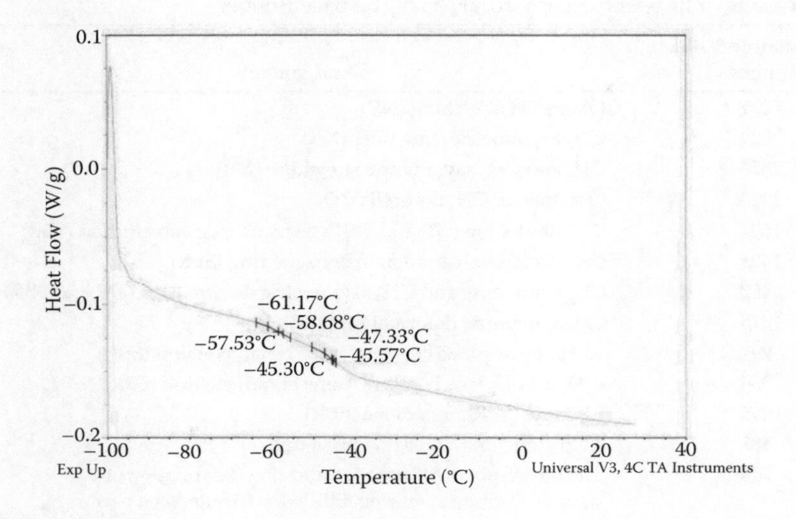

FIGURE 5.48
DSC trace of the unknown base rubber.

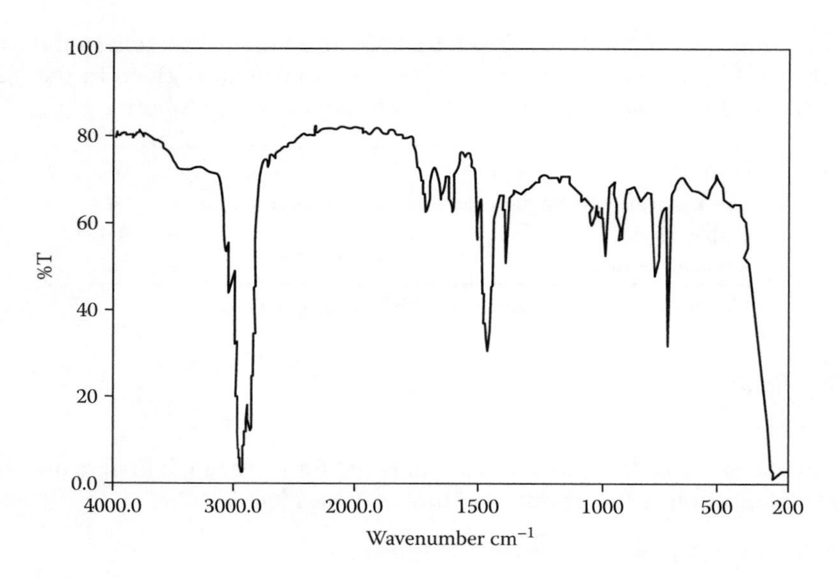

FIGURE 5.49
IR spectra of the base rubber.

5.7.3.2.5 Estimation of Probable Composition of the Compound

The TGA curves of the compound in nitrogen and oxygen atmospheres are given in Figures 5.50 and 5.51, respectively. The derivative thermogram exhibits two prominent peaks (due to decomposition of polymers) which indicate that the base rubber compound is a blend of two rubbers. The peak maximum temperatures suggest that the polymers may be NR and SBR. Based on the TGA curves in nitrogen and oxygen atmospheres,

TABLE 5.26
Results of IR Spectroscopy Analysis of the Base Rubber

Sample 3—Base Rubber	Assignment
3031	Olefin CH stretching (NR)
2929	CH_2 asymmetric stretching (NR)
2856	CH_2 and CH_3 asymmetric stretching (NR)
2723	Overtone of CH_2 umbrella (NR)
1637	C=C stretching (NR and SBR); benzene ring substitution (SBR)
1491	C-C stretching vibration in benzene ring (SBR)
1450	CH_2 symmetric and CH_3 asymmetric deformation (NR and SBR)
1375	CH_3 symmetric deformation (NR)
991	-C-H out-of-plane bending, trans butadiene unit (SBR)
966	-CH out-of-plane bending, trans butadiene unit (SBR)
908	1,2-vinyl (SBR); butadiene (SBR)
887	-C-H stretching due to isoprene unit
754	-CH out-of-plane deformation and ring deformation of monosubstituted benzene (SBR); due to butadiene part
699	-CH out-of-plane deformation and ring deformation of monosubstituted benzene (SBR)

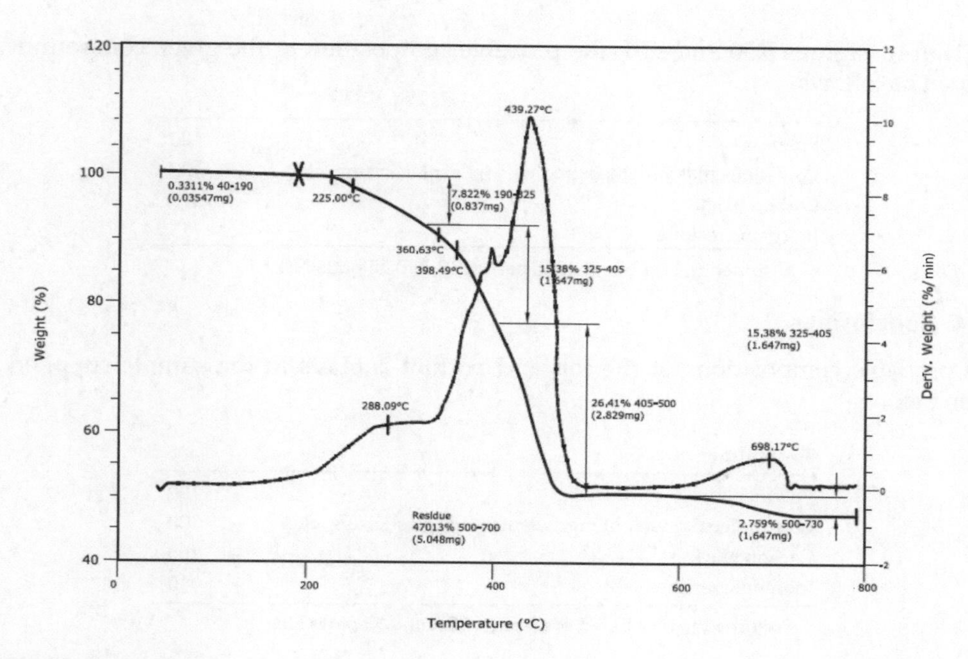

FIGURE 5.50
TGA and DTGA curves of top rubber in nitrogen.

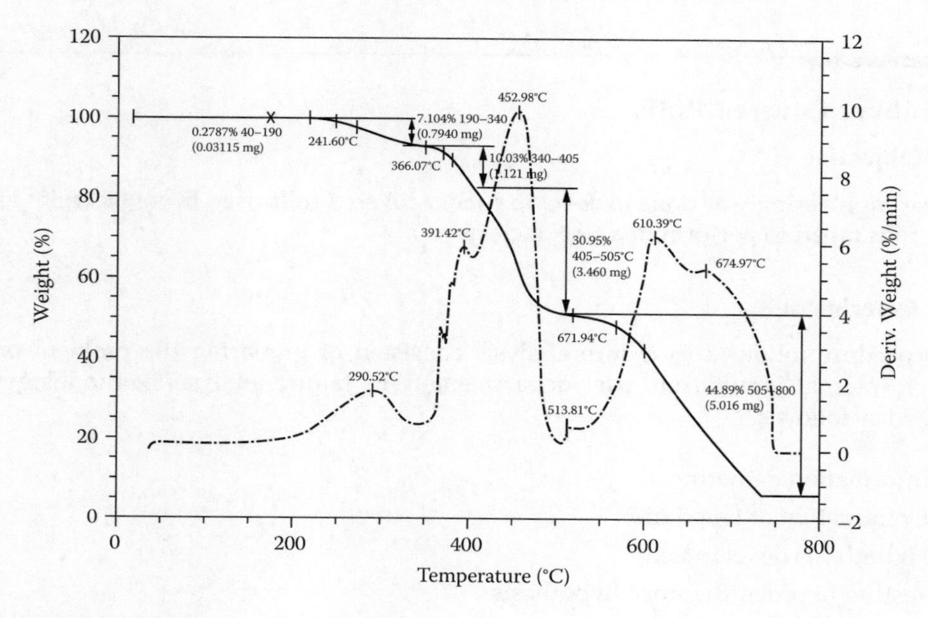

FIGURE 5.51
TGA and DTGA curves of top rubber in oxygen.

as given in Figures 5.50 and 5.51, the probable composition of the given compound was derived as follows:

Polymer[a]	–	100
Low molecular weight ingredients (oils, plasticizers, etc.)	–	20
Carbon black	–	110
Inorganic residue	–	15

[a] Assumed to be a blend of 65 parts SBR and 35 parts NR

5.7.4 Conclusions

The probable compositions of the top and bottom rubbers in the sample supplied are given below:

Top Rubber

Polymer	–	100
Low molecular weight ingredients (oils, plasticizers, etc.)	–	20
Carbon black	–	100
Inorganic residue	–	10

[a] Assumed to be a blend of 65 parts SBR and 35 parts NR

Base Rubber

Polymer	–	100
Low molecular weight ingredients (oils, plasticizers, etc.)	–	20
Carbon black	–	110
Inorganic residue	–	15

[a] Assumed to be a blend of 65 parts SBR and 35 parts NR

5.8 Rubber Covered Rolls

5.8.1 Objective

Reverse engineering was done to develop rubber covered rolls used in contact with alkali. These rolls failed to perform in a steel factory.

5.8.2 Experiment

The procedure followed in failure analysis consisted of preparing the problem profile through internal inputs from user departments. The failure analysis methodology was designed as follows:

1. Information gathering
2. Examination of failed roll
3. Hypothesis development
4. Testing to prove/disprove hypothesis
5. Conclusions
6. Simulation testing and documentation

5.8.2.1 Sample Preparation

A format was developed for gathering information on various aspects of the rubber rolls used in different sections with respect to their mechanical parameters; working conditions (mechanical and chemical); presently used polymers used for roll covering and their hardness, covering thickness, etc.; presently available service life; and the type of failures occurring. A sample copy is shown in Table 5.27.

TABLE 5.27

Information on Failure of Rubber Rolls

Line	Remarks
1. Roll reference	
a. Dimensions in mm	
• Core OD	
• Finished OD	
• Cover length	
b. Function	
c. Designed material in use	
• Hardness Shore A	
2. Working condition	
a. Drive system	
• Direct	
• Friction	
b. Line speed (rpm)	
c. RPM	
d. Line pressure (kg/cm^2)	
e. Working mode	
• Continuous	
• Intermittent	
f. Working temperature	
g. Chemical in contact	
h. Chemical	
• On surface	
• Immersed	
3. Normal life experienced	
a. Before regrinding	
b. After regrinding	
• First regrinding	
• Second regrinding	
• Third regrinding	
4. Mode of failure experienced	
• Bond failure	
• Failure from layer	
• Waviness	
• Any other	

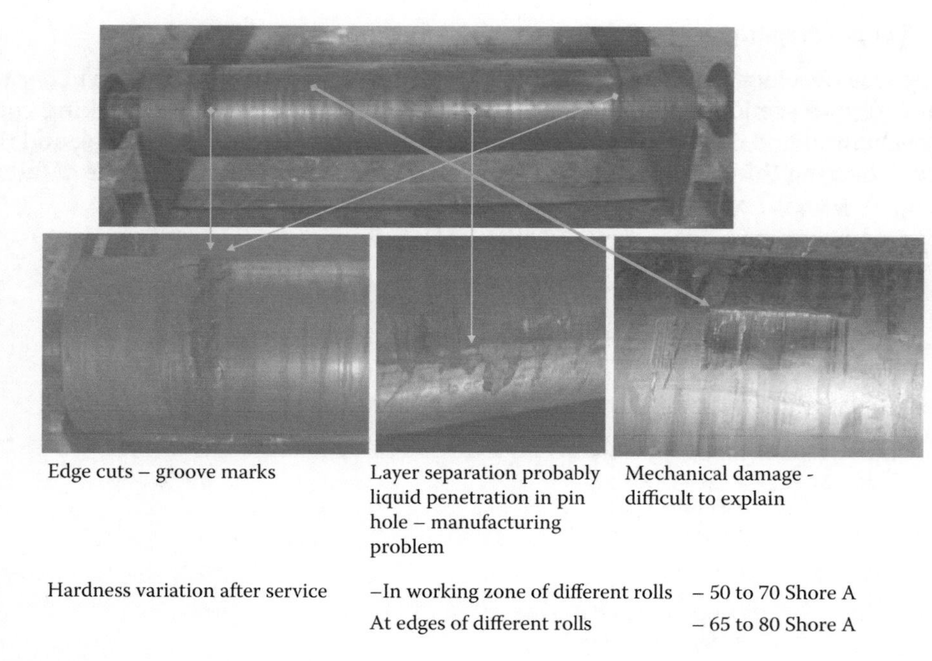

FIGURE 5.52
Defects in rubber rolls during operation in alkali rinsing area.

The collected information was studied, followed by practical observations of the working conditions of various rolls in different units and also the appearance of the defective rolls to arrive at reasons for failure. These reasons for failure were subsequently corroborated through analysis of cut pieces from various rolls at the laboratory. The type of failure observed in actual practice was recorded. A representative photograph is shown in Figure 5.52.

5.8.2.2 Analysis

The rubber samples taken from various sections were thoroughly analyzed, as described in earlier sections, to determine the polymer, filler, accelerator, processing aid, ash, etc., and a rough composition. These are not shown here to avoid repetition. The properties of the rubber vulcanizates after buffing the cut pieces from the rolls and preparing these to the required dimensions were measured and recorded.

5.8.3 Results

The formulations presented next were designed based on the requirements (see Chapter 1). The relevant properties including processing were studied. If there is a specification, the properties should be matched with the specification. If there is no specification, performance of the works in actual operation should be studied. In this particular case, there was no specification; hence the properties measured in the laboratory on cut pieces served as a rough guideline. The laboratory test results of these compounds were thoroughly examined. One such formulation is given in Table 5.28. The properties are shown in Table 5.29.

TABLE 5.28

Rubber Formulation for Alkali Section Squeeze Roll

Rubber	phr
Neoprene	100
MgO	4
Stearic acid	1
OCD	0.8
CI Resin	4.5
ISAF	42
Ivamol	5
TMT	0.45
ZnO	5
NA22	0.2
Sulfur	0.75

TABLE 5.29

Results of Properties of Squeeze Roll Rubber

After Swelling at 90°C, 7 d		Hardness (Shore A)		Physical Properties before Swelling			Properties before Swelling		DIN Abrasion	
Weight Loss (%)	Volume Loss (%)	Before Swelling	After Swelling at 90°C, 7 d	TS (MPa)	EB (%)	Modulus (MPa) at % 100 200 300	Tear Str. (kN/m)	Mooney t_5 (min) at 120°C	Before Swelling Mins. 15/20/25	After Swelling at 90°C
7.82	9.46	70–71	73	18.2	371	3.0 6.8 13.7	62.0	10.0	0.19	0.18

It was observed that these properties are much better than those of the analyzed compound. Hence, rolls were manufactured using a specially developed procedure. These were then tested in actual operation. Tables 5.30 and 5.31 display formats for information, documents, and inspection reports gathered from the factory. Finally, a service report was prepared (Table 5.32). Once the product life was satisfactory, it was assumed that the formulation designed by reverse engineering was final and frozen for a continuous manufacturing operation. By using this process, the life of the rubber rolls was enhanced ten-fold with reference to the existing life.

TABLE 5.30

Information for Documentation

Roll Number	Date of Installation	Date of Removal	Condition	Number of Days Performed	After Grinding, Number of Days Performed	Total Days Performed

TABLE 5.31
Inspection Report

Roll Number and Testing Zone	Metal to Rubber Bond Failure	Rubber to Rubber Ply/Layer Separation	Blistering/ Bulging	Surface Cracking	Surface Color Patching	Cut/ Abrasion/ Bruise Mark	Shore A Hardness	Roll Diameter (mm) Ø	Observations

TABLE 5.32
Service Report

Fresh Rolls	
Date of installation	
Date of removal (I grinding)	
Days performed	
Date of installation after I regrinding	
Date of removal (inspection)	
Days performed	
Date of reinstallation	
Date of removal (inspection)	
Days performed	
Date of reinstallation	
Date of removal (inspection)	
Days performed	
Total days in service (straight life + regrinding life)	

5.9 Rubber Part in Rubber-Metal Bonded Ring

5.9.1 Objective

The objective of this case study was identification of the chemical composition present in the rubber compounds of the rubber-metal bonded ring sample.

5.9.2 Experiment

Rubber sample was collected from the metal-rubber bonded product. The rubber part was homogenized through the closed-nip two-roll mill. The crushed sample was extracted with solvent using Soxhlet extraction for 16 h. The rubber sample was analyzed by thermogravimetric analyzer for composition analysis, viz., percent rubber, percent carbon black, and percent ash. The extraction gave the percent volatile content.

5.9.2.1 Analysis

Polymer type was analyzed by using pyrolysis FTIR. Pyrolysis was done at 500°C under an inert atmosphere. For pyrolysis, extracted rubber sample was considered to avoid any interference from the organic materials present in the rubber sample.

5.9.2.1.1 GC-MS Analysis

For identification of the volatile components, the solvent was tested using GC-MS. The organic components were determined on the basis of the atomic mass unit of the separated fraction of the materials through chromatography.

5.9.2.1.2 Ash Analysis

Ash analysis was done to identify the components of the ash. For ash analysis, a good amount of the rubber sample was taken for combustion in a muffle furnace. The ash was quantitatively measured and treated with different acids, viz., dilute hydrochloric acid followed by dilute sulfuric acid followed last by hydrofluoric acid. The metal oxides like zinc oxide, calcium oxide, and magnesium oxide, and also metal carbonates like calcium carbonate, magnesium carbonate, etc., were dissolved in dilute hydrochloric acid. Sulfuric acid generally dissolves titanium dioxide, whereas hydrofluoric acid is used for the silica-type materials. With the acid part, the analysis was done either by atomic AAS or by ICP spectrometry to determine the exact type of metal. Silica from ash was also characterized by FTIR spectroscopy. Total sulfur content in the rubber sample was determined with the help of elemental analysis.

There are several chemical test methods available to conduct the above analysis, some of which are chemical test methods that use hazardous chemicals or generate hazardous fumes during the analysis. Due to environmental regulations, it is always better to avoid these tests. It has been observed that the instrumental techniques of chemical reverse engineering have very high resolution in the results, are environmentally friendly, and also drastically reduce the testing time.

5.9.3 Results

The report obtained from the above analysis is presented in Table 5.33. The results from the above characterization methods are tabulated in Table 5.34.

TABLE 5.33

Reverse Engineering Results of Rubber Sample of Metal-Rubber Bonded Product

Test Parameters	Results	Explanation
Polymer	60.3%	It was estimated from the TGA thermogram. For analysis of the polymer content, the sample was analyzed by TGA twice—before and after extraction—as shown in Figures 5.53 and 5.54. The following principle was used to determine the polymer content of the rubber sample. The sample was first heated from 80 to 850°C at 40°C/min heating rate. Up to 600°C, the heating was performed in the presence of inert atmosphere (in N_2) and then in air/oxygen atmosphere up to 850°C. The mass loss recorded from 600 to 850°C in the presence of air/oxygen was recorded and considered as percentage of carbon black (A%), and the residue after 850°C was considered as percentage of ash content (B%) in the rubber compound. Then the rubber sample was extracted with acetone for 16 h to remove all volatile materials. After extraction, the rubber sample was dried at a temperature of 70°C in a vacuum oven. Completely dried rubber sample was further analyzed by TGA as per steps described above. This time mass loss observed in between 80°C and 600°C was considered as percentage of polymer (C%) and mass loss observed in between 600°C and 850°C as percentage of carbon black (D%) content present in after-extracted rubber sample. Actual polymer content in percent was then calculated as per the following equation:

TABLE 5.33

Reverse Engineering Results of Rubber Sample of Metal-Rubber Bonded Product (Continued)

Test Parameters	Results	Explanation
		Polymer percent = (C% × A%)/D% However, in this rubber sample, there was no carbon black present; so the polymer content was estimated as follows: From the TGA thermogram of before-extracted material, the ash content was measured (B%). Volatile matter was calculated from the extraction (V%). Then polymer content was calculated as {100 – (*B* + *V*)}. The decomposition peak obtained from the first-order derivative of the thermogram of the extracted rubber sample was also noted. Please refer to the thermograms in Figures 5.53 through 5.55.
Carbon black	Nil	Carbon black content was calculated from the TGA thermogram of the unextracted rubber sample. In this sample, the carbon black was not present. Refer to thermograms shown in Figures 5.53 and 5.54.
Ash	14.7%	Ash content was calculated from the TGA thermogram of the unextracted rubber sample. Refer to thermograms shown in Figures 5.53 and 5.54.
Volatile	25%	Volatile content was calculated by quantitative extraction of the rubber sample with acetone.
Polymer type	Acrylonitrile butadiene copolymer	Rubber sample was characterized by pyrolysis FTIR technique. The extracted and dried rubber sample was pyrolyzed under an inert atmosphere as per ASTM D3677. The pyrolyzate was directly collected on the salt plate. The salt plate along with the pyrolyzate were placed in the sample compartment of the FTIR instrument. Before placing the sample, the sample compartment was purged with nitrogen gas to remove traces of carbon dioxide. Spectrum of the sample was then characterized for the following characteristic peaks: sharp peak at 2238 cm^{-1} indicates presence of -C=N stretching. The peaks at 968 cm^{-1} and 991 cm^{-1} indicated the presence of a butadiene unit. These characteristic peaks infer the presence of acrylonitrile butadiene copolymer in the rubber sample. Refer to the pyrolyzed spectrum (Figure 5.56) of the rubber sample.
Ash analysis	HCl-soluble ash, 3.4% HF-soluble ash, 10.7%	The rubber sample without extraction was weighed and placed inside a muffle furnace at 650°C under constant supply of air. The rubber sample was burnt along with the volatile components of the sample. The leftover material was ash. The ash was then treated quantitatively as per the following scheme:

(*continued*)

TABLE 5.33

Reverse Engineering Results of Rubber Sample of Metal-Rubber Bonded Product (Continued)

Test Parameters	Results	Explanation
		Ash → Treated with HCl→ The acid solution was quantitatively filtered→ The residue was washed, dried, and weighed→ HCl-soluble ash was determined→ HCl-insoluble part was treated with concentrated H_2SO_4→ H_2SO_4 acid solution was quantitatively filtered→ The residue was washed, dried, and weighed→ H_2SO_4-soluble ash was determined→ H_2SO_4-insoluble ash was treated with HF acid→ Quantitatively the HF solution was filtered→ The residue was washed, dried, and weighed. From the above semi micro ash analysis, the following inference could be drawn about the ash type. HCl-soluble ash indicated the presence of mainly metal oxide. H_2SO_4-soluble part generally inferred the presence of titanium dioxide, and HF-soluble part of the ash reflected the presence of mainly silica. For further confirmation about the type of metal oxide, the HCl solution was analyzed by ICP spectrometry. For identifying the presence of titanium dioxide, the H_2SO_4-soluble part was analyzed by UV-Visible spectrophotometer. The presence of carbonate salt was easily detected while treating the ash with HCl. (In the presence of HCl, the ash showed effervescence.) Further confirmation about the presence of carbonate was done through the thermogravimetric analysis. In the presence of air/oxygen atmosphere, decomposition was observed in the range of 750 to 850°C at 40°C heating rate, as shown in Figure 5.53.
Total sulfur	2.99%	Total sulfur of the rubber sample was analyzed by elemental analyzer using the sample before extraction. The instrument burnt the rubber sample at high temperature in the presence of an inert gas, mainly helium. During the burning process the sulfur was converted to sulfur dioxide. This sulfur dioxide was then reduced to sulfur, which was determined by the suitable detector, preferably a TCD. There are several instruments available based on different detector systems. The most common detectors are IR detector and thermocouple detector. One can also determine the sulfur content using ion chromatography. In this rubber sample the total sulfur was directly read out by the instrument. However, in the case of carbon black filled rubber sample, total sulfur obtained from the instrument is not the actual sulfur used in the rubber compound, but also includes sulfur from carbon black, processing oils, etc. A proper precaution is needed in that case while reporting the total sulfur content of the sample.
Analysis of volatile materials in solvent extract	1. Phthalate plasticizer 2. Sulfenamide accelerator	Volatile components of the rubber sample were analyzed by using the solvent extract. Solvent was evaporated off from the extract under an inert atmosphere. Care should be taken during the evaporation phase. High-temperature heating of the extract for evaporation is not recommended, as some of the volatile components may change their chemical structures or get oxidized. As a result the identification of volatile components will not be effective, while with evaporation, a constant flow of the inert gas is always recommended. After proper drying of the extract, a little amount of methanol was added and the part of methanol dissolved residue was taken on the salt plate for FTIR analysis. The spectrum is displayed in Figure 5.57. The following characteristic wavenumbers represent functionality of the phthalate-type plasticizer:

TABLE 5.33

Reverse Engineering Results of Rubber Sample of Metal-Rubber Bonded Product (Continued)

Test Parameters	Results	Explanation
		*Strong peak at 1728 cm^{-1} due to -C=O stretching. *Peak at 1122 cm^{-1} due to -C-O stretching of -C(=O)-O-. *1072 cm^{-1} peak indicating C-O stretching of $-O-CH_2$. *742 cm^{-1} peak representing out-of-plane bending of aromatic groups. The above characteristic peaks also helped to identify the type of the phthalate. In this case, it was likely to be dioctyl phthalate type plasticizer. One should remember that for nitrile rubber or PVC type materials, plasticizer should be characterized first by FTIR technique. Otherwise, if the plasticizer is analyzed by GC-MS, the chromatography column may get contaminated with phthalate plasticizers. Another part of the methanol-dissolved residue was injected to GC with mass detector. Accelerator system analysis is one of the most difficult parts of reverse engineering. This is due to the fact that the accelerators generally lose their original chemical structure during the vulcanization process. In this case the presence of the sulfenamide accelerator was confirmed by GC with mass detector. The mass spectrum is shown in Figure 5.58. The intense peak at m/e = 135 at low voltage (14 to 16 eV) indicated the presence of marcaptobenzothiazole. Marcaptobenzothiazole is the ultimate breakdown product of most of the sulfenamide accelerators.

TABLE 5.34

Reverse Engineering Results of Rubber Sample of Metal-Rubber Bonded Product

Ingredients Identified	Results
Polymer (%)	60.3
Carbon black (%)	Nil
Total ash (%)	14.7
HCl-soluble ash (%)	3.4
HF-soluble ash (%)	10.7
HF-insoluble ash (%)	0.6
ZnO (%)	2.45
$CaCO_3$ (%)	0.46
Volatile (%)	25.0
Polymer type	Acrylonitrile butadiene rubber
Total sulfur (%)	2.99
Plasticizer type	Dioctyl phthalate
Accelerator type	Sulfenamide type

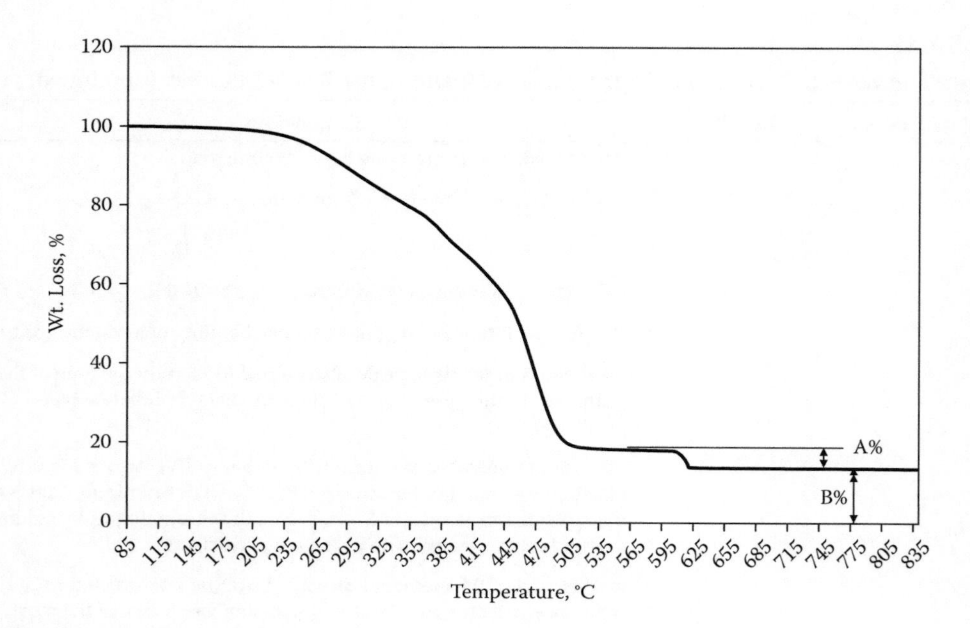

FIGURE 5.53
TGA thermogram of rubber sample before extraction.

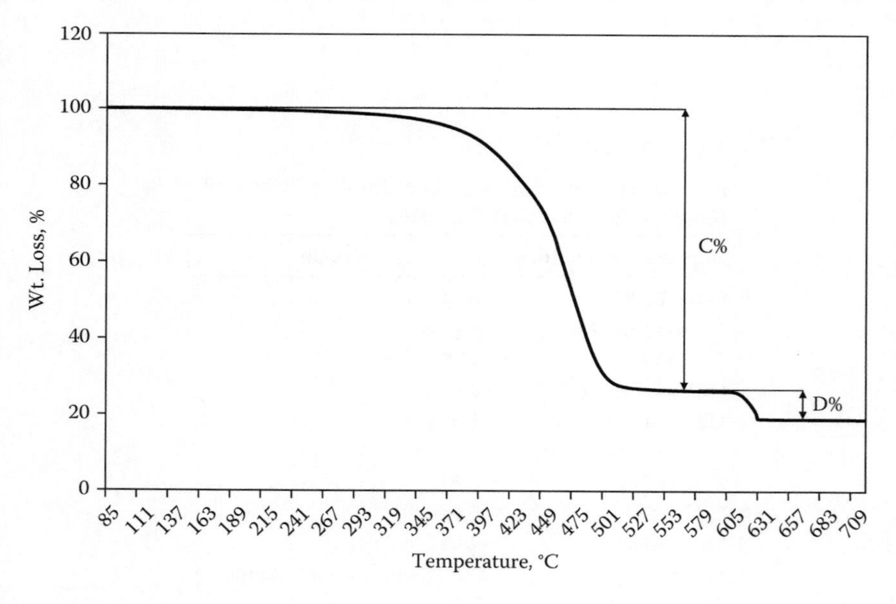

FIGURE 5.54
TGA thermogram of rubber sample after extraction.

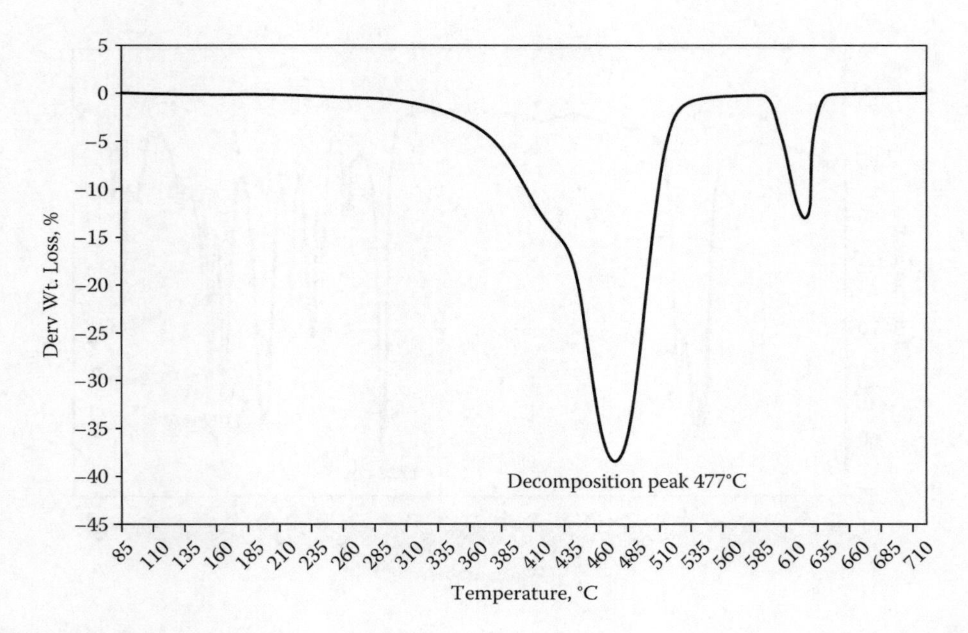

FIGURE 5.55
DTGA thermogram of rubber sample after extraction.

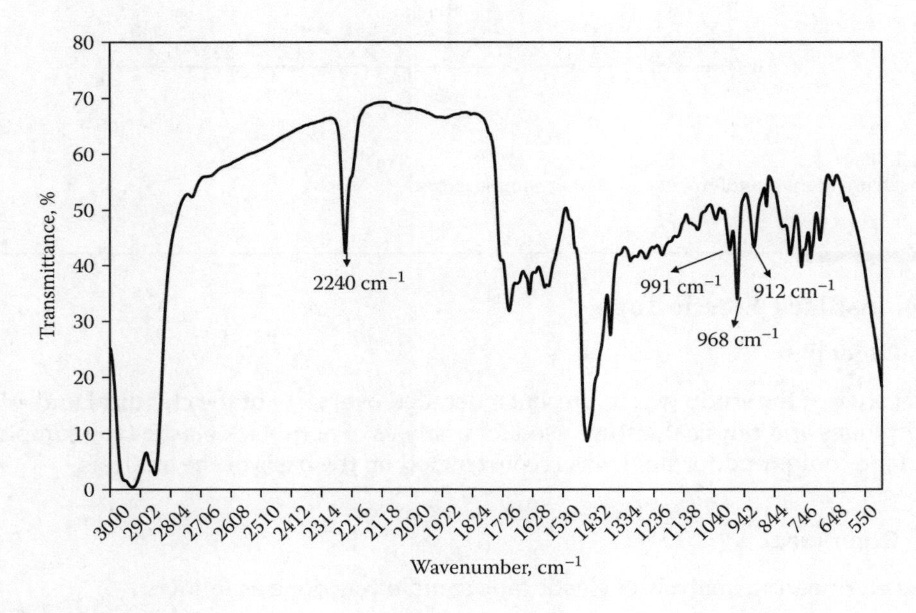

FIGURE 5.56
FTIR spectrum of rubber sample (pyrolysis spectrum).

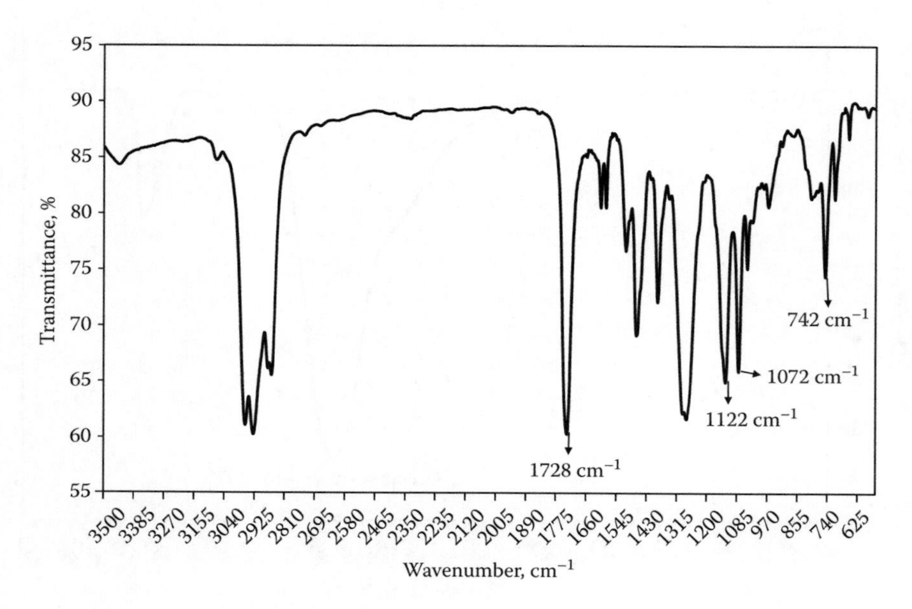

FIGURE 5.57
FTIR spectrum of toluene extract of rubber sample.

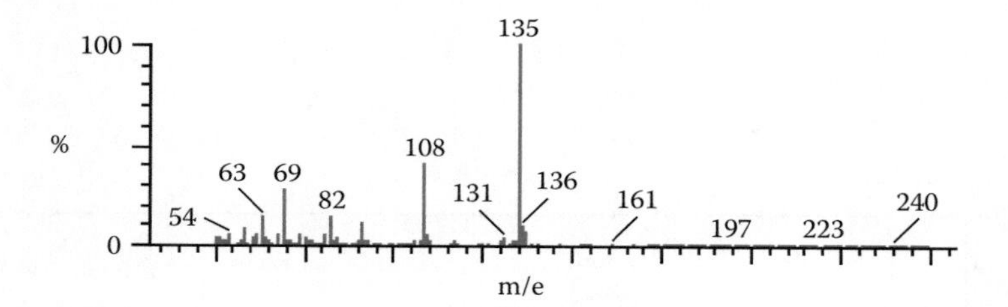

FIGURE 5.58
Mass spectrum of methanol fraction of rubber sample extract.

5.10 Non-Black Elastic Tape

5.10.1 Objective

The objective of the study was to present a detailed overview of the chemical and analytical techniques and physical testing used for analysis of non-black elastic tape sample. The elastic tape compound formula was reconstructed on the basis of the analysis.

5.10.2 Experiment

Reverse engineering analysis of elastic tape sample was done as follows:

- Extraction of rubber compound
- Polymer characterization and quantification

- Identification as well as quantification of inorganic materials in the compound
- Characterization of extract for identification of volatile components of the rubber compound
- Determination of sulfur level

5.10.2.1 Sample Preparation

Elastic tape with unknown composition was collected and passed through the closed nip of a two-roll mill to reduce the sample size.

5.10.2.2 Extraction

Extraction of the crushed elastic tape sample was carried out with Soxhlet-type extraction using acetone for 16 h at 70°C. The extracted rubber sample was heated under vacuum for complete dryness and was stored for further analysis.

TGA, FTIR, GC-MS, sulfur analysis, and ash analysis discussed in earlier chapters were used in analyzing the materials present in the elastic tape sample.

5.10.2.3 Composition Analysis

TGA was used to determine the composition with respect to polymer and ash content by heating the sample from 80 to 800°C at 40°C/min ramp rate.

5.10.2.3.1 *Analysis of Rubber*

Rubber of the elastic tape sample was analyzed by pyrolysis FTIR. The extracted sample was pyrolyzed in an inert atmosphere at 500°C. The condensate of the pyrolysis product was collected on the salt plate and analyzed by FTIR in the wavenumber range of 4000 cm^{-1} to 400 cm^{-1}.

5.10.2.3.2 *Analysis of Volatile Components*

After proper drying of the extract, the volatile components were analyzed using GC with mass detector.

5.10.2.3.3 *Analysis of Sulfur*

Analysis of sulfur was conducted with the unextracted sample using an elemental analyzer.

5.10.2.3.4 *Analysis of Ash*

For the ash analysis, a large quantity of elastic tape rubber sample was burnt inside the muffle furnace, and the total ash was measured. This ash was treated with dilute HCl and was partially dissolved in HCl. The residue of HCl was filtered and dried. The HCl-insoluble ash was fused with potassium hydrogen sulfate along with two drops of concentrated H_2SO_4. This step was followed, as the sample was a colored one. It is always preferable to follow this step if the sample is not black and the ash content is high. The fused ash was heated with a burner for 10 to 15 min and then transferred into a beaker. H_2SO_4 (6:100) was added in the beaker, and the whole mixture was boiled until dissolution. The mixture was then filtered. The filter was then treated with phosphoric acid (H_3PO_4) and hydrogen peroxide (H_2O_2). The filtrate was analyzed by UV-Visible spectroscopy at

416 nm wavelength specific for titanium dioxide (TiO_2). The absorbance was measured. The following equation was used to determine the TiO_2 percentage:

$$TiO_2 \ \text{percent} = \frac{(A_S - A_B)}{aE}$$

where:

A_S = absorbance of the sample solution

A_B = absorbance of blank solution (without ash)

a = absorptivity of TiO_2, obtained from the calibration curve with known TiO_2 concentration

E = weight (in milligrams) of the specimen.

The HCl-insoluble ash was also treated with hydrofluoric (HF) acid, and the solubility content was measured. The HCl dissolved ash was characterized by ICP spectroscopy to find out the type of metal present in the ash.

5.10.3 Results and Discussion

From the TGA thermogram (Figure 5.59) the ash content was calculated. In this figure, weight loss represented by "A" is total content of polymer and volatile components, and weight loss represented by "B" is decomposition due to the calcium carbonate content. "C" represents total residue which is also termed as ash content. Weight loss due to calcium carbonate (B) and ash content (C) of the elastic tape sample from the thermogram was calculated as 5.4% and 25.5%, respectively.

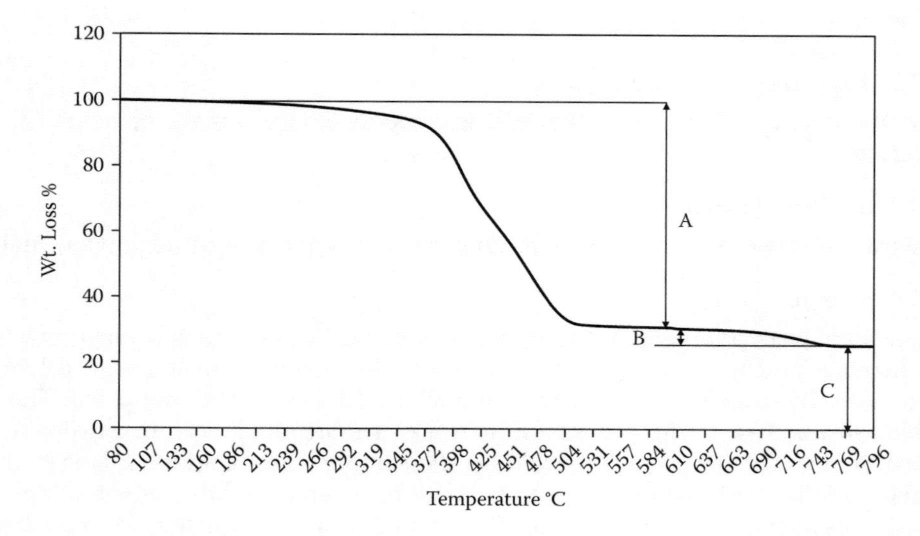

FIGURE 5.59
TGA of elastic tape.

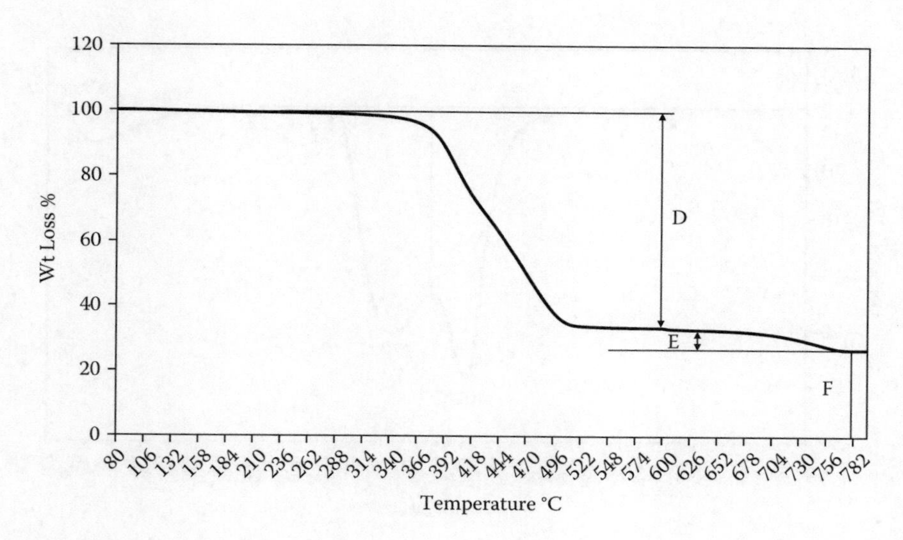

FIGURE 5.60
TGA of elastic tape after extraction.

A second TGA run of the elastic tape rubber after extraction was performed to calculate the polymer, calcium carbonate, and ash contents. The thermogram is presented in Figure 5.60. From Figure 5.60, the polymer content represented by the weight loss "D," weight loss due to calcium carbonate that is represented by "E," and residue weight "F" were calculated. The values obtained were 66.7%, 6.6%, and 26.7%, respectively.

With the help of the data, the polymer, volatile matter, and ash content were estimated. These are highlighted in Table 5.35.

TABLE 5.35
Total Compositions of the Elastic Tape

Composition	Before Extraction, %	After Extraction, %	Actual Content, %	Remarks
Polymer	Not calculated	66.7	61.9	Expression used to calculate the polymer % is: $\{D^*(B + C)\}/(E + F)$ (Refer to Figures 5.59 and 5.60.)
Weight loss due to calcium carbonate	5.4	6.6		
Ash	25.5	26.7	30.9	Calculated on the basis of the total weight loss due to calcium carbonate and the residue from the before-extraction thermogram.
Volatile matter	Not calculated	Not calculated	10.2	

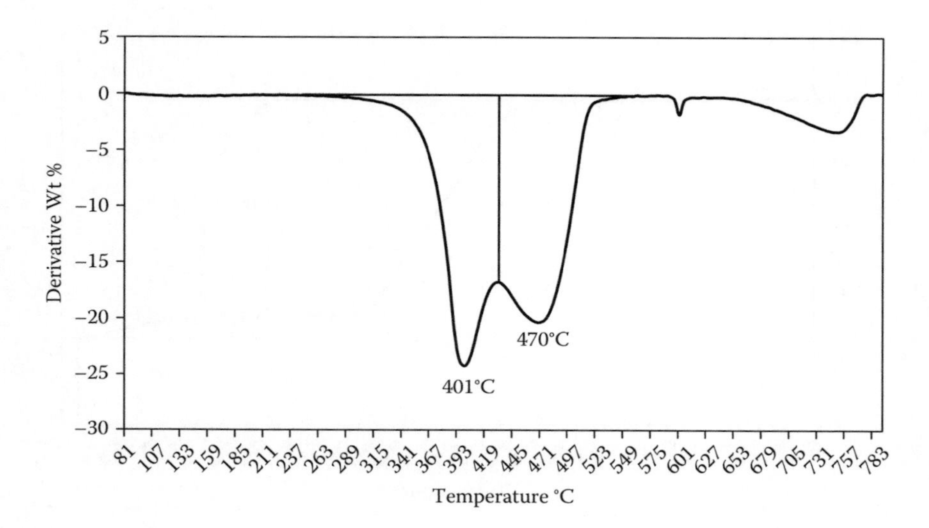

FIGURE 5.61
DTGA of elastic tape after extraction.

Derivative TGA (DTGA) thermogram of the rubber sample after extraction was also considered to characterize the presence of polymer blends. Figure 5.61 shows the DTGA thermogram of the extracted elastic tape sample. The peaks were compared with those of pristine rubbers.

The elastic tape sample contained blends of polymers as shown in Figure 5.61. DTGA was also used to estimate the blend ratio. The areas under the corresponding decomposition peaks 401°C and 470°C were estimated for blend composition. However, before calculation of the blend ratio, the polymer type was identified by pyrolysis FTIR analysis. Figure 5.62 shows the FTIR spectrum in the range of 2000 to 600 cm^{-1} for analysis of characteristic peaks of the rubber. Characteristic peak at 887 cm^{-1} represented the =C-H out-of-plane stretching of NR. Other peaks at 1450 cm^{-1} and 1377 cm^{-1} due to methyl deformation stretch of NR and 1603 cm^{-1} and 1495 cm^{-1} due to aromatic ring stretching of SBR were observed. A 700 cm^{-1} peak represented out-of-plane aromatic C-H deformation of SBR. A 760 cm^{-1} aromatic C-H deformation peak was also characteristic of SBR. The peaks at 966 cm^{-1} and 995 cm^{-1} indicated the presence of trans and vinyl double-bond out-of-plane C-H bending of the butadiene unit of SBR. The intensity of the 700 cm^{-1} aromatic C-H deformation revealed that a good amount of styrene was present in the blends.

After confirming the presence of the polymers in the blend, the blend composition was estimated using the DTG thermogram of the extracted rubber sample. From Figure 5.61, the blend components were NR 50% and SBR 50% on the basis of the peak area calculation using the thermal software support.

Figure 5.63 shows the chromatogram obtained with the elastic tape extract. On the basis of this chromatogram, the mass spectra were derived. These are shown in Figures 5.64 through 5.67, which represent different components of the extract of the elastic tape. In Figure 5.64, the m/e at 115 indicated the presence of indene fragments, which were obtained from coumarone indene resin. In Figure 5.65, the presence of m/e at 135 indicated the presence of benzothiazole fragments that came from the accelerator system used in

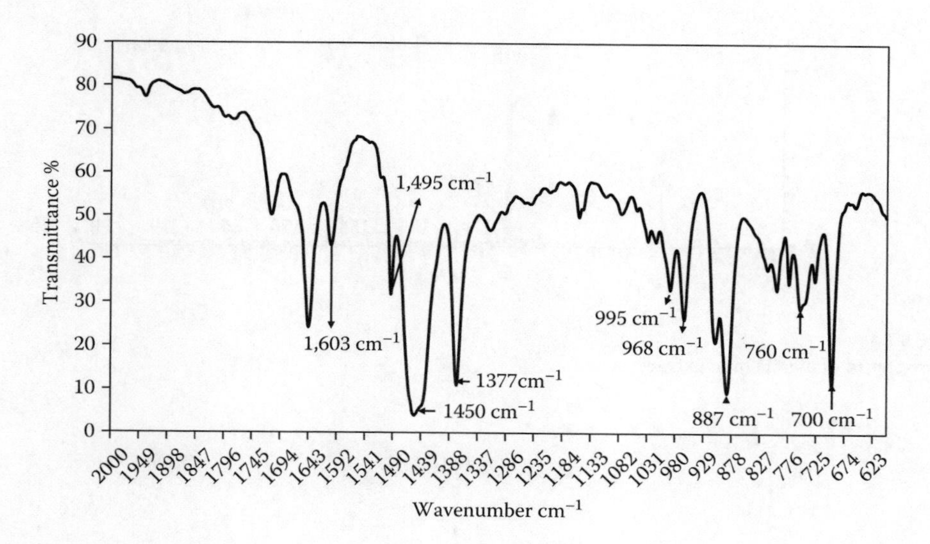

FIGURE 5.62
FTIR spectrum of elastic tape after extraction.

the elastic tape. Hence, it was assumed that a sulfenamide-type accelerator was used. The m/e peaks at 165 and 183 (Figure 5.66) indicated the presence of phenolic materials. This fraction represented the phenolic antioxidant. In Figure 5.67, the m/e peaks at 87 and 74 confirmed the presence of octadecanoic acid, which is basically the fraction of stearic acid. So, the elastic tape sample also contained stearic acid.

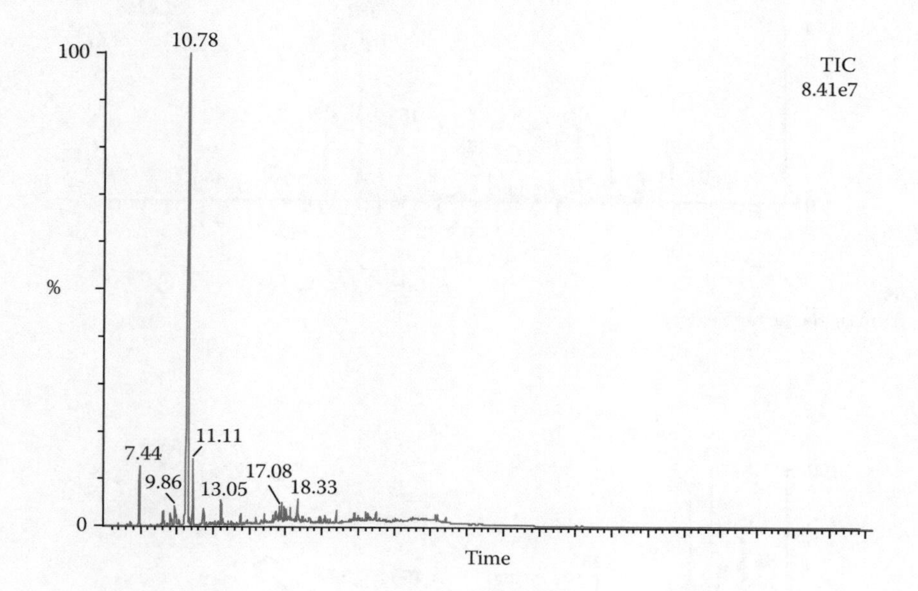

FIGURE 5.63
Chromatogram of elastic tape extract.

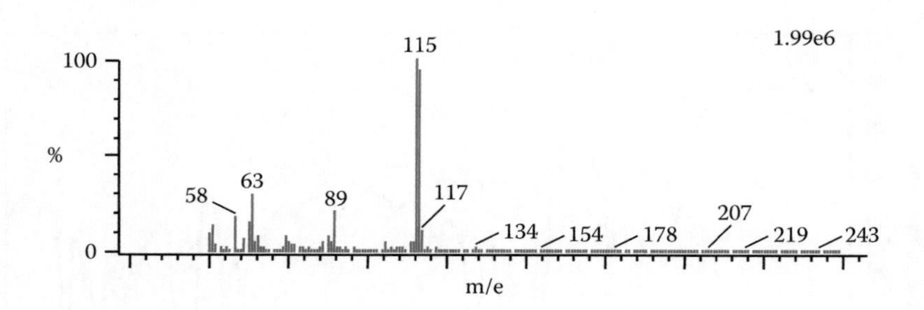

FIGURE 5.64
Mass spectrum of elastic tape extract.

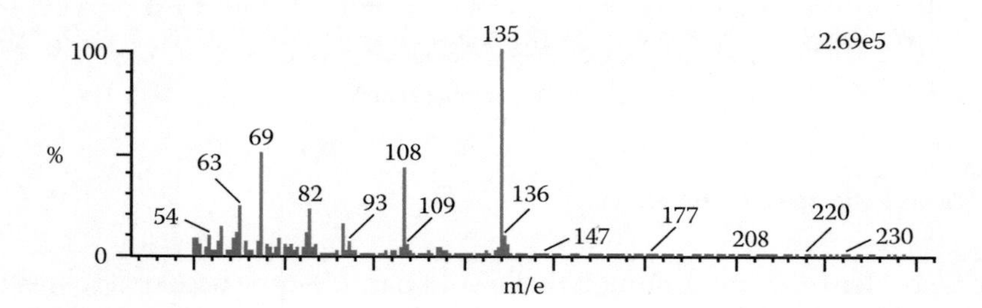

FIGURE 5.65
Mass spectrum of elastic tape extract.

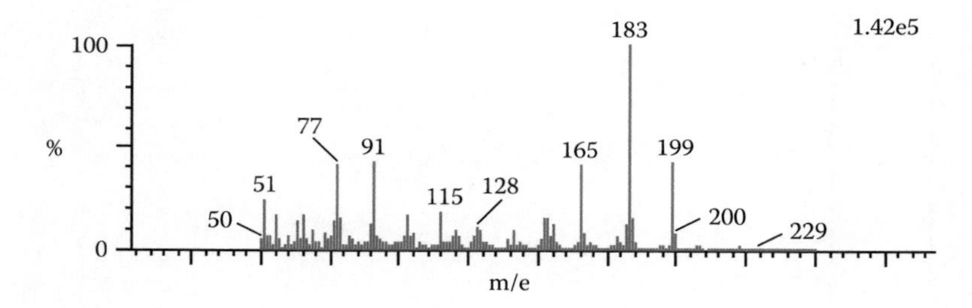

FIGURE 5.66
Mass spectrum of elastic tape extract.

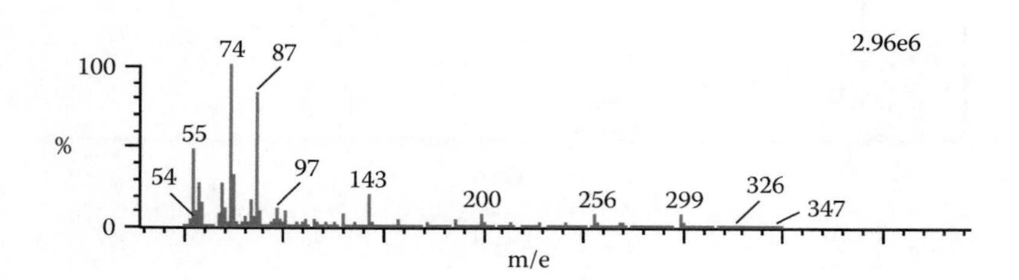

FIGURE 5.67
Mass spectrum of elastic tape extract.

TABLE 5.36

Total Chemical Composition of Elastic Tape Sample

Ingredients Identified	Results
Polymer (%)	61.9
Volatile (%)	10.2
Total ash (%)	30.9
ZnO (%)	1.3
$CaCO_3$	12.7
Silica	12.0
TiO_2	4.5
Polymer type	Blends of NR and SBR. The blend ratio is 50:50.
Total sulfur (%)	1.04
Accelerator	Sulfenamide
Antidegradants	Phenolic type
Activator organic	Stearic acid
Plasticizer type	Process oil; CI resin

Crushed sample was analyzed and the total sulfur was obtained as 1.04%. This was the total sulfur content of the sample. No correction factor was applied in this case, as this sample did not contain any carbon black. The HCl-soluble part of ash was analyzed by ICP spectroscopy. Zn and Ca were observed in HCl solution. UV-Visible study of H_2SO_4-soluble ash fraction confirmed the presence of TiO_2. Using Equation 5.1, the total content of TiO_2 was calculated. HF-soluble fraction determined the content of silica filler in the sample. These observations are summarized in Tables 5.36 to 5.37 to give the overall composition and final formulation of the elastic tape sample.

TABLE 5.37

Final Formulation of Elastic Tape Sample

Ingredients	phr
NR	50.00
SBR	50.00
ZnO	2.00
Stearic acid	1.50
Silica	19.00
$CaCO_3$	20.50
TiO_2	7.00
Phenolic antioxidant	1.50
Naphthenic oil	5.00
CI resin	8.00
Sulfur	1.70
TBBS	1.20
Retarder	0.20

5.11 Rubber Diaphragm of Audio Speaker

5.11.1 Objective

The objective of this study is to characterize a rubber diaphragm using different analytical and chemical techniques.

5.11.2 Experiment

Chemical composition of the rubber diaphragm was analyzed using the following steps:

- Extraction of rubber sample
- Polymer characterization and quantification
- Determination of carbon black content
- Semi-quantitative analysis of the inorganic materials of the rubber sample
- Qualitative analysis of the volatile components
- Determination of sulfur level

The above analyses were performed using Soxhlet apparatus, TGA, FTIR, GC-MS, and inductively coupled plasma-atomic emission spectrophotometry (ICP-AES). For each instrumental analysis, steps were followed to get proper resolution of the test results. The analytical quality assurance norm was strictly followed to avoid any ambiguity in the results.

5.11.2.1 Sample Preparation

The diaphragm sample was collected and homogenized using a two-roll mill. The homogenized sample was used for further analysis.

5.11.2.2 Extraction

Extraction of the homogenized diaphragm sample was carried out using Soxhlet-type extraction. Acetone was used as the extracting solvent. The extraction was continued for 16 h to get equilibrium extraction. The temperature of the extraction was maintained at 70°C.

After complete extraction, the rubber sample was vacuum dried at 100°C. The extract was kept for further analysis of the volatile components.

5.11.2.3 Composition Analysis

TGA was used to determine the composition of the diaphragm sample. The TGA analysis was conducted at 80 to 850°C at a 40°C/min heating rate. Two different environments were used in the TGA. First, the sample was heated from 80 to 600°C in N_2 atmosphere, and from 600 to 850°C the heating was performed in O_2 atmosphere. Between 80°C and 600°C, the decomposition of the sample was due to low boiling volatile fraction of the compound and polymer, and between 600°C and 850°C, the decomposition was due to carbon black oxidation.

5.11.2.3.1 Analysis of Rubber

Rubber of the diaphragm sample was analyzed by pyrolysis FTIR. This spectrum was captured in the region of 4000 to 600 cm^{-1} wavenumber. The fingerprint region (i.e., 1500 to 600 cm^{-1}) of the FTIR spectrum was analyzed for characterization of the rubber type.

5.11.2.3.2 Analysis of Volatile Components

Solvent extract was used to characterize the volatile components. GC with mass detector was used to characterize the volatiles. Solvent extract was dried and dissolved in methanol for injection of the extract into the GC column. Column selection was one of the important parts of GC analysis.

5.11.2.3.3 Analysis of Sulfur

Sulfur estimation was done using elemental analyzer. The homogenized diaphragm sample was used to analyze the total sulfur content.

5.11.2.3.4 Analysis of Ash

Semi-quantitative ash analysis of the diaphragm sample was performed using Inductively Coupled Plasma-Optical Emission Spectrometer (ICP-OES) instrument. The ash was generated in the muffle furnace and then treated with dilute HCl for solubilizing the ash. The HCl-soluble ash was purged in ICP-OES, and the metal content present in the solution was analyzed.

5.11.3 Results and Discussion

From the TGA thermogram (Figure 5.68), the carbon black and ash content were calculated. In Figure 5.68, weight loss represented by "A" is due to the total content of polymer and

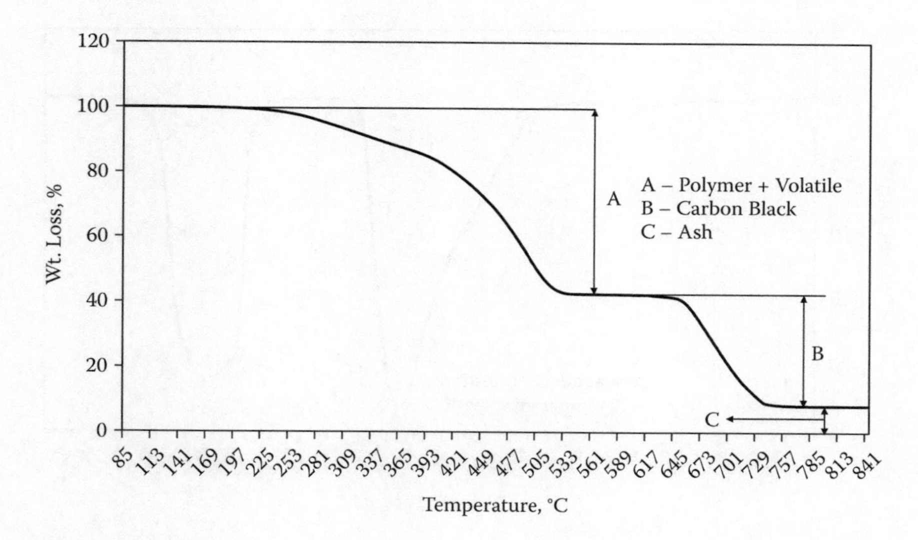

FIGURE 5.68
TGA of diaphragm rubber sample before extraction.

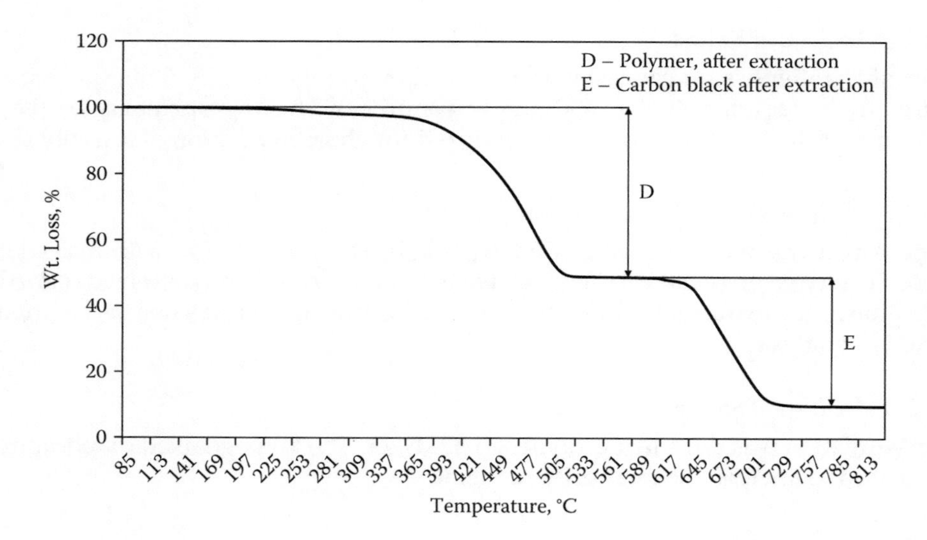

FIGURE 5.69
TGA of diaphragm sample after extraction.

volatile components, and weight loss represented by "B" is total carbon black content. "C" represents total residue which is also termed as ash content. Carbon black and ash contents of diaphragm samples from the thermogram were calculated as 34.2% and 9%, respectively.

Figure 5.69 shows the TGA thermogram of a diaphragm rubber sample after extraction. Polymer content is represented as "D" and carbon black content as "E." From these two thermograms (Figures 5.68 and 5.69), the total polymer, carbon black, and ash contents were analyzed. Figure 5.70 shows the DTGA thermogram indicating peak decomposition

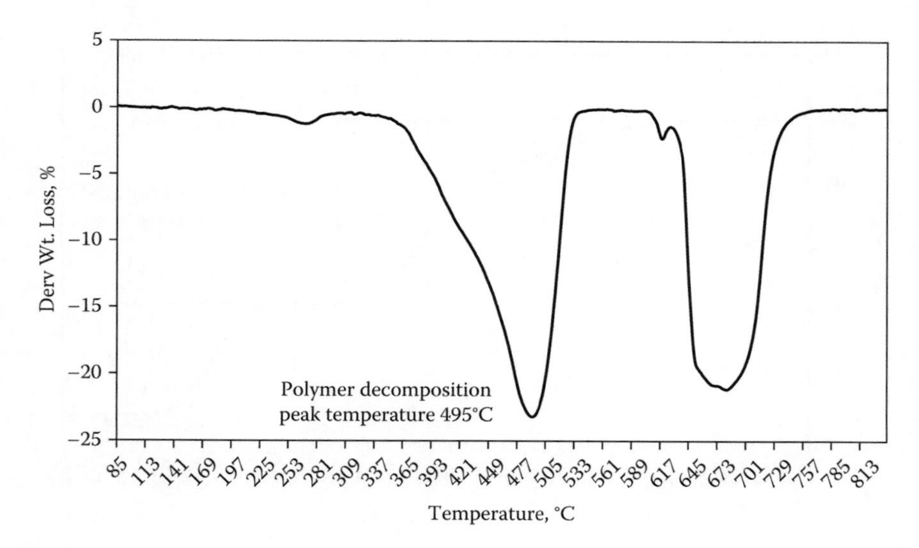

FIGURE 5.70
DTGA of diaphragm sample after extraction.

TABLE 5.38
Total Compositions of the Diaphragm Sample

Composition	Actual Content (%)
Polymer	43.3
Carbon black	34.2
Ash	9.0
Volatile matter	13.5

temperature of the rubber. This thermogram also confirmed that the diaphragm sample contained only a single polymer.

The total volatile component was calculated by subtracting the sum of total carbon black, ash, and polymer from 100. This is also known as the macro composition of the diaphragm sample. The estimated composition is shown in Table 5.38.

Rubber type was analyzed by pyrolysis FTIR. Pyrolysis of the extracted diaphragm rubber sample was performed as per ASTM D3677. The sample was pyrolyzed at 550°C under inert atmosphere. The pyrolyzed condensate was collected in a vial. The condensate was diluted with toluene and placed on the salt plate. After proper drying of the salt plate, the same was placed inside the sample compartment of the FTIR system, and the spectrum was collected from 4000 to 600 cm^{-1} wavenumber. The peak analysis was performed in the fingerprint region of the spectrum (i.e., 1500 to 650 cm^{-1}).

Figure 5.71 shows the FTIR spectrum from 2500 to 600 cm^{-1} for analysis of characteristic peaks of the rubber. The peaks at 699 cm^{-1}, 906 cm^{-1}, 990 cm^{-1}, and 966 cm^{-1} were observed.

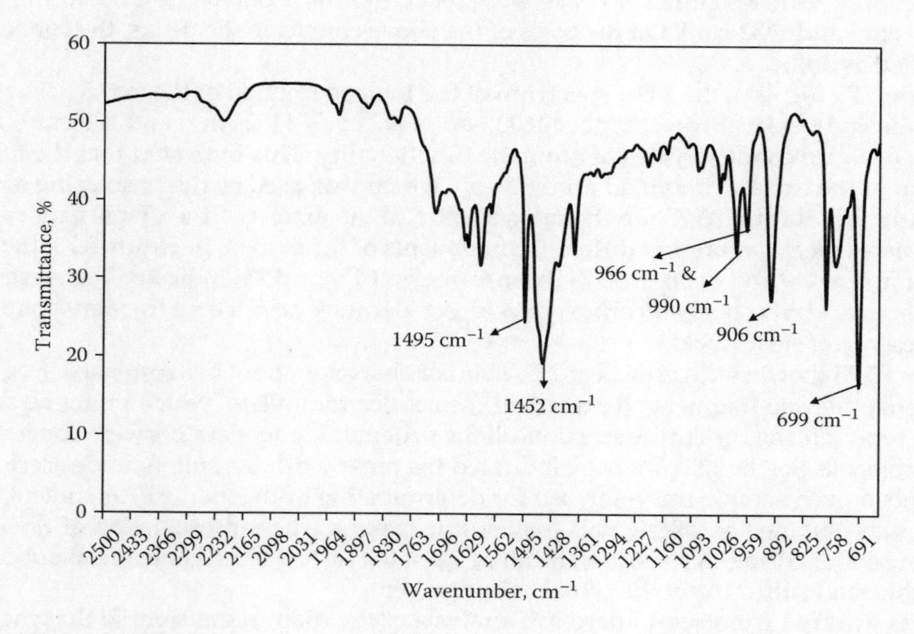

FIGURE 5.71
FTIR spectrum of pyrolyzed diaphragm sample after extraction.

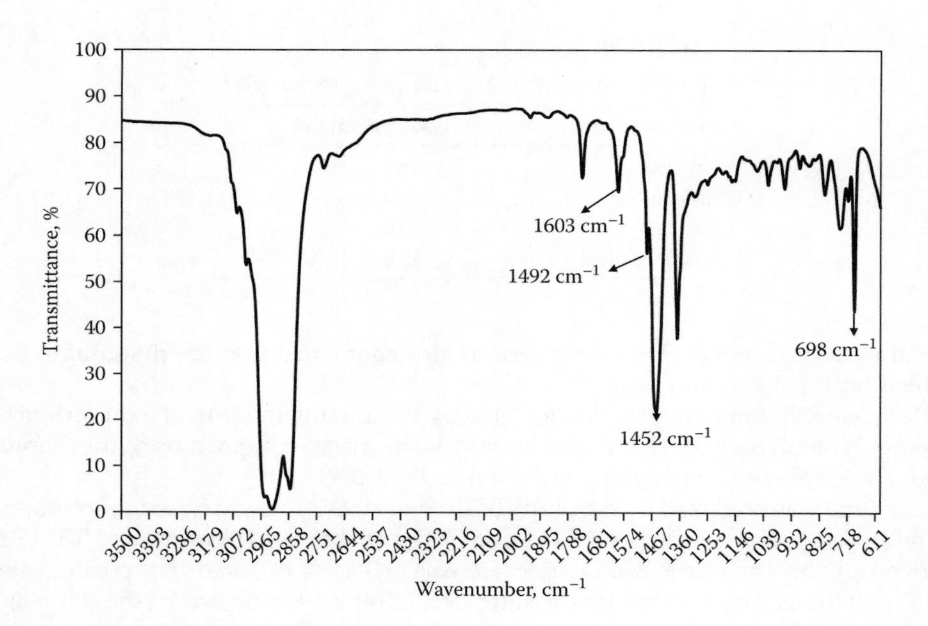

FIGURE 5.72
FTIR spectrum of diaphragm sample extract—toluene fraction.

At 699 cm^{-1} the out-of-plane aromatic C-H deformation peak was represented. The 906 cm^{-1}, 990 cm^{-1}, and 966 cm^{-1} peaks indicated the trans and vinyl double bonds and out-of-plane C-H bending. Other significant peaks were related to the aromatic ring breathing mode at 1495 cm^{-1} and 1452 cm^{-1}. On the basis of the above characteristic peaks, the rubber was identified as SBR.

Figure 5.72 displays the FTIR spectrum of the toluene fraction of the extract. The spectrum showed peaks at wavenumber 1603 cm^{-1}, 1492 cm^{-1}, 1452 cm^{-1}, and 698 cm^{-1}. All of these peaks are characteristics of aromatic functionality. This indicated that the toluene fraction of the extract contained aromatic oil, which was used as the processing aid.

Figures 5.73 through 5.75 are the mass spectra of the extract of the diaphragm sample. These mass spectra represent different components of the extract. In Figure 5.73, the presence of m/e at 149 and in Figure 5.74 the m/e peaks at 157 and 214 indicated the presence of long-chain acid, viz., benzodicarboxylic acid, octadecanoic acid. These fractions confirmed the presence of stearic acid.

Figure 5.75 shows the m/e peak at 135. This is a characteristic of benzothiazole fragment. The benzothiazole fragments represented the accelerator system, which might be sulfenamide type. During the curing reaction, all the sulfenamide accelerators were converted to benzothiozole. So, the 135 m/e peak indicated the presence of the sulfenamide accelerator.

Homogenized sample was analyzed for determination of the total sulfur content. Total sulfur was obtained as 1.38%. This total sulfur is not a true representation of the actual sulfur used for curing. A correction factor of 1% of the carbon black content was subtracted from this total sulfur to get the actual sulfur percent.

It was inferred from semi micro ash analysis of the diaphragm sample that the HCl-soluble part of the ash was basically ZnO. The HCl-insoluble ash was further treated

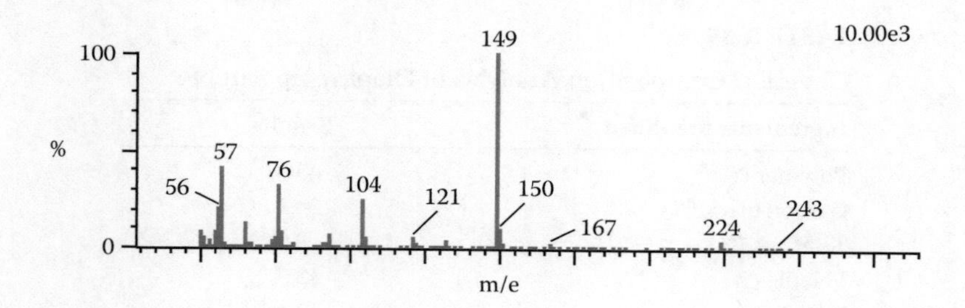

FIGURE 5.73
Mass spectrum of diaphragm sample extract.

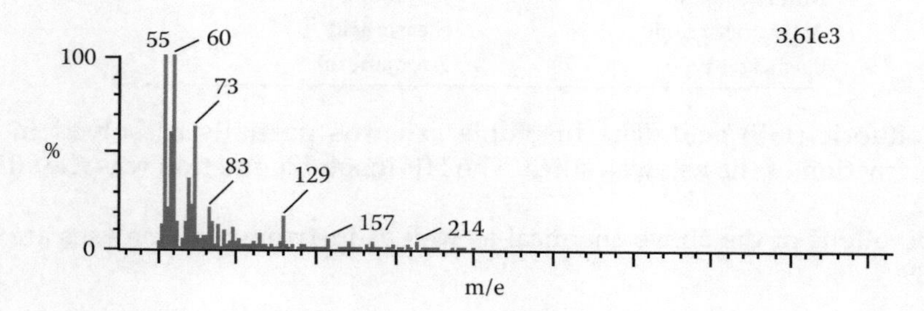

FIGURE 5.74
Mass spectrum of diaphragm sample extract.

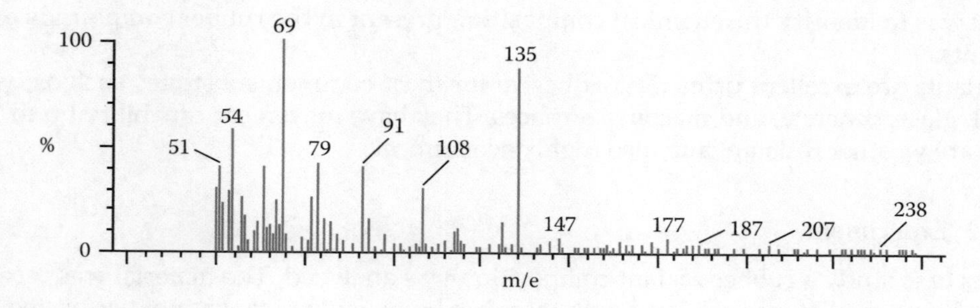

FIGURE 5.75
Mass spectrum of diaphragm sample extract.

TABLE 5.39

Chemical Composition Analysis of Diaphragm Sample

Ingredients Identified	Results
Polymer (%)	43.3
Carbon black (%)	34.2
Total ash (%)	9.0
Volatile (%)	13.5
ZnO (%)	3.4
Silica (%)	3.0
Silicate (%)	2.6
Polymer type	SBR
Total sulfur (%) after correction	1.04 (considering 1% correction factor of total carbon black)
Accelerator	Sulfenamide
Antidegradants	Not found
Activator organic	Stearic acid
Plasticizer type	Aromatic oil

with hydrofluoric (HF) acid. The insoluble ash was partially dissolved in HF. This HF-soluble fraction of the ash was silica. The HF-insoluble fraction was the silicate type of material.

The observations of the above chemical as well as instrumental analyses are tabulated in Table 5.39.

5.12 Rubber Sealant

5.12.1 Objective

Rubber sealants and caulks are designed for the toughest applications including roof sealants, foam roof sealants, metal roof sealants, metal siding, log homes/cabins, extreme environments, and all types of window sealants. The objective of this case study was to identify the chemical composition present in the rubber compounds of the sealants.

Sealants are excellent primerless adhesive for most common substrates such as wood, metal, glass, concrete, and masonry surfaces. They have movement capability up to 10%. They are weather resistant and also highly economical.

5.12.2 Experiment

In this case study a rubber sealant composition was analyzed. The material was received as a yellow tape. The material was soft; therefore homogenization was not performed. The sample of the sealant tape was used as it is.

5.12.2.1 Analysis

Polymer type was analyzed by using pyrolysis FTIR. Pyrolysis was done at 500°C under an inert atmosphere. For pyrolysis, extracted rubber sample was considered to avoid any interference from the organic materials present in the rubber sample.

For identification of the volatile components, the solvent was analyzed by using GC-MS. The organic components were analyzed on the basis of the atomic mass unit of the separated fraction of the materials.

For ash analysis, a good amount of rubber sample was taken in a muffle furnace. The ash was quantitatively measured and treated with different acids, *viz.*, dilute hydrochloric acid followed by dilute sulfuric acid and last by hydrofluoric acid. With the acid dissolved part, the analysis was done either by AAS or by ICP spectrometry.

Total sulfur content in the rubber sample was determined with the help of elemental analysis.

5.12.3 Results

The analysis report is presented in Table 5.40. With the above characterizations, formulation of the sealant can be derived as tabulated in Table 5.41.

TABLE 5.40
Reverse Engineering Results of Rubber Sealant

Test Parameters	Results	Explanation
Polymer	35.5%	Polymer content was estimated from TGA. One part of the experiments was done with unextracted sample, and the second analysis was performed with extracted sample. See Figures 5.76 through 5.78.
Carbon black	Nil	The sample was colored and contained no carbon black. So, carbon black was not estimated.
Ash	55.4%	Ash content was calculated from the TGA thermogram of the non-extracted sealant sample. Weight loss C, in Figure 5.76 shows the amount of ash in the rubber sealant sample.
		Refer to the thermogram presented in Figure 5.76.
Volatile content	9.1%	Volatile content was measured by quantitative extraction of the rubber sealant sample. In this case the extraction was conducted using acetone as the extracting solvent.
Polymer type	Butyl rubber	For identification of the polymer in the rubber sealant sample, the extracted rubber sealant sample was pyrolyzed under an inert atmosphere. The pyrolyzed condensate was collected on the salt plate and was analyzed by FTIR technique following ASTM D3677 standard.
		The spectrum of the sample was then characterized on the basis of its characteristic peaks. The following peaks in the fingerprint region of the spectrum were of interest:

(*Continued*)

TABLE 5.40

Reverse Engineering Results of Rubber Sealant (*Continued*)

Test Parameters	Results	Explanation
		3. =C-H stretching at 888 cm^{-1}—typical for isoprene unit 4. splitted peaks of C-H stretch at 1387 cm^{-1} and 1366 cm^{-1}—typical characteristics of the butyl unit Refer to the FTIR spectrum for the pyrolyzed rubber sample (Figure 5.79).
Ash analysis	HCl-soluble ash, 43.4% HCl-insoluble ash, 12.0% H_2SO_4-soluble ash, 4.0% HF-soluble ash, 8.0%	The unextracted rubber sample was weighed and placed inside the muffle furnace at 650°C under constant supply of air. The rubber sample was burnt along with the volatile components of the sample. The leftover material was ash. The ash was then treated quantitatively as per the following scheme: Ash→ Treated with HCl→ Acid solution was quantitatively filtered→ The residue was washed, dried, and weighed→ HCl-soluble part of ash was determined→ HCl-insoluble part was treated with concentrated H_2SO_4→ The H_2SO_4 acid solution was quantitatively filtered→ The residue was washed, dried, and weighed→ H_2SO_4-soluble ash was determined→ H_2SO_4-insoluble ash was treated with HF acid→ Quantitatively the HF solution was filtered→ The residue was washed, dried, and weighed. From the above semi micro ash analysis, different materials of the ash were analyzed.
Total sulfur	Nil	Total sulfur of the rubber sample was analyzed using the unextracted sample by elemental analyzer.
Analysis of volatile materials in solvent extract	1. Polybutene 2. Hindered phenol	Volatile components of the rubber sample were analyzed by using the solvent extract. The extract was dried under controlled conditions to avoid any oxidation as well as degradation of the extract material. The residue of the extract was dissolved in toluene as well as methanol. The toluene fraction was analyzed by FTIR, and the methanol fraction was injected to GC with mass detector. The FTIR spectrum is shown in Figure 5.80. The overall GC chromatogram is shown in Figure 5.81. Polybutene was identified through the FTIR of toluene fraction. The characteristic peaks were observed at wavenumbers 1366 cm^{-1}, 1388 cm^{-1}, and 3000 cm^{-1}. The sharp peaks at 1366 cm^{-1} and 1388 cm^{-1} were an indication of the isobutylene group. 300 cm^{-1} represented the alkyl -CH stretch. These are representations of isobutene. In Figure 5.82, the mass spectrum of the methanol fraction is shown. The peak at m/e = 107 at low voltage (14 to 16 eV) indicated the presence of hindered phenol.

TABLE 5.41

Reverse Engineering Results

Ingredients Identified	Results
Polymer (%)	35.5
Carbon black (%)	Nil
Total ash (%)	55.4
HCl-soluble ash (%)	43.4
H_2SO_4-soluble ash (%)	4.0. This is basically TiO_2.
HF-soluble ash (%)	8.0. This is basically silica.
ZnO (%)	1.24
Ca as $CaCO_3$ (%)	38.9
Volatile (%)	9.1
Polymer type	Butyl rubber
Total sulfur (%), after correction with respect to carbon black	Nil
Plasticizer type	Polybutene
Accelerator type	Hindered phenol

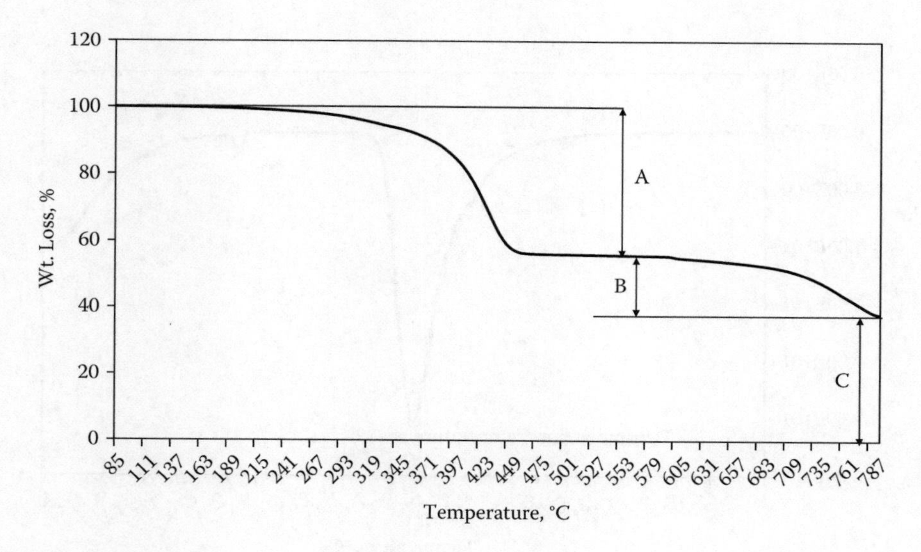

FIGURE 5.76
TGA thermogram of rubber sealant sample before extraction.

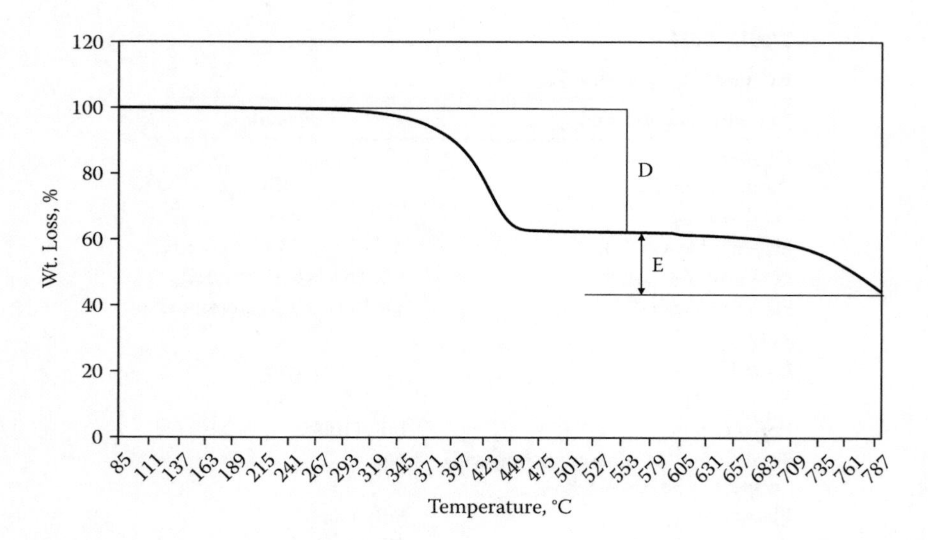

FIGURE 5.77
TGA thermogram of rubber sealant sample after extraction.

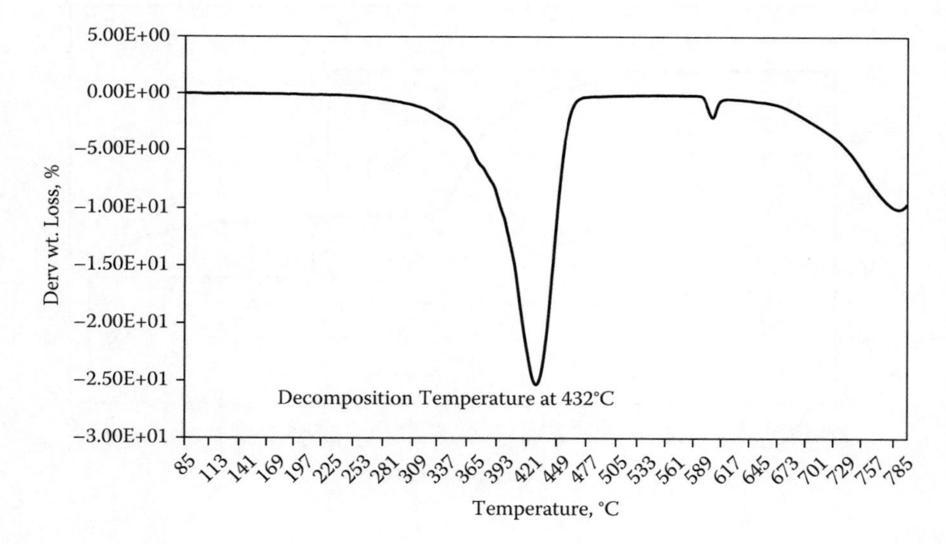

FIGURE 5.78
DTGA thermogram of rubber sealant sample after extraction.

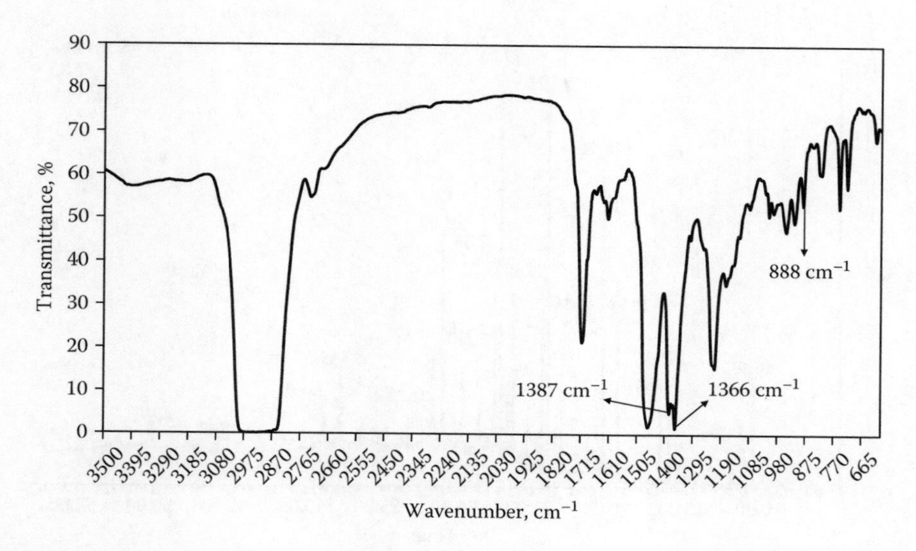

FIGURE 5.79
FTIR spectrum of pyrolyzed rubber sealant sample.

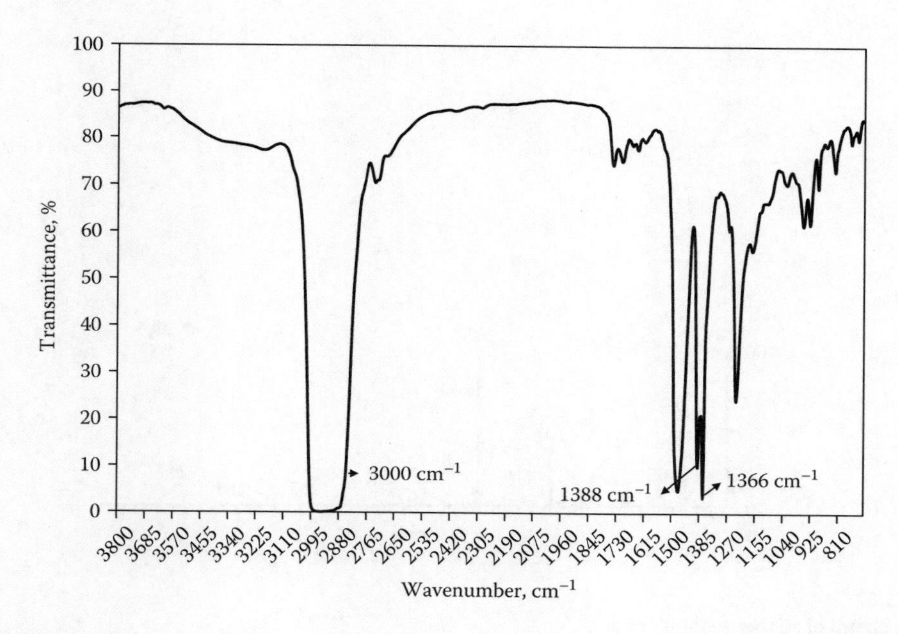

FIGURE 5.80
FTIR spectrum of rubber sealant extract—toluene fraction.

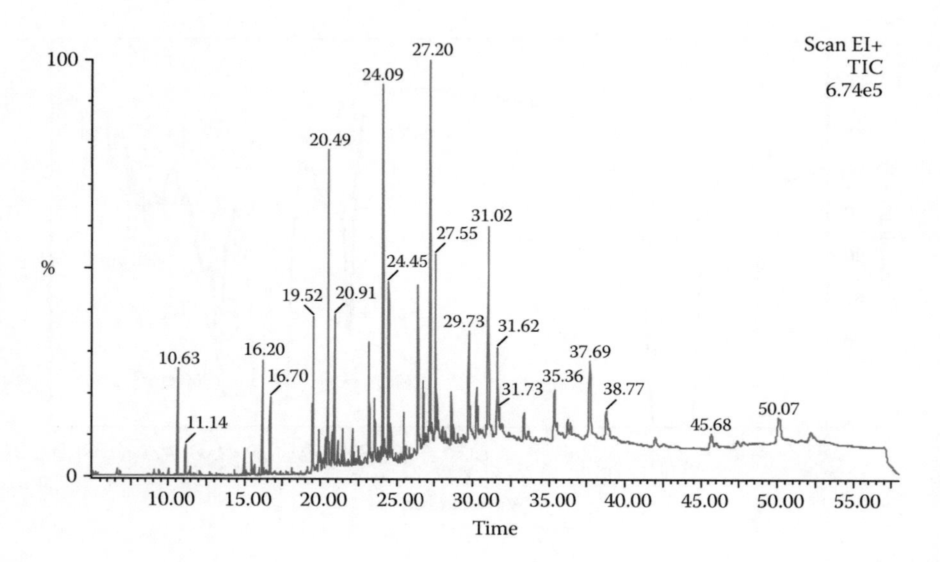

FIGURE 5.81
Chromatogram of rubber sealant extract—methanol fraction.

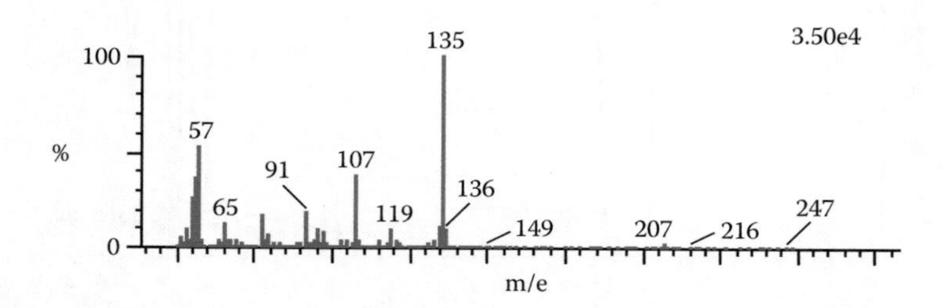

FIGURE 5.82
Mass spectrum of rubber sealant extract.

Bibliography

Abouzahr, S., Wilkes, G.L. *J. Appl. Poly. Sci.* 1984, 29, 2695.

Ahn, S.J., Lee, K.H., Kim, B.K., Jeong, H.M. *J. Appl. Poly. Sci.* 2000, 78, 1861.

Alex, R., De, P.P. *Kaut. Gumm. Kunst.* 1990, 43, 1002.

Alex, R., De, P.P., De, S.K. *Polym. Commun.* 1990, 31, 3, 118.

Alex, R., De, P.P., De, S.K. *Kaut. Gummi. Kunst.* 1991, 44, 333.

Aluas, M., Filip, C. *Solid State Nucl. Magn. Reson.* 2005, 27, 165.

Amraee, L.A., Kathab, A.A., Aghafarajollah, S. *Rubber Chem. Technol.* 1995, 69, 130.

Anadhan, S., De, P.P., De, S.K., Bhowmick, A.K. *J. Appl. Polym. Sci.* 2003, 88, 1976.

Anderson, D.A., Freeman, E.S. *J. Polym. Sci.* 1961, 54, 253.

Annual Book of ASTM Standards: Volume 05.01.

Annual Book of ASTM Standards: Volume 06.03.

Annual Book of ASTM Standards: Volume 07.01.

Annual Book of ASTM Standards: Volume 09.01.

Aronne, A., Napolitano, R., Pirozzi, B. *Eur. Polym. J.* 1986, 22, 703.

Arvanitoyannis, I., Kolokuris, I., Blanshard, J.M.V., Robinson, C. *J. Appl. Polym. Sci.* 1993, 48, 987.

ASTM D 1418, Standard practice for rubber and rubber latices, Nomenclature.

ASTM D 1566, Standard terminology relating to rubber.

ASTM D 1765, Standard classification system for carbon blacks used in rubber products.

ASTM D 2227, Standard specification for natural rubber (NR) technical grades.

ASTM D 297, Standard test methods for rubber products—chemical analysis.

ASTM D 3849, Standard test method for carbon black—morphological characterization of carbon black using electron microscopy.

ASTM D 4676, Standard classification for rubber compounding materials—Antidegradants.

ASTM D 5538, Standard practice for thermoplastics elastomers—Terminology and abbreviations.

ASTM D1054, Resilience using a Goodyear-Healey rebound pendulum.

ASTM D1795, Standard classification system for carbon blacks used in rubber products.

ASTM D3053, Standard terminology relating to carbon black.

ASTM D3182, Standard practice for rubber—Materials, equipment and procedures for mixing standard compounds and preparing standard vulcanized sheets.

ASTM D3183, Preparation of pieces for test purposes from products.

ASTM D3184, Standard methods for evaluation of and test formulations for natural rubber.

ASTM D3677, Standard test methods for rubber identification by infrared spectroscopy.

ASTM D5992, Guide for dynamic testing of vulcanized rubber and rubber like materials using vibratory methods.

ASTM D624, Tear strength of conventional vulcanized rubber and thermoplastic elastomers.

ASTM D945, Rubber properties in compression or shear (mechanical oscillograph).

Audebert, R. *Ann. Chim. Paris Ser.* 1968, 14, 49.

Ayala, J.A., Hess, W.M., Dotson, A.O., Joyce, G. *Rubber Chem. Technol.* 1990, 63, 747.

Ayala, J.A., Hess, W.M., Kistler, F.D., Joyce, G. ACS Rubber Div. Meeting, Washington, DC, 1990 Paper 42.

Azzurri, F., Flore, S.A., Alfonso, G.C., Calleja, F.J.B. *Macromolecules* 2002, 35, 9069.

Baader, T. *Kaut. u. Gummi* 7, 1954, 15WT.

Barbin, W.W., Rodgers, M.B. In *Science and Technology of Rubber*, 2nd ed., Mark, J.E., Erman, B., Eirich, F.R., Eds., Academic Press, New York, 1994, pp. 419–468.

Barlow, F.W. *Rubber Compounding, Principles, Materials and Techniques*, Marcel Dekker, New York, 1993.

Bart, J.C.J. *J. Anal. Appl. Pyrolysis* 2001, 3, 58.

Bart, J.C.J. In *Additive in Polymers, Industrial Analysis and Application*, Wiley, Chichester, England, 2005, p. 29–48.

Bartlett, C.J. *J. Elastomers and Plastics* 1978, 10, 369.

Bashir, H., Linares, A., Acosta, J.L.J. *Polym. Sci.: Polym. Phys.* 2001, 39, 1017.

Bateman, L., Moore, C.G., Porter, M., Saville, B. In *Chemistry and Physics of Rubber-Like Substances*, Bateman, L. Ed., Wiley, New York, 1963.

Baxter, R.A. In *Thermal Analysis*, Schwenker, Jr., R.F., Garn, P.D. Eds., Academic Press, New York, 1969, Volume 1, p. 65.

Bayer, F.L., *Adv. Chem. Ser.* 1983, 203, 693.

Bell, C.L.M., Gillham, J.K., Benci, J.A. SPE Third Annual Tech. Conf. Antec 74, California, 1974.

Beltran, M., Garcia, J.C., Marcilla, A. *Eur. Polym. J.* 1997, 33, 453.

Bennett, F.N.B. *Polym. Test.* 1980, 1, 2.

Bentley, D.J. *Paper Film Foil Conv.* 1997, 71, 24.

Berger, M., Buckley, D.J. *J. Polym. Sci.* 1963, 1, 2945.

Bernard, D., Baker, C.S.L., Wallace, I.R., *NR Technol.*, 1985, 16, 19.

Berry, J.P., Sambrook, R.W. Rubbercon 1977, 77, Paper 28.

Bhattacharjee, S., Bender, H., Padliya, D. *Rub. Chem. Technol.* 2003, 76, 1057.

Bhattacharya, S., Bhowmick, A.K., Avasthi, B.N. *Ind. Eng. Chem. Res.* 1991, 30, 108–6.

Bhattacharya, S.K., Chakraborty, I., Bhowmick, A.K., Dutta, B. *A rubber formulation for rubber covering rolls used in steel industry.* Indian Pat. Appl. (2007), CODEN:INXXBO IN 2005KO01188 A 20070504 AN 2007:515231.

Bhattacharya, M., Bhowmick, A.K. *Wear* 2010, 269, 152.

Bhaumik, T.K., Gupta, B.R., Bhowmick, A.K. *Wear* 1988, 128, 167.

Bhowmick, A.K. *J. Mater. Sci.* 1986, 5(10), 1042.

Bhowmick, A.K., Rampalli, S., Gallagher, K., and McIntyre, D., *J. Appl. Polym. Sci.*, 1987, 33, 1125.

Bhowmick, A.K., White, J.R., *J. Mater. Sci.* 2002, 37, 5141.

Bhowmick, A.K. *Current Topics of Elastomer Research*, CRC Press, Taylor and Francis, Boca Raton, FL, 2008.

Bhowmick, A.K., Hall, M.M., Benary, H.A. *Rubber Products Manufacturing Technology*, Marcel Dekker, New York, 1994.

Bhowmick, A.K. *J. Mater. Sci.* 1986, 21(11), 3927.

Bhowmick, A.K. *Kaut. Gummi Kunst.* 1984, 37, 191.

Bhowmick, A.K., Biswas, A., Krishnamurthy, R., Preschilla, N., Gupta, S. *Thermoplastic elastomer composition, method of making and articles thereof,* US Pat. 201393-1 (GP2 - 0448).

Bhowmick, A.K., De, S.K. Eds. *Fractography of Rubbery Materials*, Elsevier, New York, 1991.

Bhowmick, A.K., De, S.K. Eds. *Thermoplastic Elastomers from Rubber-Plastic Blends*, Ellis Horwood, New York, 1990.

Bhowmick, A.K., De, S.K. *Rub. Chem. Technol.* 1979, 52, 985.

Bhowmick, A.K., De, S.K. *Rub. Chem. Technol.* 1980, 53, 107.

Bhowmick, A.K., Gent, A.N., Pulford, C.T.R. *Rub. Chem. Technol. 1983,* 56, 226.

Bhowmick, A.K., Heslop, J., White, J.R. *Polym. Degrad. Stab.* 2001, 74, 513.

Bhowmick, A.K., Inoue, T. *J. Appl. Polym. Sci.* 1993, 49, 1893.

Bhowmick, A.K., McIntyre, D. Eds. Polymer and composite characterization: Special issue. *J. Macromol. Sci., Chem.* 1989.

Bhowmick, A.K., Mukhopadhyay, R., De, S.K. *Rub. Chem. Technol.* 1979, 52, 725.

Bhowmick, A.K., Stephens, H.L. *Handbook of Elastomers*, 2nd ed., Marcel Dekker, New York, 2001.

Bhowmick, A.K., Banerjee, D. *Improved rubber-covered conveyor belt consuming reduced energy for driving the same,* Indian Pat. Appl. (2007), CODEN:INXXBO IN 2005KO00800 A 20060908 AN 2007:398411.

Bigg, D.C.H., Hill, R.J. *J. Appl. Polym. Sci.* 1976, 20, 565.

Binder, J.L. *J. Polym. Sci. Part A* 1965, 3, 1587.

Blackledge, R.D. *J. Forensic Sci.* 1980, 25, 583.

Blackledge, R.D. *J. Forensic Sci.* 1981, 26, 3, 554.

Blackley, D.C. *Synthetic Rubbers: Their Chemistry and Technology*, Applied Science, New York, 1983, p. 279.

Blazek, A. *Thermal Analysis,* Van Nostrand Reinhold, London, 1973.
Blazso, M. *J. Anal. Appl. Pyrol.* 1997, 39, 1.
Blazsó, M., Czégény, Z.S., Csoma, C.S. *J. Anal. Appl. Pyrolysis* 2002, 64, 249.
Bloechle, D.P. *J. Elast. Plast.* 1978, 10, 377.
Blow, C.M. In *Rubber Technology and Manufacture,* Blow, C.M. Ed., Newnes-Butterworth, London, 1977, p. 29.
Blume, A., Luginsland, H.D., Meon, W., Uhrlandt, S. In *Rubber Compounding, Chemistry and Applications,* Rodgers, B. Ed. Marcel Dekker, New York, 2004.
Bohnert, T., Stanhope, B., Gruszecki, K., Pitman, S., Elsworth, V. Antec 2000, *Conference proceedings,* Orlando, FL, 2000, p. 648.
Bonfils, S., Laigneau, J.C., de Livonniere, H., Saint, B.J. *Kaut. u. Gummi Kunst.* 1999, 52, 32.
Boonstra, B.B., Cochrane, H., Dannenberg, E.M. *Rubber Chem. Technol.* 1975, 48, 558.
Borchardt, J.H., Daniels, F. *J. Am. Chem. Soc.* 1956, 79, 41.
Bove, J.L., Dalven, P. *The Science of the Total Environment* 1984, 36, 313.
Brandna, P., Zima, J. *J. Anal. App. Pyro.* 1991, 21, 207.
Brandolini, A.J., Hills, D.D. *NMR Spectra of Polymers and Polymer Additives.* Marcel Dekker, New York, 2000.
Brantley, H.L., Day, G.L. ACS Rubber Div. Meeting, New York, 1986 Paper 33.
Brazier, D.W., Nickel, G.H. *Rubber Chem. Technol.* 1975, 48, 661.
Brazier, D.W., Schwartz, N.V. *Rubber Chem. Technol.* 1978, 51, 1060.
Brennan, W.P. *Thermal Analysis Application Study,* PerkinElmer Corporation, Waltham, MA, Order number TAAS 22, 1977, 22, p. 16.
Briscoe, B.J., Sebastian, K.S., Adams, M.J. *J. Phys. D* 27 1994, 6, 156.
Bristow, G.M.J. *Nat. Rubber Res.* 1990, 5, 82.
Bristow, G.M. *NR Technol.* 1983, 13, Part 3.
Brown, R.P. *Polym. Test* 2, 1980, 1.
Brown, R.P. *RAPRA Members J.* 1975, 3, 2.
Brown, R.P., Boult, B.G. *RAPRA Members J.* 1975.
Brown, R.P., Hands, D. *RAPRA Members J.* 1973.
Brown, R.P., Hughes, R.C. *RAPRA Bull.* 1972.
Brown, R.P., Hughes, R.C., Norman, R.H. *RAPRA Members J.* 1974.
Brown, R.P., Jones, W.L. *RAPRA Bulletin,* 1972.
Brown, R.P. *Physical Testing of Rubber,* Applied Science, London, 1979.
Bryant, C.L. In *Rubber Technology and Manufacture,* Blow, C.M., Ed., Newnes-Butterworth, London, 1977, p. 119.
Brydson, J.A. *Rubber Materials and Their Compounds,* Elsevier Applied Science, London, 1988, pp. 125–126.
BS 903:Part A14, Determination of modulus in shear or adhesion to rigid plates—Quadruple test method.
BS 903:Part A3.2, Determination of tear strength—Small (Delft) test pieces.
BS 903-A64, Method for the preparation, mixing and vulcanization of rubber test mixes.
BS ISO 23529, General procedures for preparing and conditioning test pieces for physical test methods.
Buchberger, W., Stiftinger, M. In *Advances in Polymer Science,* Vol. 248, Mass Spectrometry of Polymers—New Techniques, Hakkarainen, M., Ed., Springer-Verlag, Berlin, Heidelberg, Germany, 2012, p. 39.
Buist, J.M. Int. Rubber Conf. 96, Manchester, UK, 1996, Paper No. 1.
Bulgin, D., Wratten, R. *Proc. 2nd Int. Congress on Rheology,* Butterworth, London, 1953.
Burfield, D.R., Lim, K.L. *Macromolecules,* 1983, 16, 1170.
Burhin, H.G. *Kaut. u Gummi Kunst.* 1994, 47, 262.
Burhin, H., Spreutels, W., Sezna, J. ACS Rubber Div. Conference, Detroit, 1989, Paper No. 74.
Burns, J.R., Turnbull, D. *J. Polym. Sci. Part A Polym. Physics* 1968, 6, 775.
Busfield, J.J.C., Thomas, A.G. *Rubber Chem. Technol.* 1999, 72, 876.

Byers, J.T. In *Rubber Technology*, Morton, M., Ed., Van Nostrand Reinhold, New York, 1987.
Cantor, A.S. *J. Appl. Poly. Sci.* 2000, 77, 826.
Caprin, J.C., Macander, R.F. In *Rubber Technology*, Morton, M., Ed., Van Nostrand Reinhold, New York, 1987.
Causin, V., Marega, C., Marigo, A., Ferrara, G., Idiyatullina, G., Fantinel, F. *Polymer* 2006, 47, 4773.
Chakraborty, S., Kar, S., Ameta, R., Dasgupta, S., Mukhopadhyay, R., Bandyopadhyay, S. *Rubber World* 2010.
Champaneria, R.K., Harris, B., Lotfipour, M., Packham, D.E., Turner, D.M. *Plast. Rubber Process. Appln.* 1987, 8, 185.
Chandra, A.K., Mukhopadhyay, R., Bhowmick, A.K. *J. Elast.Plast.*, 1997, 29, 34.
Chang, L.L., Woo, E.M., Liu, H.L. *Polymer* 2004, 45, 6909.
Chang, T-L., Mead, T.E. *Anal. Chem.* 1971, 43, 534.
Charlton, D.H., Yang, J., The, K.K. *Rubber Chem. Technol.* 1994, 67, 481.
Chau, K.W., Yang, Y.C., Geil, P.H. *J. Mater. Sci.* 1986, 21, 3002.
Cheng, T.-L., Su, A.-C. *Polymer*, 1995, 36, 73.
Cheng-Yu Wang, F. *J. Chromatogr. A* 2000, 891, 325.
Cheng-Yu Wang, F. *J. Chromatogr. A* 2000, 883, 199.
Cheng-Yu Wang, F., Buzanowski, W.C. *J. Chromatogr. A* 2000, 891, 313.
Cheng-Yu Wang, F. *J. Chromatogr. A* 2000, 886, 225.
Chih-An Hu, *J. Anal. Chem.* 1977, 49, 537.
Chiou, J.S., Barlow, J.W., Paul, D.R. *J. Appl. Polym. Sci.* 1985, 30, 3911.
Chiou, J.S., Barlow, J.W., Paul, D.R. *J. Appl. Polym. Sci.* 1985, 30, 2633.
Chiu, J. In *Applied Polymer Analysis and Characterization: Recent Developments in Techniques, Instrumentation and Problem Solving*, Jr., Mitchell, J., Ed., Hanser, Munich, Germany, 1987, Chapter IIG.
Chiu, J. *Anal. Chem.* 1968, 40, 1516.
Chiu, J. In *Analytical Calorimetry*, Volume 5, Johnson, J.F., Gill, P.S., Eds., Plenum Press, New York, 1984, p. 197.
Chiu, J. *Thermochimica Acta* 1970, 1, 231.
Choi, S.S. *J. Anal. Appl. Pyrol.* 2001, 57, 249.
Choi, S.S. *Korea Polym. J.* 2001, 9, 45.
Choi, S.S. *Rubber Chem. Technol.* 1996, 69, 325.
Choi, S.-S., Han, D.-H. *J. Anal. Appl. Pyrolysis* 2007, 80, 53.
Choudhury, A., Bhowmick, A.K., Ong, C., Soddemann, M. *Polym. Engg. Sci.* 2010, 50, 1389.
Choudhury, A., Bhowmick, A.K., Ong, C., Soddemann, M. *J. Nanosci. Tech.* 2010, 10, 5056.
Choudhury, A., Ong, C., Bhowmick, A.K. *J. Appl. Polym. Sci.*, 2010, 116, 1428.
Ciampelli, F., Morero, D., Cambini, M. *Macromol. Chem.* 1963, 6, 250.
Cianetti, E., Pecci, G. *Ind. Gomma* 1969, 13, 47.
Clamroth, R. *Polym. Test* 1984, 2, 4.
Clarson, S.J., Dodgson, K., Semlyen, J.A. *Polymer* 1985, 26, 930.
Clarson, S.J., Semlyen, J.A., Dodgson, K. *Polymer* 1991, 32, 2823.
Cody, C.A., Dicarlo, L., Faulseit, B.K. In Proceedings—10th North American Thermal Analysis Society Conference, 1980, Boston, MA, p. 137.
Cody, C.A., Dicarlo, L., Faulseit, B.K. *Am. Lab.* 1981, 13, 93.
Coleman, M.M., Painter, P.C. *Infrared-Absorption Spectroscopy*, vol. 8, 2nd ed., Wiley, New York, 1986, p. 69.
Coleman, M.M., Painter, P.C. *Appl. Spect. Rev.* 1984, 20, 255.
Colletti, T.A., Renshaw, J.T., Schaefer, R.E. *J. Vinyl Additive Technol.* 1998, 4, 233.
Colombian Chemicals, *The Nature of Carbon Black*, November 1990, p. 16.
Compounders pocketbook, Chemicals for Rubber Industry, Flexsys, Kingsport, TN.
Constitutive Models for Rubber II Conference, Hanover, 2001, Proceedings, Balkema, Leiden, The Netherlands.
Constitutive Models for Rubber III Conference, London, 2003, Proceedings, Balkema, Leiden, The Netherlands.

Coots, R.J., Kolycheck, E.G., *Rubber World* 1993, 208, 19.
Cotton, G.R., Murphy, L.J. *Rubber Chem. Technol.* 1988, 61, 609.
Couchman, P.R., Karasz, F.E. *Macromolecules* 1978, 11, 117.
Coveney, V.A., Jamil, S., Johnson, D.E. Constitutive Models for Rubber III Conference, 2003, Proceedings.
Cowie, J.M.G., McEwen, I.J. *Polymer* 1973, 14, 423.
Coz, D., Baranwal, K.C., In *Rubber Compound Analysis and Formula Reconstruction, in Elastomer Technology: Special Topics*, Baranwal, K.C., Stephens, H., Eds. Rubber Division, American Chemical Society, Akron, OH, 2003, chap. 18.
Crane, D., Ness, P.E. *ACS Rubber Div. Meeting*, Detroit, 1989, Paper No. 126.
Cukor, P., Lanning, E.W. *J. Chromatographic Sci.* 1971, 9, 487.
Dannis, M.L. *J. Appl. Polym. Sci.* 1962, 6, 283.
Danusso, F., Levi, M., Gianotti, G., Turri, S. *Polymer*, 1993, 34, 3687.
Datta, S., *Rubber Compounding, Chemistry and Applications* Rodgers, B., Ed., Marcel Dekker, New York, 2004.
Davies, C.K.L., De, D.K., Thomas, A.G. *Rubber Chem. Technol.* 1994, 67, 716.
Davis, B. *Rubber & Plastics News* 1994, 24, 44.
Day, G.L., Futamura, S. ACS Rubber Div. Meeting, New York, 1986, Paper No, 22.
de Luca, M.A., Machado, T.E., Notti, N.B., Jacobi, M.M. *J. Appl. Poly. Sci.* 2004, 92, 798.
De Sarkar, M., De, P.P., Bhowmick, A.K. *J. Appl. Poly. Sci.* 1999, 71, 1581.
De Sarkar, M., De, P.P., Bhownick, A.K., *Polymer*, 2000, 41, 907.
De, R.C., Auriernma, F., Corradi, M., Caliano, L., Talarico, G. *Macromolecules* 2008, 41, 8712.
Dean, G.O., Duncan, J.C., Johnson, A.F. *Polym. Test.* 4, 1984, No. 2–4.
DelVecchio, R.J. *Understanding Design of Experiments*, Hanser, Cincinnati, OH, 1997.
Demarest, C. *Plast. News* 18, 1989, 20, 16.
Deuri, A.S., Bhowmick, A.K. *J. Appl. Polym. Sci.* 1988, 35, 327.
Di Lorenzo, M.L., Righetti, M.C. *Polymer* 2008, 49, 1323.
Dick, J., Pawlowski, H. *ACS Rubber Div. Meeting*, Nashville, TN, 1992 Paper No. 37.
Dick, J., Pawlowski, H. *Polym. Test.* 1995, 14, 45.
Dick, J., Pawlowski, H. *Polym. Test.* 1996, 15, 207.
Dick, J., Pawlowski, H., Scheers, E. *ACS Rubber Div. Meeting*, Orlando, FL, 1993, Paper No. 98.
Dinesh Kumar, K., Tsou, A.H., Bhowmick, A.K. *Macromolecules* 2010, 43, 4184.
Dinsmore, H.L., Smith, D.C. *Rubber Chem. Technol.* 1949, 22, 572.
Doi, Y., Yano, A., Soga, K., Burfield, D.R. *Macromolecules* 1986, 19, 2409.
Donempudi, S., Yaseen, M. *Polym. Eng. Sci.* 1999, 39, 399.
Duck, E.W. Proceedings of the 12th Annual Meeting of International Institute of Synthetic Rubber Producers, Kyoto, Japan, 1974, p. 17.
Duerden, F. *Plast. Rubber In.* 1986, 11, 22.
DuPont Dow Elastomers, Nordel Engineering Properties & Applications, E-13 193.
DuPont Dow Elastomers, Viton, H-42586 (2/1995).
Duval, C. *Inorganic Thermogravimetric Analysis*, 1st ed., Elsevier, Amsterdam, The Netherlands, 1953.
Dyszel, S.M. In *Analytical Calorimetry*, Volume 5, Johnson, J.F., Gill, P.S., Eds. Plenum Press, New York, 1984, p. 277.
Dyszel, S.M. *Thermochimica Act* 1983, 61, 169.
Eagles, A.E., Fletcher, W. ASTM Sp. Publication 553 Symposium Philadelphia, 1973.
Edwards, B.C. *J. Polym. Sci., Polym. Phy.* 1975, 13, 1387.
Ennor, J.L. *J. IRI*, 1968.
Fahmy, I., Abdul-Wahab, M.S., Shalash, R.J. *Polym. Test.* 1990, 9, 127.
Famulok, T., Roch, P. Proceedings of the Institute of Materials International Rubber Conference, IRC 96, Manchester, UK, 1996, Paper No. 3.
Ferry, J.D. *Viscoelastic Properties of Polymers*, Wiley, New York, 1970.
Fetterman, M.Q. *Rubber World* 1986, 194, 38.
Finney, Papa, J., ACS Rubber Div. 151st Meeting, Anaheim, CA, Spring 1997, Paper No. 62.

Fischer, S.G., Chiu, J. *Thermochimica Acta* 1983, 65, 9.
Flory, P.J. *Principles of Polymer Chemistry*, Cornell University Press, Ithaca, NY, 1953, Chapter 12.
Flynn, J.H., *Thermal Analysis*, 2nd ed., Epse, S., Ed., Wiley, New York, 1986, p. 690.
Fox, T.G. *Bull. Am. Phys. Soc.* 1956, 1, 123.
Freakley, P.K., Payne, A.R. *Theory and Practice of Engineering with Rubber*, Applied Science, London, 1978.
Freeman, E.S., Carroll, B. *J. Phys. Chem.* 1958, 62, 394.
Freeman, W.J. *J. Anal. Appl. Pyrol.* 1979, 1, 3.
Freeman, W.J. *Characterization of Polymers*, 2nd ed., Wiley, New York, 1986, p. 291.
Friedersdorf, C.B., Duvdevani, I. *ACS Rubber Div. Meeting*, Chicago, 1994, Paper No. 4.
Fujiwara, Y. *Polymer Bulletin* 1985, 13, 253.
Fusco, J.V., Housb, P. In *Rubber Technology*, Morton, M., Reinhold, V.N., Eds., New York, 1987, p. 289.
Gächter, R., Müller, H. *Plastics Additives Handbook*, 4th ed., Hansa, Munich, Germany, 1993.
Ganguly, A., Bhowmick, A.K., Li, Y. *Macromolecules* 2008, 41, 6246.
García, J.C., Marcilla, A. *Polymer* 1998, 39, 431.
Garcia, J.C., Marcilla, A., Beltran, M. *Polymer* 1998, 39, 2261.
Garcia, J.C., Marcilla, A. *Polymer* 1998, 39, 3507.
Garcia, J.C., Marcilla, A. *Polymer* 1998, 39, 431.
Garn, P.D. *Thermoanalytical Methods of Investigation*, Academic Press, New York, 1965.
Garner, D.P., DiSano, M.T. *Polym. Mater. Sci. Eng.* 1996, 75, 301.
Garvey, B.S., Whitlock, M.H., Freeze, J.A. *Ind. Eng. Chem.* 1942, 34, 1309.
Gent, A.N. *Rubber Chem. Technol.* 1994, 67, 549.
Gent, A.N., Kim, H.J. *Rubber Chem. Technol.* 1990, 63, 613.
Gent, A.N., Thompson, T.T., Ramsier, R.D. *Rubber Chem. Technol.* 2003, 76, 779.
Gent, A.N. *Engineering with Rubber*, Hanser, Berlin, 2001.
Gent, A.N., Zhang, L.-Q. *J. Polym. Sci., Part B: Polym. Phys.* 2001, 39, 811.
George, S., Varughese, K.T., Thomas, S. *Polymer* 2000, 41, 5485.
Gerspacher, M., O'Farrell, C.P., Nikiel, L., Yang, H.H., Le, Mehaute, F. ACS Rubber Div. Meeting, Montreal, Canada, 1996, Paper No. 6.
Ghafouri, S.N., Freakley, P.K. *Polym. Test.* 1992, 11, 101.
Ghebremeskel, G.N., Hendrix, C. *Rubber Plastics News*, 1998, 28.
Ghebremeskel, G.N., Shield, S.R. *Rubber World* 2003, 26, 227.
Ghebremeskel, G.N., Sekinger, J.K., Hoffpauir, J.L.,Hendrix, C. *Rubber Chem. Technol.* 1996, 69, 874.
Ghosh, S., Khastgir, D., Bhowmick, A.K., Mukunda, P.G. *Polym. Deg. Stab.* 2000, 67, 427.
Gibbons, W.S., Kusy, R.P. *Polymer* 1998, 39, 6755.
Giddings, J.C. *J. Chem. Educ.* 1973, 50, 667.
Giddings, J.C., Meyers, M.N., Yang, F.J.F., Smith, L.K. *Mass Analysis of Particles and Macromolecules by Field-Flow Fractionation*, M. Kerker, New York, 1976, p. 381.
Gillham, J.K., Roller, M.B. *Polym. Eng. Sci.* 1971, 11, 295.
Gillham, J.K. *Polym. Eng. Sci.* 1967, 7, 225.
Gillham, J.K. In *Thermoanalysis of Fibers and Fiber-Forming Polymers*, Jr., Schwenker, R.F., Ed., Applied Polymer Symposium, Interscience, New York, 1966, 2, 45.
Gillham, J.K. *Polym. Eng. Sci.* 1979, 19, 676.
Glossary of Terms—Glossary of Technical Rubber Terms, published by BF Goodrich Company.
Gmelin, E., Sarge, St.M. *Pure. Appl. Chem.* 1995, 67, 11, 1789.
Goel, S.K., Beckman, E.J. *Polymer* 1993, 34, 1410.
Gonzalez-Roa, G., Ramos-deValle, L.F., Sanchez-Adame, M. *J. Vinyl Technol.*, 1991, 13, 160.
Gonzalez-Vila, F.J., Lankes, U., Ludemann, H.D. *J. Anal. Appl. Pyrol.* 2001, 349, 58.
Gordon, M., Taylor, J.S. *J. Appl. Chem.* 1952, 2, 493.
Gordon, M., Taylor, J.S. *Rubber Chem. Technol.* 1953, 26, 323.
Gorl, U., Hunsche, A. ACS Rubber Div. 150th Meeting, Washington, DC, 1996. Paper No. 76.
Goyert, W., Hespe, H. *Kunststoffe* 1978, 68, 12, 819.
Graf, H.J., Hellberg, G., Lauhus, W.P., Werner, H. *Kunststoffe Germ. Plast.* 1989, 79, 7.

Grassie, N. *Developments in Polymer Degradation*, Applied Science, London, 1998.

Graves, D.F. In *Rubber in Kent and Riegels Handbook of Industrial Chemistry and Biotechnology*, Part 1, Kent, J.A., Ed., Springer Science + Business Media, New York, 2007, p. 689.

Gregory, M.J. *Polym. Test.* 4, 1984, No. 2–4.

Gross, S.M., Flowers, D., Roberts, G., Kiserow, D.J., DeSimone, J.M. *Macromolecules* 1999, 32, 3167.

Gross, S.M., Goodner, M.D., Roberts, G.W., Kiserow, D.J., DeSimone, J.M. Abstracts of Papers of the American Chemical Society, 1999, 218, U426.

Gul, V.E., Sdobnikova, O.A., Khanchich, O.A., Peshekhonova, A.L., Samoilova, L.G. *Int. Polym. Sci. Technol.* 1996, 23, 85.

Hakkarainen, M. In *Advances in Polymer Science*, Vol. 211, *Chromatography for Sustainable Polymeric Materials*, Albertsson A.-C., Hakkarainen, M., Eds., Springer-Verlag, Berlin, Heidelberg, Germany, 2008, p. 23.

Hall, M.M., Thomas, A.G.J. *IRI* 1973, 65.

Han, M., White, J.L., Nakajima, N., Brzowskowski, R. *Kaut. u Gummi Kunst.* 1990, 43, 60.

Hands, D., Norman, R.H., Stevens, P. *Rubber Conference*, Birmingham, AL, 1984.

Hands, D., Horsfall, F. Paper 11, *ACS Rubber Div. Meeting*, Houston, TX, 1983.

Harkin-Jones, E.M.A. Rotational molding of reactive plastics, PhD Thesis, University of Belfast, 1992.

Hashim, A.S., Azahare, B., Ikeda, Y., Kohjiya, S. *Rubber Chem. Technol.* 1998, 71, 289.

Hassan, M., Chandra, A.K., Mukherjee, R., Bhowmick, A.K. *Rubber World*, 1992, 207, 25.

Hatakeyama, T., Quinn, F.X. *Thermal Analysis: Fundamentals and Applications to Polymer Science*, 2nd ed., Wiley, New York, 2000, p. 13.

Hatakeyama, T., Quinn, F.X. *Thermal Analysis: Fundamentals and Applications to Polymer Science*, 2nd ed., Wiley, New York, 2000.

Hawley, S. *Polym. Test.* 1997, 16, 327.

Heme, H., Hine, D.J., Wright, M. *RABRM Res. Report* 1952, 70.

Hepburn, C. Rubber Compounding Ingredients. Need, Theory and Innovation Part I, Rapra Review Reports 1995, 75, 7.

Herrera, M., Matuschek, M., Kettrup, A. *J. Anal. Appl. Pyrolysis* 2003, 70, 35.

Hertz, Jr. D.L., In *Handbook of Elastomers*, Bhowmick A.K., Stephens, H.L., Eds., Marcel Dekker, New York, 1988, p. 464.

Hesketh, T.R., van Bogart, J.W.C., Cooper, S.L. *Polym. Eng. Sci.*, 1980, 20, 190.

Hess, W. M., Kemp, W. K. *Rubber Chem. Technol.* 1983, 56, 390.

Hess, W.M., Chirico, V.E. *Rubber Chem. Technol.* 1977, 50, 301.

Hess, W.M., Scott, C.E., Callan, J.E. *Rubber Chem. Technol.* 1967, 40, 371.

Himmelsbach, M., Buchberger, W., Reingruber, E., *Polym. Degrad. Stab.* 2009, 94, 1213.

Hitt, D.J., Gilbert, M. *Polym. Test.* 1994, 13, 219.

Hoffman, D.J., Garcia, L.G. *J. Macromol. Sci. Phys.* 1981, B20, 3, 335.

Hoffmann, D.J., Collins, E.A. *Rubber Chem. Technol.* 1979, 52, 676.

Hofmann, W. *Rubber Technology Handbook*, Carl Hanser Verlag, Munich, Germany, 1989.

Holden, G., Legge, N.R., Quirk, R.P., Schroeder, H.E. *Thermoplastic Elastomers*, 2nd ed., Hanser, Munich, Germany, 1996.

Holownia, B.P., James, E.H. *Rubber Chem. Technol.* 1993, 66, 749.

Hong, S.W. In *Rubber Compounding, Chemistry and Applications*, Rodgers, B., Ed., Marcel Dekker, New York, 2004.

Hsu, J.-M., Yong, D.-L., Huang, S.K. *J. Polym. Res.* 1999, 6, 67.

Hueske, E.E., Clodfelter, R.W. *J. Forensic Sci.* 1977, 22, 636.

Hummel, D.O. *Macromol. Symp.* 1997, 119, 65.

Humpidge, R.T., Mattews, D., Morrel, S.H., Pyne, J.R. *Rubber Technol.* 1973, 46, 148.

Irwin, W.J. *J. Anal. Appl. Pyrol.* 1979, 1, 89.

Isayev, A.I., Cao, D., Dinzburg, B. ACS Rubber Div. Meeting, Cleveland, OH, 1995, 17–20, Paper No. 47.

Ishiaku, U.S., Shaharum, A., Ismail, H., Mohd. Ishak, Z.A. *Polym. Intl.* 1998, 45, 83.

ISO 1827, Determination of modulus in shear or adhesion to rigid plates—Quadruple test method.
ISO 2285, Determination of tension set under constant elongation, and of tension set, elongation and creep under constant tensile load.
ISO 23529, General procedures for preparing and conditioning test pieces for physical test methods.
ISO 2393. Rubber test mixes—Preparation, mixing and vulcanisation—Equipment and procedures.
ISO 34 Part 1, Determination of tear strength—Trouser, angle and crescent test pieces.
ISO 34 Part 2, Determination of tear strength—Small (Delft) test pieces.
ISO 37, Determination of tensile stress strain properties.
ISO 4662, Determination of rebound resilience of vulcanisates.
ISO 4664-1, Determination of dynamic properties—General guidance.
Jacobsen, S., Fritz, H.G. *Polym. Eng. Sci.* 1999, 39, 1303.
James, D.I., Gilder, J.S. *RAPRA Members Report.* 1984, 87.
Janowska, G., Slusarski, L. *J. Therm. Anal. Calorimetry* 2001, 65, 205.
Jansson, K.D., Zawodny, C.P., Wampler, T.P. *J. Anal. Appl. Pyrolysis* 2007, 79, 353.
Jha, A., Bhowmick, A.K. *J. Appl. Polym. Sci.* 1999, 74, 1490.
Jha, A., Bhowmick, A.K. *Polym. Deg. Stab.* 1998, 62, 575.
Jiang, S.D., Duan, Y.X., Li, L., Yan, D.D., Chen, E.Q., Yan, S. *Polymer* 2004 45, 6365.
Jimenez, A., Lopez, J., Iannoni, A., Kenny, J.M. *J. Appl. Polym. Sci.* 2001, 81, 1881.
Johnson, B.B., Chiu, J. *Thermochimica Acta* 1981, 50, 57.
Jr. Fox, T.G., Flory, P.J. *J. Appl. Phys.* 1950, 21, 581.
Jr. Gibson, E.K., Johnson, S.M. *Thermochimica Acta* 1992, 4, 49.
Jr. Gorman, W.B. inventor; *Bridgestone/Firestone Inc., assignee;* US 5,191,1993.
Kader, M.A., Bhowmick, A.K. *J. Appl. Polym. Sci.* 2003, 895, 1442.
Kainradl, P. *Rubber Chem. Technol.* 1956, 29, 1082.
Kalay, G., Kalay, C.R. *J. Appl. Polym. Sci.* 2003, 88, 814.
Kandyrin, L.B., Al'tzitser, U., Anfimov, B.N., Kuleznev, V.N. *Kauch. i. Rezina* 1977, 4, 100.
Kar, K.K., Bhowmick, A.K. *Polymer* 1999, 40, 683.
Kar, S., Maji, P.K., Bhowmick, A.K. *J. Mater. Sci.*, 2010, 45, 64.
Kar, S., Bhowmick, A.K. *J. Nanosci. Nanotechnol.* 2009, 9(5), 3144.
Kerby, R.E., Tiba, A., Culbertson, B.M., Schricker, S. Knobloch, L. *J. Macromol. Sci. A* 1999, A36, 227.
Kern, W.J., Futamura, S. ACS Rubber Div. Meeting, Quebec, 1987, Paper No. 78.
Khlok, D., Deslandes, Y., Prud'homme, J. *Macromolecules* 1976, 9, 809.
Kirchhoff, J., Mewes, D. *Kaut. u Gummi Kunst.* 2002, 55, 373.
Koopmann, R. *Polym. Test.* 5, 1985, 5.
Koopmann, R., Kramer, H. Conference Internationale du Caoutchouc 60, Paris, 1982, Paper No. 12.
Kopp, S., Wittmann, J.C., Lotz, B. *Polymer* 1994, 35, 908.
Kopp, S., Wittmann, J.C., Lotz, B. *Polymer* 1994, 35, 916.
Kow, C., Morton, M., Fetters, L.J., Hadjichristidis, N. *Rubber Chem. Technol.* 1982, 55, 245.
Kramer, H. *ACS Rubber Div. Meeting,* Washington, DC, 1990, Paper No. 5.
Kramer, H. *Polym. Test.* 2, 1981, 2, 7.
Kramer, H. *Rubber Plast. News* 1992, 21, l6.
Krause, S., Lu, Z-H., Iskandar, M. *Macromolecules* 1982, 15, 1076.
Krishnamoorti, R., Graessley, W.W., Fetters, L.J., Garner, R.T., Lohse, D.J. *Macromolecules* 1998, 31, 2312.
Kucherskii, A.M., Kaporovskii, B.M. *Polym. Test.* 1997, 16, 481.
Kumar, N., Chandra, A., Mukhopadhyay, R. *In. J. Polym. Sci.* 1996, 34, 91.
Kumar, R. *Am. Lab.* 1999, 32.
Kusch, P., Knupp, G., Morrisson, A., Bregg, R.K., Ed. In *Horizon in Polymer Research,* Nova Science, New York, 2005, p. 141.
Kusch, P., Fink, W., Schroeder-Obst, D., Obst, V. *Mass Spectrometr. Int. J.* 2008, 84, 76.
Kusch, P., Knupp, G. *J. Polym. Environ.* 2004, 12, 2, 83.
Kusch, P., Knupp, G. *LC·GC Ausgabe in deutscher Sprache* 2007, 28.
Kusch, P., Knupp, G. *Nachrichten aus der Chemie* 2009, 57, 6, 682.

Kusch, P., Obst, V., Schroeder-Obst, D., Knupp, G., Fink, W. *LC·GC Ausgabe in deutscher Sprache* 2008, 5.
Kuzma, L.J. *Rubber Technology*, 3rd ed., Morton M., Ed., Van Nostrand Reinhold, New York, 1987, p. 235.
Lacabanne, C. Contribution à l'étude des Propriétés Diélectriques des Polysulfonamides, University of Toulouse, France. PhD Thesis, 1974.
Lai, M.-K., Wang, J.-Y., Tsiang, R.C.-C. *Polymer*, 2005, 46, 2558.
Laird, J.L., Edmondson, M.S., Reidel, J.A. *Rubber World* 1997, 217, 42.
Lake, G.J., Yeoh, O.H. *Polym. Sci. Polym. Phys.* 1987, 25, 157.
Lake, G.J., Yeoh, O.H. *Rubber Chem. Technol.* 1980, 53, 1.
Lallandrelli, L.C., et al. *Makromol. Chem.* 1992, 193, 669.
Langer, H.J. In *Treatise on Analytical Chemistry*, Part 1, 2nd ed., Kolthoff, I.M., Elving, P.J., Sandell, E.B., Eds., Wiley, New York, 1980.
Lattimer, R.P., Harris, R.E., Schulten, H. *Rubber Chem. Technol.* 1985, 58, 577.
Laufer, Z., Diamant, Y., Gill, M., Fortuna, G. *Int. J. Polym. Mater.* 1978, 6, 159.
Lavergne, C., Lacabanne, C. *IEEE Electrical Insulation Magazine*, 1993, 9, 5.
Lazcano, S., Marco, C., Fatou, J.G., Bello, A. *Eur. Poly. J.* 1988, 24, 991.
Leblanc, J.L. *Eur. Rubber J.* 1998, 162, 20.
Leblanc, J.L. *Plast. Rubber Comp. Process. Appl.* 1994, 21, 81.
Leblanc, J.L. *Plast. Rubber Process. Appl.* 1982, 1.
Leblanc, J.L. *J. Appl. Polym. Sci.* 2000, 78, 1541.
LeBras, J., Papirer, E. *Rubber Chem. Technol.* 1979, 52, 43.
Legge, N.R. *Elastomerics* 1991, 123, 9, 14.
Li, B., Hu, G.H., Cao, G.P., Liu, T., Zhao, L., Yuan, W.K. *J. Appl. Polym. Sci.* 2006,102, 3212.
Li, B., Zhu, X.Y., Hu, G.H., Liu, T., Cao, G.P., Zhao, L., et al. *Polym. Eng. Sci.* 2008, 48, 1608.
Li, L., Liu, T., Zhao, L., Yuan, W.K. *Macromolecules* 2009 42, 2286.
Li, Q.F., Tian, M., Kim, D.G., Wu, D.Z., Jin, R.G. *J. Appl. Polym. Sci.* 2002, 83, 1600.
Liang, J.-Z. *Polym. Test.* 2004, 23, 1, 77.
Liebman, S.A., Levy, E.J. *Adv. Chem. Ser.* 1983, 203, 617.
Lilaonitkul, A., West, J.C., Cooper, S.L. *J. Macromol. Sci. Part B: Physics* 1976, 4, 563.
Lindley, P.B. *Nat. Rubber Tech. Bull.* 1970, 8, 3.
Liu, K., Jakab, E., Zmierczak, W., Shabtai, J.S., Meuzelaar, H.L.C. *ACS Preprints, Division of Fuel Chemistry*, 1994, 39, 576.
Lodding, W., *Gas Effluent Analysis*, Marcel Dekker, New York, 1967.
Lombardi, G. *For Better Thermal Analysis*, 2nd ed., International Confederation for Thermal Analysis, Institute of Mineralogy, Rome, Italy, 1980, p. 18.
Lotti, C., Canevarolo, S.V. *Polym. Test.* 1998, 17, 523.
Lu, K.B., Yang, D. *Polym. Bull.* 2007, 58, 731.
Ma, W.M., Yu, J., He, J.S. *J. Polym. Sci. Part B–Polym. Phy,.* 2007, 45, 1755.
Ma, W.M., Yu, J., He, J.S. *Macromolecules* 2004, 37, 6912.
Ma, X., Sauer, J.A., Hara, M. *J. Polym. Sci., Polym. Phys.* 1997, 35, 1291.
MacKinney, P.V. *J. Appl. Polym. Sci.* 1965, 9, 3359.
Macosko, C.W. *Rheology Principles, Measurements and Applications*, Wiley-VCH, New York, 1994.
Magil, I.R., Demin, S. *Rubber World* 1999, 221, 24.
Maiti, M., Sadhu, S., Bhowmick, A.K. *J. Appl. Polym. Sci.* 2005, 96, 443.
Maiti, M., Patel, J., Naskar, K., Bhowmick, A.K. *J. Appl. Polym. Sci.* 2006, 5463, 102.
Malaysian Rubber Producers' Research Association, The Natural Rubber Formulary and Property Index, Imprint of Luton, England, 1984.
Male, F.J. *ACS Rubber Div. Meeting*, Chicago, 1994, Paper No. 11.
Male, F.J. *ACS Rubber Div. Meeting*, Montreal, Canada, 1996, Paper No. 28.
Manoj, N.R., De, P.P., De, S.K. *J. Appl. Polym. Sci.* 1993, 49, 133.
Manoj, N.R., De, S.K., De, P.P. *Rubber Chem. Technol.* 1993, 66, 550.
Marchionni, G., Ajroldi, G., Righetti, M.C., Pezzin, G. *Polym. Commun.* 1991, 32, 71.
Marcilla, A., Garcia, J.C., Beltran, M. *Eur. Polym. J.* 1997, 33, 753.

Marcilla, A., Garcia, J.C. *Eur. Polym. J.* 1997, 33, 349.
Mareanukroh, M., Eby, R.K., Scavuzzo, R.J., Hamed, G.R., Preuschen, J. *Rubber Chem. Technol.* 2000, 73, p. 912.
Mark, J.E., Erman, B., Eirich, F.R. *The Science and Technology of Rubber*, Elsevier, New York, 2005.
Marsh, P.A., Cabot, J.A., Medalia, A.I. *Rubber Chem. Technol.* 1968, 41, 344.
Marshall, A.S., Petrie, S.E.B. *J. Appl. Phys.* 1975, 46, 4223.
Matuana, L.M., Balatinecz, J.J., Park, C.B. ANTEC 97. Volume III. Conference proceedings, SPE, Toronto, April 27 to May 2, 1997, p. 3580.
Matuana, L.M., Park, C.B., Balatinecz, J.J. *J. Vinyl Additive Technol.* 1997, 3, 265.
Maurer, J.J. *J. Macromol. Sci. Chem.* 1974, 8, 73.
Maurer, J.J. *Rubber Chem. Technol.* 1969, 42, 110.
McCaffrey, C., Church, E.C., Jones, F.E. Profile of Carbon Blacks in Styrene-Butadiene Rubber, Technical Report RG-129, Cabot Corporation, Boston.
McCrum, N.G., Reed, B.E., Williams, G. *Anelastic and Dielectric Effects in Polymeric Solids*, Wiley, London, 1967.
McDuff, K. Green strength—A review, *RAPRA Technol.* 1978.
McIntire, D., Stephens, H.L., Bhowmick, A.K. In *Handbook of Elastomers*, Bhowmick, A.K., Stephens, H.L., Eds., Marcel Dekker, New York, 1988, p. 1.
McNeill, I. Thermal degradation. In *Comprehensive Polymer Science*, Pergamon, New York, 1989.
Mezynski, S.M., Rodgers, M.B., Radial Medium Truck Tire Performance and Materials, ACS Rubber Div. Meeting, Akron Rubber Group, 1989.
Mitra, S., Chattopadhyay, S., Bhowmick, A.K. *Nanoscale Res. Lett.* 2009, 4, 420.
Miyoshi, T., Hayashi, S., Imashiro, F., Kaito, A. *Macromolecules* 2002, 35, 2624.
Moghe, S.R. *Rubber Chem. Technol.* 1976, 49, 247.
Möhler, H., Stegmayer, A., Kaisersberger, E. *Kaut. u. Gum. Kunst.* 1991, 44, 4, 369.
Mol, G.J. *Thermochimica Acta* 1974, 10, 259.
Moldoveanu, S.C. *Analytical Pyrolysis of Synthetic Organic Polymers*, Elsevier, Amsterdam, The Netherlands, 2005.
Morita, E. *Rubber Chem. Technol.* 1980, 53, 393.
Morley, J.F., Scott, J.R. *J. Rubber Res.* 1946, 15, 10.
Morton, M. *Rubber Technology*, Kluwer, Dordrecht, The Netherlands, 1999.
Moskala, E.J., Pecorini, T.J. *Polym. Eng. Sci.* 1994, 34, 1387.
Moyal, J.E., Fletcher, W.P. *J. Sci. Instrum.* 1945, 22, 167.
Mukhopadhyay, R., Bhowmick, A.K., De, S.K. *Polymer* 1976, 19, 1176.
Mukhopadhyay, S., De, P.P., De, S.K. *J. Appl. Polym. Sci.* 1991, 43, 2, 347.
Mukhopadhyay, S., De, S.K. *J. Mater. Sci.* 1990, 25, 4027.
Mukhopadhyay, S., De, S.K. *Polymer* 1991, 32, 1223.
Mullins, L. *RABRM Res. Memo. R.* 1948, 342.
Mullins, L. *Trans. IRI* 1948, 23, 280.
Murayama, T. *J. Appl. Polym. Sci.* 1975, 19, 3221.
Murayama, T. *J. Appl. Polym. Sci.* 1976, 20, 2593.
Murthy, V.M., Bhowmick, A.K., De, S.K. *J. Mater. Sci.* 1982, 17, 709.
Nagayama, N., Yoyoyama, M. *Mol. Crystals Liquid Crystals* 1999, 327, 19.
Nagdi, K. *Rubber as an Engineering Material: Guidelines for Users*, Hanser, Munich, Germany, 1993, p. 134.
Nagdi, K. *Thermoplastic Elastomers (TPEs), in Rubber as an Engineering Material: Guideline for Users*, Hanser, Munich, Germany, 1993.
Nagode, J.B., Roland, C.M. *Polymer* 1991, 32, 505.
Nakafuku, C., Miyaki, T. *Polymer* 1983, 24, 141.
Nakajima, N., Daniels, C.A. *J. Appl. Polym. Sci.* 1980, 25, 2019.
Nakajima, N., Harrel, E.R. *Adv. Polym. Technol.* 1986, 6, 409.
Nakajima, N., Harrell, E.R. *Rubber Chem. Technol.* 1979, 52, 962.
Nakajima, N., Isner, J.D., Harrell, E.R., Daniels, C.A. *Polym. J.* 1981, 13, 955.

Nakajima, N., Varkey, J.P. *J. Appl. Polym. Sci.* 1998, 69, 1727.

Nakamura, K., Aoike, T., Usaka, K., Kanamoto, T. *Macromolecules* 1999, 32, 4975.

Nanasawa, A., Takayama, S., Takeda, K. *J. Appl. Polym. Sci.* 1997, 66, 19–28.

Natta, G., Pino, P., Corradini, P., Danusso, F., Mantica, E., Mazzanti, G., et al. *J. Am. Chem. Soc.* 1955, 77, 1708.

Neag, C.M., Ibar, J.P., Saffell, J.R., Denning, P. *J. Coatings Technol.* 1993, 65, 37.

Neideck, K., Franzel, W., Grau, P. *J. Macromolecular Sci. B* 1999, B38, 669.

Neogi, C., Basu, S.P., Bhowmick, A.K. *J. Elast. Plast.* 1992, 24, 96.

Niemiec, S. *Polym. Test* 1, 1980, 3, 201.

Niziolek, A., Nelsen, J.G., Jones, R. *Kaut. u. Gum. Kunst.* 2000, 53, 358.

Noordermeer, J.W.M., America, R.J.H., Visser, G.W. Standards and quality in the rubber industry, London, Proceedings, 1992, p. 67.

Nordel Hydrocarbon Rubber, DuPont de Nemours and Co., Wilmington, DE, Code No: MLD A-55375, 1967.

Nordsiek, K.H. *Kaut. u. Gum. Kunst.* 1985, 38, 178.

Norem, S.D., O'Neill, M.J., Gray, A.P. *Thermochimica Acta* 1970, 1, 29.

Norman, R.H. *Plast. Rubber Int. 5*, 1980, 6, 243.

Norman, R.H. *Polym. Test.* 1980, 1, 4.

Oberster, E., Bouton, T.E., Valaites, J.K. *Die Angewandte Mackromolekulare Chemie* 1973, 29/30, 291.

Oertan-Lamontagne, M.C., Parthum, K.A., Seitz, W.R., Tomellini, S.A. *Appl. Spectroscopy* 1994, 48, 1539.

Ohtani, H., Ueda, S., Tsukahara, Y., Watanabe, C., Tsuge, S. *J. Anal. Appl. Pyrol.* 1993, 25, 1.

Oregui, X., Santamaría, A., Hiriart, J.M., Etcheveste, J. *J. Vinyl Technol.* 1985, 7, 98.

Ou, Y.C., Yu, Z.Z., Vidal, A., Donnet, J.B. *Rubber Chem. Technol.* 1994, 67, 834.

Ozawa, T.J. *Thermal Anal.* 1970, 2, 301.

Painter, P.C., Coleman, M.M., Koenig, J.L. *The Theory of Vibrational Spectroscopy and Its Application to Polymeric Materials*, Wiley, New York, 1982.

Palatinol 711P. Technical Data Sheet. BASF 1996.

Parameswaran, V., Shukla, A. *J. Mater. Sci.* 1998, 33, 3303.

Park, S., Choe, S. *Macromol. Res.* 2005, 13, 297.

Parker, J.R., Waddell, W.H., *J. Elastom. Plast.* 1996, 28, 140.

Pavia, D.L., Lampman, G.M., Kriz, G.S. *Introduction to Spectroscopy*, 2nd ed., Harcourt Brace College, Philadelphia, 1996.

Pawlowski, H., Dick, J. *ACS Rubber Div. Meeting*, Pittsburgh, 1994, paper 32.

Pawlowski, H., Dick, J. *Rubber World* 1992, 206, 35.

Payne, A.R. RABRM Res. Memo. R411, 1958.

Payne, A.R. RABRM Res. Report 76, 1955.

Payne, A.R. *Engineer* 1959, 207, 328.

Payne, A.R., Scott, J.R. *Engineering Design with Rubber*, MacLaren and Sons, London, 1960.

Penzel, E., Rieger, J., Schneider, H.A. *Polymer* 1997, 38, 325.

Perera, M.C.S., Ishiaku, U.S., Ishak, Z.A.M. *Eur. Polym. J.* 2001, 37, 167.

Perera, M.C.S., Ishiaku, U.S., Ishak, Z.A.M. *Polym. Degr. Stab.* 2000, 68, 393.

Phair, M., Wampler, T. ACS Rubber Div. 150th Meeting, Washington, DC, American Chemical Society, 1996, Paper No. 69.

Pica, D., Barket, R., Rice, P., Ma, C.C. *Rubber World* 1979, 180, 4.

Pinoit, D., Prud'homme, R.E. *Polymer* 2002, 43, 2321.

Piper, G.H., Scott, J.R. *J. Sci. Instrum.* 1945, 22, 206.

Poirier, N., Derenne, S., Rouzaud, J.N., Largeau, C., Mariotti, A., Balesdent, J., Maquet, J. *Org. Geochem.* 2000, 31, 813.

Prime, R.B. In *Thermal Characterisation of Polymeric Materials,* Turi, E.A., Ed., Academic Press, New York, 1981, Chapter 5.

Pron, A., Nicolau, Y., Genoud, F., Nechtschein, M. *J. Appl. Polym. Sci.* 1997, 63, 971.

Pruneda, F., Sunol, J.J., Andreu-Mateu, F., Colom, X. *J. Therm. Anal. Calorimetry* 2005, 80, 1, 187.

Radhakrishnan, S.R., Saini, D.R. *J. Appl. Poly. Sci.* 1994, 52, 1577.
Raemakers, K.G.H., Bart, J.C.J. *Thermochima Acta*, 1997, 295, 1.
Ramesh, P., De, S.K. *J. Mater. Sci.* 1991, 26, 2846.
Ramey, K.C., Hayes, M.W., Alteneau, A.G. *Macromolecules* 1973, 6, 795.
Ramos-de Valle, L., Gilbert, M. *J. Vinyl Technol.* 1990, 12, 222.
Ramos-de, V.L., Gilbert, M. *Plast. Rubber Composites Process. Appl.* 1991, 15, 207.
Raos, P. *Kaut. u. Gummi Kunst.* 1992, 45, 957.
Reig, F.B., Gimeno Adelantado, J.V., Peris Martinez, V., Moya Moreno, M.C.M., Domenech Carbo, M.T. *J. Mol. Struct.* 1999, 529, 480.
Reiner, E., Moran, T.F. *Adv. Chem. Ser.* 1983, 203, 703.
Reiter, S.M., Buchberger, W., Klampfl, C.W. *Anal. Bioanal. Chem.* 2011, 400, 2317.
Richwine, J.R. *Elastomerics* 1991, 123, 21.
Roberts-Austen, N.C. *Proceedings of the Royal Society of London* 1891, 49, 347.
Rodgers, M.B., Waddell, W.H., Klingensmith, W., *Kirk-Othmer Encyclopedia of Chemical Technology*, 5th ed., Wiley, New York, 2004.
Rogalewicz, R., Batko, K., Voelkel A. *J. Environ. Monit.* 2006, 8, 750.
Rogalewicz, R., Voelkel, A., Kownacki, I. *J. Environ. Monit.* 2006, 8, 377.
Roland, C.M., Santangelo, P.G., Baram, Z., Runt, J. *Macromolecules* 1994, 27, 5382.
Roland, C.M. *Macromolecules* 1987, 20, 2557.
Roovers, J., Toporowski, P.M. *Macromolecules* 1992, 25, 1096.
Roy Chowdhury, N., Bhowmick, A.K. *J. Mater. Sci.* 1988, 23, 2187.
Roy, A., Bhowmick, A.K., De, S., K. *J. Appl. Polym. Sci.* 1993, 39, 263.
Ruiz, C.S.B., Machado, L.D.B., Vanin, J.A., Volponi, J.E. *J. Therm. Anal. Cal.* 2002, 67, 335.
Saad, A.L.G., Hussien, L.I., Ahmed, M.G.M., Hassan, A.M. *J. Appl. Polym. Sci.* 1998, 69, 685.
Sadeghi, G.M.M., Morshedian, J., Barikani, M. *Polym. Int.* 2003, 7, 1083.
Sadeghi, G.M.M., Morshedian, J., Barikani, M. *Polym. Test.* 2003, 22, 168.
Sadhu, S., Dey, R.S., Bhowmick, A.K. *Polym. Polym. Composites* 2008, 16, 283.
SAE Standard 52236.
Sahoo, S., Bhowmick, A.K. *J. Appl. Polym. Sci.* 2007, 106, 3077.
Sahoo, S., Bhowmick, A.K. *Rubber Chem. Technol.* 2007, 80, 826.
Sahoo, S., Maiti, M., Ganguly, A., George, J.J., Bhowmick, A.K. *J. Appl. Polym. Sci.* 2007, 105, 2407.
Santee, E.R., Chang, R., Morton, M. *J. Polym. Sci., Polym. Ed.* 1973, 449.
Sarkar, A., Bhowmick, A.K., Chakravarty, S.N. *Polym. Test.* 1988, 8, 415.
Sarkar, M.D., De, P.P., Bhowmick, A.K. *Rubber Chem. Technol.* 1997, 70, 855.
Schnabel, W. *Polymer Degradation Principles and Applications*, Hanser, New York, 1981.
Schneider, N.S., Sprouse, J.F., Hagmauer, G.L., Gillham, J.K. *Polym. Eng. Sci.* 1979, 19, 304.
Schneider, W.A., Huybrechts, F., Nordesik, K.H. *Kaut. u. Gum. Kunst.* 1989, 44, 528.
Mark, J.E., Erman, B., Eirich, F.R., Eds., *Science and Technology of Rubber*, 2nd ed., Academic Press, San Diego, CA, 1994.
Scott, J.R. *RABRM Lab. Circular* 1940, 172.
Scott, J.R. *J. Rubber Res.* 1944, 4, 13.
Scott, J.R. *J. Rubber Res.* 1945, 18.
Scott, J.R. *RABRM Lab. Circular* 1935, 134.
Scott, J.R. *Trans. IRI* 1931, 7, 169.
Scott, J.R. *Trans. IRI* 1935, 10, 481.
Sen, A., Bhattacharya, A.S., De, P.P., Bhowmick, A.K. *J. Thermal Anal.* 1993, 39, 887.
Sen, A., Mukherjee, B., Bhattachayya, A.S., De, P.P., Bhowmick, A.K. *J. Appl. Polym. Sci.* 1991, 43, 1673.
Sen, A., Mukherjee, B., Bhattachayya, A.S., De, P.P., Bhowmick, A.K. *Die Angewandte Chemie* 1991, 191, 15.
Sena, J.A. ACS Rubber Div. Meeting, Indianapolis, 1984, Paper No. 22.
Sengupta, R., Sabharwal, S., Tikku, V.K., Somani, A.K., Chaki, T.K., Bhowmick, A.K. *J. Appl. Polym. Sci.* 2006, 99, 1633.

Seung-Hwan, L., Shiraishi, N. *J. Appl. Polym. Sci.* 2001, 81, 243.
Sezna, J.A. ACS Rubber Div. Meeting, Louisville, KY, 1996, Paper No. 3.
Shield, S.R., Ghebremeskel, G.N. *Rubber Chem. Technol.* 2001, 74, 803.
Shield, S.R., Ghebremeskel, G.N. *Rubber World* 2000, 24, 223.
Shield, S.R., Ghebremeskel, G.N. *J. Appl. Polym. Sci.* 2003, 88, 1653.
Shiga, S., Oka, N. *Int. Polym. Sci. Technol.* 1995, 22, T43.
Shin, S.M., Shin, D.K., Lee, D.C. *Polym. Bull.* 1998, 40, 599.
Shivakumar, E., Das, C.K., Pandey, K.N., Alam, S., Mathur, G.N. *Macromol. Res.* 2005, 1, 81.
Shmakova, N.A., Slovokhotova, N.A., Shukhov, F.F. *Intl. Polym. Sci. Technol.* 1995, 22, 50.
Shoemaker, K. *Rubber Technol. Int.* 1996, p. 150.
Šics, I., Ezquerra, T.A., Baltá Calleja, F.J., Tupureina, V., Kalninš, M. *J. Macromol. Sci., Part B: Physics*, 2000, B39, 5–6, 761.
Simha, R., Boyer, R.F. *J. Chem. Phys.* 1962, 37, 1003.
Simpson, B.D. *The Vanderbilt Rubber Handbook*, 12th ed., Babbit, R., Ed., R.T. Vanderbilt, Norwalk, CT, 1978.
Sircar, A.K., Lamond, T.G. *Rubber Chem. Technol.* 1975, 48, 301.
Sircar, A.K., Galaska, M.L., Rodrigues, S., Chartoff, R.P. *Rubber Chem. Technol.* 1999, 72, 513.
Sircar, A.K., Lamond, T.G. *Rubber Chem. Technol.* 1972, 45, 329.
Sisson, J. *Rubber Plast. News* 1997, 26, 6.
Smith, D.E. *Thermochimica Acta*, 1976, 14, 370.
Smith, R.W., Bryg, V. *Rubber Chem. Technol.* 2006, 79, 520.
Smothers, W.J., Chiang, Y. *Handbook of Differential Thermal Analysis*, Revised edition, Chemical Publishing, New York, 1966.
Sobeih, K.L., Baron, M., Gonzales-Rodrigues, J. *J. Chromatogr.* 2008, 1186, 1–2, 51.
Spathis, G., Maggana, C. *Polymer* 1997, 38, 2371.
Spetz, G. *Polym. Test.* 1995, 14, 13.
Spinka, O. Polymer Testing Conference, Rapra, Shawbury, 1996, Paper No. 9.
Stacer, R.G., Yanyo, L.C., Kelley, F.N. ACS Rubber Div. Meeting, Denver, CO, 1984, Paper No. 27.
SAE International, Standard method for determining continuous upper temperature resistance of elastomers, 1992.
Stern, H.J. *History in Rubber Technology and Manufacture*, Blow, C.M., Ed., Newnes-Butterworth, London, 1977, p. 7.
Stevenson, A., Malek, K.A. *Rubber Chem. Technol.* 1994, 67, 743.
Stewart, L.N. In *Proceedings of the 3rd Toronto Symposium on Thermal Analysis*, Toronto, Canada, 1969, p. 205.
Subramaniam, A. In *Rubber Technology*, 3rd ed., Morton, M., Ed., Van Nostrand Reinhold, New York, 1987, p. 184.
Swarin, S.J., Wims, A.M. *Rubber Chem. Technol.* 1974, 47, 1193.
Sweet, G.C. Special purpose elastomers, In *Developments in Rubber Technology-1*, Whelan, A., Lee, K.S., Eds., Applied Science, Barking, UK, 1979, p. 86.
Sykes, G.F., Burks, H.D., Nelson, J.B. In *Proceedings of the Natural SAMPE Symposium Exhibition*, 1977, 22, 350.
Tai, H. *Polym. Eng. Sci.* 1999, 39, 1320.
Takayanagi, M. Memoirs of the Faculty of Engineering, Kyushu University, 1963, 23, 41.
Tanaka, Y., Takeuchi, Y. *J. Polym. Sci. Part A-2* 1971, 9, 43.
Tangorra, G. *Rubber Chem. Technol.* 1961, 34, 347.
Tareev, B.M. *Physics of Dielectric Materials*, Mir, Moscow, Russia, 1975.
Taylor, R.H. *ASTM Bull.* 1943.
Taylor, R.H., Fielding, J.H., Mooney, M. *Rubber Age N.Y.* 1947, 61, 567.
Thavamani, P., Bhowmick, A.K. *Rubber Chem. Technol.* 1992, 65, 31.
Thavamani, P., Bhowmick, A.K. *Rubber Comp. Proc. Appl.* 1993, 20, 239.
Thomas, W., Kotz, D.A., Tabb, D.L. *Rubber World* 1995, 211, 34.
Thorburn, B., Hoshi, Y. *Rubber World* 1992, 206, 17.

Thorn, A.D., Robinson, R.A. In *Rubber Products Manufacturing Technology*, Bhowmick, A.K., Hall, M.A., Benarey, H.A., Eds., Marcel Dekker, New York, 1994.
Tienpont, B., David, F., Vanwalleghem, F., Sandra, P. *J. Chromatogr. A* 2001, 911, 2, 235.
Trask, C.A., Roland, C.M. *Macromolecules* 1989, 22, 256.
Treloar, L.R.G. *The Physics of Rubber Elasticity*, Oxford University Press, New York, 1975.
Tsuge, S., Ohtani, H. *Basic and Databook of Pyrolysis Chromatography for Polymer*, Techno System, Tokyo, 1995.
Tsuge, S., Ohtani, H. *Polym. Deg. Stab.* 1997, 58, 109.
Uniplex FRP-45. Flame retardant plasticizer. Unitex Chemical Corporation, Greensboro, NC.
Unseld, K., Albohr, O., Herrmann, V., Fuchs, H.B. *Kaut. und Gummi Kunst.* 2000, 53, 52.
Van Steenwinckel, D., Hendrickx, E., Samyn, C., Engels, C., Persoons, A. *J. Mater. Chem.* 2000, 10, 2692.
Varughese, K.T., De, P.P., Sanyal, S.K., De, S.K. *J. Appl. Poly. Sci.* 1989, 37, 2537.
Varughese, K.T., Nando, G.B., De. P.P., De, S.K. *J. Mater Sci.*, 1988, 23, 3894.
Vitali, M., Pizzoli, M. *Macrom. Chem. Phys.* 1998, 199, 429.
Voet, A., Marawski, J.C. ACS Div. of Rubber Chem. 1974, Paper No. 42, Preprint 12.
Voet, A., Morawski, J.C., Donnet, J.B. *Rubber Chem. Technol.* 1977, 50, 342.
Volponi, J.E., Mei, L.H.I., Rosa, D.S. *Polym. Test.* 2004, 23, 461.
Vroomen, G.L.M., Visser, G.W., Gehring, J. *Rubber World* 1991, 205, 23.
Waddell, W.H., Rodgers, M.B., Tracey, D.S. ACS Rubber Div. Meeting, Grand Rapids, MI, 2004 Paper A.
Waddell, W.H., Bhukuni, R.S., Barbin, W.W., Sandstrom, P.H. In *The Vanderbilt Rubber Handbook*, 13th ed., Ohm, R.F., Ed., R.T. Vanderbilt Company, Norwalk, CT, 1990.
Waddell, W.H., Rodgers, M.B., Tracey, D.S. Rubber Div. Meeting, Grand Rapids, MI, 2004, Paper H.
Wagner, M.P. In *Rubber Technology*, Morton, M., Ed., Van Nostrand Reinhold, New York, 1987, p. 95.
Wagner, M.P. *Rubber Chem. Technol.* 1976, 55, 703.
Wampler, T.P., Ed. *Applied Pyrolysis Handbook*, 2nd ed., CRC Press, Boca Raton, FL, 2007.
Wampler, W.A., Carlson, T.F., Jones, W.R. In *Rubber Compounding, Chemistry and Applications*, Rodgers, B., Ed., Marcel Dekker, New York, 2004.
Wang, M.J., Patterson, W.J. *Proceedings of the International Rubber Conference*, Manchester, UK, 1996, Paper No. 43.
Wang, G., Chen, Y. *Polym. Test.* 1991, 10, 315.
Wang, C.B., Cooper, S.L. *J. Polym. Sci. Polym. Phys.* 1983, 21, 1, 11.
Waters, D.N., Paddy, J.L. *Anal. Chem.* 1988, 60, 53.
Watson, E.S., O'Neill, M.J., Justin, J., Brenner, N. *Anal. Chem.* 1964, 36, 1233.
Wendlandt, W.W., Gallagher, P.K. In *Thermal Characterisation of Polymeric Materials*, Turi, E.A., Ed., Academic Press, New York, 1981, Chapter 1.
Wendlandt, W.W. *Thermal Analysis*, 3rd ed., Wiley, New York, 1986, p. 1.
Wendlandt, W.W. *Handbook of Commercial Scientific Instruments*, Volume 2, Marcel Dekker, New York, 1974, p. 144.
Wendlandt, W.W. *Thermal Methods of Analysis,* Volume 19, 2nd ed., Wiley Interscience, New York, 1974,
Wendlandt, W.W., Collins, L.W. In *Benchmark Papers in Analytical Chemistry*, Volume 2, Dowden, Hutchinson and Ross, Stroudsburg, PA, 1976.
Werner Hoffmann, *Rubber Technology Handbook*, Hanser, Munich, Vienna, New York, 1989, p. 171.
Wharlow, R.W. *RABRM Res. Memo. R 375*, 1951.
White, D.M., Garland, D.S., Beyer, L., Yoshikawa, K. *J. Anal. Appl. Pyrolysis* 2004, 71, 107.
White, J.L. *Rubber Chem. Technol.* 1977, 50, 163.
White, J.L. *Rubber Processing: Technology, Materials and Principles*, Hanser, Munich, Germany, 1995, p. 13.
White, J.L. *Rubber Processing: Technology, Materials and Principles*, Hanser, Munich, Germany, 1995, p. 4.
White, J.L., Soos, I. *Rubber Chem. Technol.* 1993, 66, 3, 435.
Whorlow, R.W. *Rheological Techniques*, Ellis Harwood, Chichester, UK, 1992.

Wickson, E.J. *Handbook of PVC Formulating*, Wiley Interscience, New York, 1993, p. 674.
Wilkes, G.L., Wildnauer, R. *J. App. Phys.* 1975, 46, 4148.
Williams, I. *Ind. Eng. Chem.* 1924, 16, 362.
Williams, M.L., Landel, R.F., Ferry, J.D. *J. Am. Chem. Soc.* 1955, 77, 3701.
Winsor, D.L., Scheinbeim, J.I., Newman, B.A. *J. Polym. Sci., Polym. Phys.* 1996, 34, 2967.
Wolf, C.J., Grayson, M.A., Fanter, D.L. *Anal. Chem.* 1980, 52, 348.
Wolff, S. ACS Rubber Div. Meeting, Las Vegas, NV, 1990.
Wolff, S., Gorl, U. The Influence of Modified Carbon Blacks on Viscoelastic Compound Properties, International Rubber Conference, 1991.
Wolff, S. *Rubber Chem. Technol.* 1983, 55, 967.
Wolff, S., Wang, M.J. *Rubber Chem. Technol.* 1992, 65, 329.
Wolff, S., Wang, M.J., Tan, E.H. *Rubber Chem. Technol.* 1993, 66, 163.
Wood, L.A. *J. Polym. Sci.* 1958, 28, 319.
Worfield, R.W., Cuevas, E., Bamet, F.R. *J. Appl. Polym. Sci.* 1968, 12, 1147.
Worldwide Rubber Statistics, 2003, International Institute of Synthetic Rubber Producers, Houston, TX, 2003.
Wunderlich, B. *Thermal Analysis of Polymeric Materials*, Springer-Verlag, Berlin, Germany, 2005, p. 279.
Best Materials discount warehouse, www.bestmaterials.com.
Currell, G. Statistical Tables and Formula Sheet, May 2008, www.calcscience.uwe.ac.uk/w2/am/Resources/Statistical%20Tables.pdf.
Continental Carbon, www.continentalcarbon.com.
Statistics Formula Sheet, www.personal.maths.surrey.ac.uk/st/K.Young/form_sheet.pdf.
www.reagenssi.kemia.helsinki.fi
www.water800.com
www.wilderness.net/toolboxes/documents/vum/Basic%20statistics%20vocabulary.pdf
Zorge, www.zorge.com.
Xie, H.-Q., Pan, Y.-J., Guo, J.-S. *J. Macromol. Sci., Phys.* 2003, B42, 2, 257.
Xinping, H., Kok Siong Siow, *Polymer* 2000, 41, 8689.
Yamada, K., Funayama, Y. *Rubber Chem. Technol.* 1990, 63, 669.
Yang, X., Cai, J., Kong, X., Dong, W., Li, G., Zhou, E. *Macromol. Chem. Phys.* 2001, 202, 1166.
Yeoh, H. Fundamental Characteristics and Properties of Natural Rubber, in Education Symposium No. 35, Philadelphia, May 1995, ACS Rubber Div., p6.
Yeoh, O.H. *ACS Rubber Div Meeting*, Nashville, TN, 1998, Paper No. 2.
Yeoh, O.H. *J. Rubber Res. Inst. Malaysia*, 1984, 32, 227.
Yeoh, O.H. *Plast. Rubber Process. Appln.* 1984, 4, 141.
Yeoh, O.H. *Rubber Chem. Technol.* 1997, 70, 175.
Yeoh, O.H., Pinter, G.A., Banks, H.T. *ACS Rubber Div. Meeting*, Cincinnati, OH, 2000, Paper No. 84.
Yi, Y.X., Zoller, P. *J. Poly. Sci., Part B: Polym. Phys.* 1993, 31, 779.
Yokouchi, M., Kobayashi, Y. *J. Appl. Polym. Sci.* 1981, 26, 4307.
Yoshida, J. *Chem. Abs.* 1952, 46, 3786.
Yu, X., Nagarajan, M.R., Li, C., Gibson, P.E., Cooper, S.L. *J. Polym. Sci., Part B: Polym. Phys.* 1986, 24, 2681.
Yu, X., Song, C., Li, C., Cooper, S.L. *J. Appl. Poly. Sci.* 1992, 44, 409.
Yuen, H.K., Mappes, G.W., Grote, W.A. *Thermochimica Acta* 1982, 52, 1–3, 143.
Zhang, B., Tan, B. *Eur. Polym. J.* 1998, 34, 571–575.
Zhang, L., Zhang, G., Carlisle, G.O., Crowder, G.A. *J. Mater. Sci., Mater. Electronics* 2000, 11, 229.
Zitomer, F. *Anal. Chem.* 1968, 40, 1091.

Appendix A: Statistical Aspects of Chemical Analysis

A.1 Basic Statistics Vocabulary

Population: The set of all measurements of interest to the sample collector.

Parameter: A characteristic of the population.

Statistic: A characteristic of a sample drawn from a population.

Accuracy: Closeness of a measured or computed value to its true value. How close the sample estimates are to the actual population. This is affected by nonsampling errors (errors from method of measurement, improperly designed or executed sampling plan).

Precision: Closeness of repeated measurements of the same quantity to each other. Consistency of measurements, or reliability. Restricted to errors of sampling, malfunctioning equipment, or observer bias.

- A biased but sensitive scale may give inaccurate but precise weight readings.
- An insensitive scale may give an accurate reading, by chance, but it would be an imprecise reading since a number of readings would likely all yield different weights.
- The bullseye of a dartboard is accurate. However, a person who throws consistently off to the left will produce a group of darts clustered together (precise throws, but inaccurate).
- Unless there is bias in a measuring instrument, precision will lead to accuracy.

Measures of central tendency: The three commonly used measures are: mean, median, and mode. The *mean* is the average value of a set of measurements (the sum of all the measurements divided by the total number of measurements). The *median* of a set of measurements is the middle value when the measurements are arranged from lowest to highest. The *mode* is the measurement that occurs most often.

Measures of spread (measures of variability)

Range: The difference between the largest and smallest values. A large range indicates a large amount of variability.

The most common measure is the *sample standard deviation* (s). This is a measure of how far a typical observation deviates from the sample average. Since this is a measure of spread about the *mean,* it should be used only when the mean is used as the measure of center. The standard deviation (s) is the square root of the *variance* S^2. The variance will be large if the observations are widely spread about their mean, and small if the observations are all close to the mean. Because the variance involves squaring the deviations, it doesn't have the same unit of measurement as the original observations. Lengths measured in centimeters have a variance measured in squared centimeters. Taking the square root remedies this, so the *s* measures dispersion about the mean in the original scale. The standard deviation equals 0 only when there is no spread (all observations have the same value). Otherwise, *s* is greater than 0.

Sampling error: A standard deviation among estimates rather than among individual observations. Regardless of the sample size (generally, the larger the sample the more likely it is that the measured use will represent the whole population), there is always some difference between the sample statistics and the population parameters being estimated. The most appropriate sample size depends on how much sampling error you accept.

Standard error: A mathematical way of expressing sampling error. If it is small, the measure of reliability is good. Standard error is most commonly computed for the mean. It indicates the amount of variation among means taken from many samples. How far do they deviate from the population mean? The greater the variation in values from sample to sample, the larger the standard error and the less reliable the statistic for estimating parameters. Standard error is affected by sample standard deviation (*s*) and by sample size. If *s* is constant but the sample size increases, the standard error decreases.

Confidence interval: The range + or – from the sample mean.

Confidence level: Usually 95%. The probability that the interval contains the true mean is 95%. A confidence level of 67% may be sufficient for broad policy decisions, while site-specific or allocation decisions may require a 95% confidence level.

- To increase the confidence limit (99%), the confidence interval must become wider. However, as confidence increases, the statement becomes more vague.
- Use is estimated to be 123 + or –10% at the 67% confidence level. Use is between 111 and 13567% of the time.

Correlation: The manner in which two or more variables fluctuate with reference to each other. Do they vary together? Also tells us the strength of the association (measures intensity of association observed between two variables, and whether the correlation could be expected to chance alone). One variable may be the cause of the other, but we neither know nor assume this. Both variables must be random.

- Positive correlation—as education increases, income increases
- Negative correlation—as education increases, number of children decreases
- Correlation coefficient—+1 (perfect positive association) to –1 (perfect negative correlation). There is no dispersion about the line of relationship.

A.2 Measures of Central Tendency

A.2.1 Mean, Median, and Mode

One of the most basic purposes of statistics is simply to enable us to make sense of large numbers.

A.2.1.1 Mean

The *mean* is one of the most useful and widely used techniques for measuring the central tendency, which is also widely known as the *average*. Calculation of mean is done by adding a set of data and dividing by the number of data. The equation is represented as:

$$\overline{X} = \frac{\sum X}{N}$$

where:

$\overline{X}$ (called the X-bar) is the symbol for the mean

Σ (the Greek letter *sigma*) is the symbol for summation

X is the symbol for the data

N is the symbol for the number of data

So if we know or suspect that our data may have some extreme scores that would distort the mean, what measure can we use to give us a better measure of central tendency? One such measure is the median.

A.2.1.2 Median

If the analysis data are normally distributed, the preferred measure of central tendency is the mean. However, if data are not normally distributed, and some extreme data are available in the dataset, the mean of the data will get distorted. In such a case the median is a better measure of central tendency.

The median is the point in the distribution above which and below which 50% of the data lie. In other words, arrange the data in order from highest to lowest (or lowest to highest) and find the middle-most score—that's the median.

Calculations for the median are simple, arranging the data from high to low, counting up to the mid-category, and then dividing it as necessary. The statistical formula for calculating the median is as follows:

$$Mdn = 1 + \left\{ \frac{\frac{N}{2} - \Sigma f_0}{f_w} \right\} i$$

where:

Mdn is the median

L is the lower limit of the interval containing the median

N is the total number of scores

Σf_0 is the sum of the frequencies or number of scores up to the interval containing the median

f_w is the frequency or number of scores within the interval containing the median

i is the size or range of the interval

A.2.1.3 Mode

The third and last of the measures of central tendency is the mode. The mode is the most frequently occurring data or value in the dataset.

The other part of the central tendency is dispersion.

A.2.2 Standard Deviation and the Normal Curve

A.2.2.1 A Measure of Dispersion: Standard Deviation

It is always preferable to know not only what the central tendency is in a set of data or values (i.e., the mean, the median, or the mode), but also how bunched up or spread out the data or values are. The most widely used indicator of dispersion is the standard deviation, which is based on the deviation of each score from the mean.

The standard deviation provides a measure of just how spread out the data or values are: a high standard deviation means the data are widely spread; a low standard deviation means they're bunched up closely on either side of the mean.

The calculation formula for the standard deviation is:

$$\sigma = \sqrt{\frac{\sum d^2}{N}}$$

where:

σ (little sigma) is the standard deviation
d^2 is a datum's deviation from the mean squared
N is the number of cases

The standard deviation is important, because regardless of the mean, it makes a great deal of difference whether the distribution is spread out over a broad range or bunched up closely around the mean.

A.2.2.2 The Normal Curve

Along with the standard deviation, one associated statistical terminology is the normal curve. Why is it called the normal curve? The reason is that many measurable things are distributed in the shape of this curve. In Figure A.1, a set of data is shown, which are normally distributed. The range is from 0 to 200, the mean and median are 100, and the standard deviation is 20. In a normal curve, the standard deviation indicates precisely how the data are distributed. In the normal curve, the percentage of data is marked off by standard deviations on either side of the mean. In the range between 80 and 20 (that's one standard deviation on either side of the mean), there are 68.26% of cases. In other words, in a normal distribution, roughly two-thirds of the data lie between one standard deviation on either side of the mean. Two standard deviations on either side of the mean will include 95.44% of the data; three standard deviations will encompass 98.74% of the data; and so on.

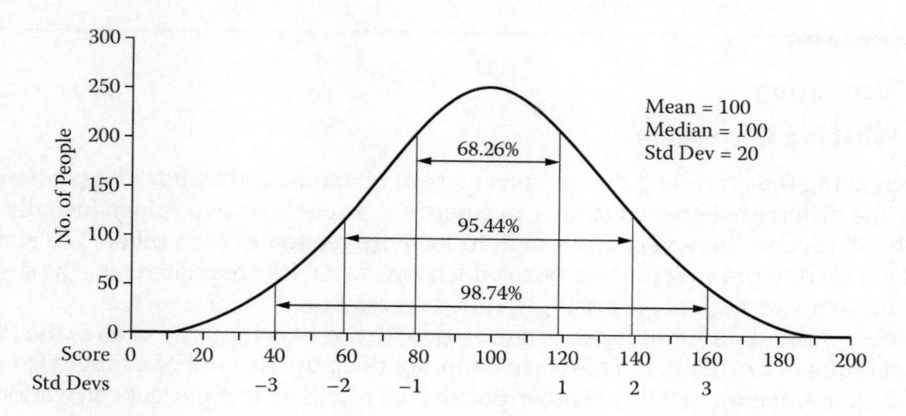

FIGURE A.1
Normal curve showing the percent of cases lying within 1, 2, and 3 standard deviations from the mean.

This is valuable to understand in its own right and will become useful when determining estimates of the significance of difference between means.

A.2.3 Testing the Difference between Means: The *t*-Test

This is one of the most important parts of basic statistics. One of the most widely used statistical methods for testing the difference between means is called the *t*-test.

The formula for the *t*-test is:

$$t = \frac{X_1 - X_2}{\sqrt{\frac{(n_1 - 1)s_1^2 + (n_2 - 1)s_2^2}{n_1 + n_2 - 2}\left[\frac{n_1 + n_2}{n_1 n_2}\right]}}$$

where:

$\overline{X}_1$ is the mean for Group 1
$\overline{X}_2$ is the mean for Group 2
n_1 is the number of people in Group 1
n_2 is the number of people in Group 2
s_1^2 is the variance for Group 1
s_2^2 is the variance for Group 2

The only thing in this formula is the symbol s^2, which stands for the variance. The variance is the same as the standard deviation without the square root (i.e., it's nothing more than the sum of the deviations of all the data from the mean divided by $n - 1$).

The formula above is for testing the significance of difference between two independent samples

A.3 Correlation

A.3.1 What is a Correlation?

After knowing the central tendency—mean, median, mode, and standard deviation—and testing the difference between means, one needs to know how two things (usually called "variables" because they vary from high to low) are related to each other. The statistical technique for determining the degree to which two variables are related (i.e., the degree to which they co-vary) is, not surprisingly, called correlation.

There are several different types of correlation; the most commonly used is the Pearson Product Moment Correlation. This correlation, signified by the symbol r, ranges from –1.00 to +1.00. A correlation of 1.00, whether positive or negative, is a perfect correlation. This means that as data on one of the two variables increase or decrease, the data on the other variable increase or decrease by the same magnitude. A correlation of 0 means there is no relationship between the two variables (i.e., when scores on one of the variables go up, scores on the other variable may go up, down, or whatever).

Thus, a correlation of 0.8 or 0.9 is regarded as a high correlation (i.e., there is a very close relationship between scores on one of the variables with the scores on the other). Correlations of 0.2 or 0.3 are regarded as low correlations (i.e., there is some relationship between the two variables, but it's a weak one).

A.3.2 Computing the Pearson Product Moment Correlation

The formula for the Pearson product moment correlation is:

$$r_{xy} = \frac{n\Sigma XY - \Sigma X \Sigma Y}{\sqrt{[n\Sigma XY^2 - (\Sigma X)^2][n\Sigma XY^2 - (\Sigma Y)^2]}}$$

where:

r_{xy} is the correlation coefficient between X and Y

n is the size of the sample

X is the individual's score on the X variable

Y is the individual's score on the Y variable

XY is the product of each X score times its corresponding Y score

X^2 is the individual X score squared

Y^2 is the individual Y score squared

Appendix B: Material Properties

B.1 Comparison of Elastomer Properties

Rubber	Elastomer Grade	Glass Transition T_g °C	Ozone Resistance	Resistance to Tear Propagation[3]	Tensile Strength[3]	Resistance to Abrasion[3]	Approximaate Figures for Low Temperature TR
NR (IR)	Natural rubber (synthetic polyisoprene)	–72	L	VH	VH	M/H	–45
SBR	Styrene-butadiene rubber	–50	L	H	H	H	–28
BR	Cis-1,4 polybutadiene rubber	–112	L	L/M	M	VH	–72
IIR	Butyl rubber	–66	M	M/H	M	M	–38
BIIR	Brominated butyl rubber	–66	M	M/H	M	M	–38
CIIR	Chlorinated butyl rubber	–66	M	M/H	M	M	–38
EPDM,S	Ethylene-propylene-terpolymer, sulfur crosslinked	–55	H	M	M	M	–35
EPDM,P	Ethylene-propylene-co-polymer, peroxide crosslinked	–55	H	L	L/M	L	–35
CR	Polychloroprene	–45	M	H	H	M/H	–25
NBR							
Nitrile Rubber							
• Low ACN content		–45	L	M	M/H	H	–28
• Medium ACN content		–34	L	M	M/H	H	–20
• High ACN content		–20	L	M	M/H	H	–10
H-NBR	Hydrogenated nitrile rubber	–30	H	M	M/H	H	–18
X-NBR	Carboxyl groups containing NBR	–30	L	M	H	VH	–18
EVM	Ethylene-vinylacetate copolymer	–30	H	M	M/H	M	–18
ACM	Polyacrylate rubber	–22 to –40	H	M	M	M	–10 to –20

Rubber	Elastomer Grade	Glass Transition T_g °C	Ozone Resistance	Resistance to Tear Propagation[3]	Tensile Strength[3]	Resistance to Abrasion[3]	Approximaate Figures for Low Temperature TR
EAM	Ethylene-acrylate rubber	–40	H	M	M	M	–20
CM	Chlorinated polyethylene	–25	H	M	M/H	M	–12
CSM	Sulfonated polyethylene	–25	H	M/H	M/H	M	–10
AU	Polyester urethane rubber	–35	H	H	H	H	–22
EU	Polyether-urethane rubber	–55	H	H	H	H	–35
MVQ	Dimethyl polysiloxane containing vinyl	–120	VH	L	L	L	–85
FVMQ	Fluorosilicone rubber	–70	VH	L	L	L	–45
FKM	Fluororubber	–18 to –50	VH	M	M/H	M	–10 to –35
CO	Polyepichloro-hydrin	–26	H	M	M	M	–10
ECO	Epichloro-hydrin ethylenoxide copolymer	–45	H	M	M	M	–25
OT	Polyethylene glycol	–50	H	L	L	L	–30

Elastomer	Compression Set at –20°C	Compression Set at +120°C	Heat Resistance after 5 h	Heat Resistance after 70 h	Working Temperature	Compression Set at Room Temperature
NR (IR)	15	70	150	120	100	8
SBR			195	130	110	
BR			170	100	90	
IIR	12	60	200	160	150	10
BIIR	12	60	200	160	150	10
CIIR	12	60	200	160	150	10
EPDM-S	20	50	200	170	140	8
EP(D)M,P	20	10	220	180	150	4
CR	50	30	180	130	125	10
NBR						
• Low ACN content	40	45	170	140	125	8
• Medium ACN content	45	50	180	145	125	8
• High ACN content	45	55	190	150	125	8
H-NBR		30	230	180	160	
X-NBR		60	170	140	120	
EVM	95	4	200	160	160	40
ACM	25	10	240	180	170	5
EAM			240		175	
CM			180	160	150	
CSM			200	140	150	
AU	25	70	170	100	75	7
EU	25	70	170	100	75	7
MVQ	10	3	>300	275	225	2
FVMQ		30	>300	220	215	
FKM	50	20	>300	280	250	18
CO		20	240	170	150	
ECO		20	220	150	135	
OT			170	120	100	

Note: L, low; M, medium; H, high; VH, very high.

B.2 Selection of Rubber (w.r.t. Oil and Temperature Resistance)

See Figure B.1.

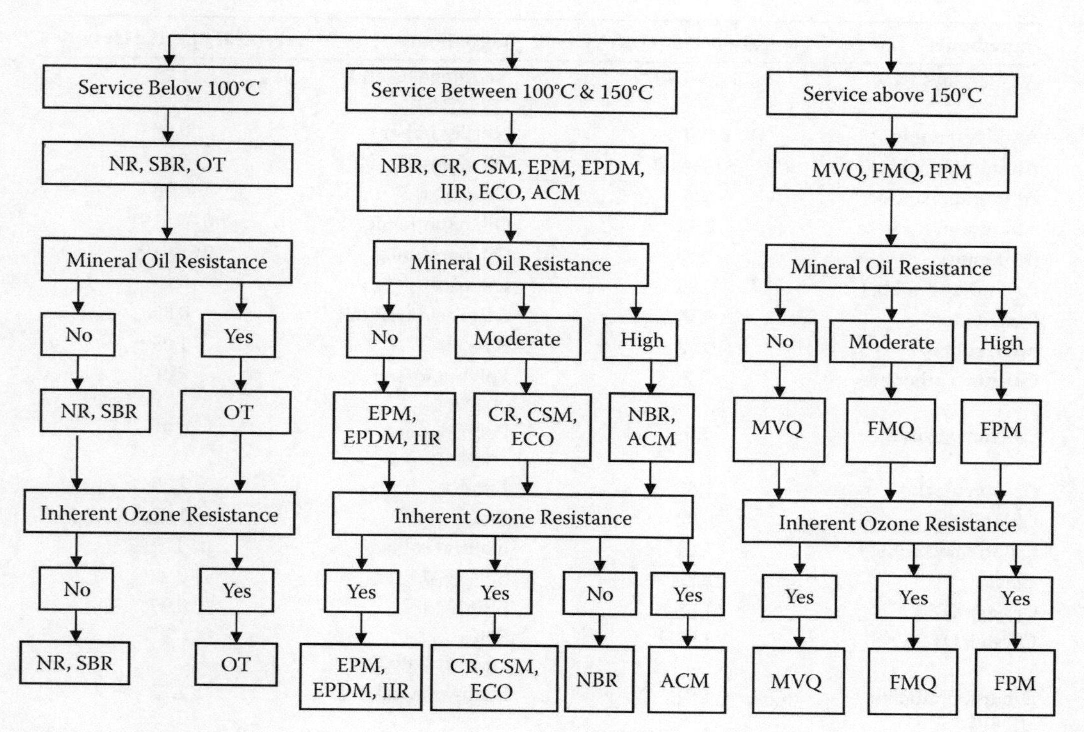

FIGURE B.1
Selection of rubber (w.r.t. oil and temperature resistance).

B.3 Specific Gravity of Compounding Materials

Ingredients	Typical Specific Gravity	Ingredients	Typical Specific Gravity
Accelerators	1.00–1.40	Neoprene (G and W types)	1.23
Antidegradants	0.9–1.1	Nitrile rubber	0.98
Aluminum oxide	3.14–4.00	Oil, aromatic	0.96
Aluminum silicate	2.6	Oil, castor	0.96
Aluminum hydrate	2.42	Oil, naphthenic	0.89–0.91
Bentonite	2.5	Oil, paraffinic	0.86–0.90
Bromobutyl rubber	0.93	Oil, pine	0.93
Iron oxide	5.15	Oleic acid (red oil)	0.89
Butyl rubber	0.92	Pine tar	1.08
Calcium carbonate	2.7	Polybutadiene rubber	0.9
Calcium stearate	2.1	Polyisoprene rubber	0.91
Carbon black	1.85	Pumice	2.35
Chlorobutyl	0.92	Rosin	1.08
Chlorinated rubber	1.64	Natural rubber	0.91–0.93
Clays	2.6	SBR 1500	0.92
Crystex OT 20	1.61	SBR 1712	0.95
Crystex OT 33 AS	1.53	Silica (precipitated)	2
Dibutyl phthalate (DBP)	1.04	Soapstone (talc)	2.72
Diethylene glycol	1.12	Sodium bicarbonate	2.2
EPDM	0.86	Stearic acid	0.85
Factice	1.04–1.08	Soluble sulfur	2
TMQ	1.1	Insoluble sulfur OT20	1.61
Hexamethylene tetramine	1.02	Insoluble sulfur OT33	1.43
Hypalon 20	1.1	Insoluble sulfur OT34AS	1.53
Iron oxide	5.14	Titanium dioxide, pure	3.88
Kaolin	2.6	Ultramarine blue	2.35
Litharge	9.35	Whiting, precipitated	2.62
Magnesium carbonate	2.2	Zinc oxide	5.57
Mica (alumina potassium silicate)	2.82	Zinc stearate	1.05
Mineral rubber	1.04	Zinc yellow	3.3

B.4 Solubility Parameter of Various Polymers, Solvents, and Oils

Material	Solubility Parameter $(cal/cm^3)^{1/2}$	Material	Solubility Parameter $(cal/cm^3)^{1/2}$
Elastomers		*Solvents*	
1,2 BR	7.91	Acetone	9.75
1,4 BR	8.15	Acetonitrile	11.95
CR	8.85	Benzene	9.05
FPM	8.15	*n*-Butane	6.7
FVMQ	8.8	Chloroform	9.33
IIR	7.84	Cyclohexane	8.2
NBR (18% ACN)	8.7	Cyclohexanol	10.69
NBR (29% ACN)	9.38	Cyclohexanone	10
NBR (34% ACN)	9.92	*n*-Decane	7.73
NBR (39% ACN)	10.3	Dichloroethane	9.42
NR	8.15	Dichloromethane	9.73
PBR	8.2	Diethyl ether	7.61
PE	7.9	Dimethyl ether	8.8
EPDM	8	Ethanol	12.9
SBR (12.5% Sty)	8.22	Ethyl acetate	9.08
SBR (15% Sty)	8.35	Ethyl benzene	8.08
SBR (23% Sty)	8.43	Ethyl chloride	8.5
SBR (40% Sty)	8.58	MEK	9.3
TM	9	*n*-Heptane	7.45
Q	7.3	*n*-Hexadecane	8
PVC	9.57	Isopentane	6.75
IIR	7.6	Methanol	14.48
CSPE	9.1	Methyl acetate	9.63
PS	8.83	*n*-Octane	7.55
Solvents		*n*-Propanol	11.92
n-pentane	7.05	Carbon tetra chloride	8.62
Carbon disulfide	10	*O*-Xylene	9
Toluene	8.9		
P-Xylene	8.8		
Oils and Plasticizers			
Paraffinic oil	7.24–7.5	Napthenic oil	7.72–8.2
Aromatic oil	8.00–9.2	DOP	8.9
DBP	9.4	PetJelly	7.5

B.5 Relative Properties of Rubber

	NR/IR	SBR	BR	EPDM	IIR	CR	NBR	T	Q	ECO	CSM	FKM	ACM	CM	U
Tensile strength (gum)	E	P	P	P	F	E	P	P	P	F	G	G-E	P	F	—
Tensile strength (reinforced)	E	G-E	G	G-E	G	G-E	E	F	F	G	G-E	G-E	F-G	G	G-E
Tear resistance (cold)	G	F-G	G	G	G	G	F	G	F	F	F	F	P	G	G
Tear resistance (hot)	G	F	F	E	F	G	P	P	P	—	—	—	—	—	G
Abrasion resistance	E	G-E	E	E	F-G	E	E	F	F	G	E	G	F	E	E
Atmospheric aging	P	P	P-F	E	G-E	E	F	E	E	E	E	E	E	E	E
Oxidation resistance	G	G	G	E	G-E	G	F	E	E	VG	E	E	E	E	E
Heat resistance	G	G	G	G	G-E	G	G	F	E	VG	E	E	G	E	F-G
Low temperature flexibility	E	G	E	G	F	F	F	G	E	F-G	F	G	P	G	E
Compression set	F-G	P-F	G	G	F	P-F	G	P-F	E	G	F	VG	G	VG	G
Impermeability	F	F	P-F	P-F	E	G	E	E	F	E	P	P	P	P	G
Flame resistance	P	P	P	P	P	E	P	P	F	F-G	G	G	P	G	P
Alkali resistance	G	G	G	G	V	G	F	G	F	VG	VG	F	P-F	VG	G
Acid resistance (dilute)	G	F-G	G	G	E	E	G	G	G	G	E	E	F	VG	G
Acid resistance (conc.)	F-G	F-G	G	G	E	G	F	P	F	F	E	E	F	VG	F
Dielectric properties	G-E	G	G	E	G-E	P-F	P	G	E	P-G	G	E	F	G-E	G-E
Animal and vegetable oils	P-G	P-G	P-G	P	E	G	E	E	G	E	E	E	VG		G-E
Oil and gasoline	P	P	P	P	P	G	E	E	F	E	G	E	G	G	E
Solvent resistance: Aliphatic hydrocarbons	P	P	P	P	P	G	E	E	G	G	G	E	E	G	G
Aromatic hydrocarbons	P	P	P	P	P	F	F	E	G	G	F-G	E	F	F	P
Chlorinated solvents	P	P	P	P	P	P	P	E	F	—	—	—	—	—	F
Oxygenated solvents	G	G	G	G	G	P	P	G	G	P	P	P	P	P	F

P, poor; F, fair; G, good; VG, very good; E, excellent.

IR, polyisoprene; CR, neoprene; T, polysulfide; Q, silicone; ECO, epichlorohydrin; CSM, sulfonated polyethylene; FKM, flurorubber; ACM, polyacrylate; CM, chlorinated polyethylene; U, polyurethane.

Note: Silicone rubbers are affected by aromatic hydrocarbons; good resistance can be obtained by using flurosilicone rubbers.

B.6 Calculation of Specific Gravity

To determine the specific gravity of a compound:

1. Express the weight of each compounding ingredient in kilograms.
2. Divide the weight of each ingredient by its specific gravity, add the results, and divide the sum into the weight of the compound.

For example:

Ingredient	Weight (kg)	Specific Gravity	Volume (L)
Rubber	100	0.92	108.696
ZnO	5	5.57	0.898
St. acid	2	0.85	2.253
Carbon black	45	1.85	24.324
TMQ	2	1.1	1.818
TBBS	0.6	1.28	0.469
Sulfur	2.5	2.0	1.250
Total	157.1		139.808

Specific gravity of compound = 157.1/139.808 = 1.124.

B.7 Infrared (IR) Absorptions for Representative Functional Groups

Functional Group	Molecular Motion	Wavenumber (cm^{-1})
Alkanes	C-H stretch	2950–2800
	CH_2 bend	~1465
	CH_3 bend	~1375
	CH_2 bend (4 or more)	~720
Alkenes	= CH stretch	3100–3010
	C = C stretch (isolated)	1690–1630
	C = C stretch (conjugated)	1640–1610
	C-H in-plane bend	1430–1290
	C-H bend (monosubstituted)	~990 and ~910
	C-H bend (disubstituted-E)	~970
	C-H bend (disubstituted-1,1)	~890
	C-H bend (disubstituted-Z)	~700
	C-H bend (trisubstituted)	~815
Alkynes	Acetylenic C-H stretch	~3300
	C,C triple bond stretch	~2150
	Acetylenic C-H bend	650–600

Functional Group	Molecular Motion	Wavenumber (cm^{-1})
Aromatics	C-H stretch	3020–3000
	C=C stretch	~1600 and ~1475
	C-H bend (mono)	770–730 and 715–685
	C-H bend (ortho)	770–735
	C-H bend (meta)	~880 and ~780 and ~690
	C-H bend (para)	850–800
Alcohols	O-H stretch	~3650 or 3400–3300
	C-O stretch	1260–1000
Ethers	C-O-C stretch (dialkyl)	1300–1000
	C-O-C stretch (diaryl)	~1250 and ~1120
Aldehydes	C-H aldehyde stretch	~2850 and ~2750
	C=O stretch	~1725
Ketones	C=O stretch	~1715
	C-C stretch	1300–1100
Carboxylic acids	O-H stretch	3400–2400
	C=O stretch	1730–1700
	C-O stretch	1320–1210
	O-H bend	1440–1400
Esters	C=O stretch	1750–1735
	C-C(O)-C stretch (acetates)	1260–1230
	C-C(O)-C stretch (all others)	1210–1160
Acid chlorides	C=O stretch	1810–1775
	C-Cl stretch	730–550
Anhydrides	C=O stretch	1830–1800 and 1775–1740
	C-O stretch	1300–900
Amines	N-H stretch (1 per N-H bond)	3500–3300
	N-H bend	1640–1500
	C-N stretch (alkyl)	1200–1025
	C-N stretch (aryl)	1360–1250
	N-H bend (oop)	~800
Amides	N-H stretch	3500–3180
	C=O stretch	1680–1630
	N-H bend	1640–1550
	N-H bend (1°)	1570–1515
Alkyl halides	C-F stretch	1400–1000
	C-Cl stretch	785–540
	C-Br stretch	650–510
	C-I stretch	600–485
Nitriles	C,N triple bond stretch	~2250
Isocyanates	-N=C=O stretch	~2270
Isothiocyanates	-N=C=S stretch	~2125
Imines	R_2C = N-R stretch	1690–1640

Functional Group	Molecular Motion	Wavenumber (cm^{-1})
Nitro groups	$-NO_2$ (aliphatic)	1600–1530 and 1390–1300
	$-NO_2$ (aromatic)	1550–1490 and 1355–1315
Mercaptans	S-H stretch	~2550
Sulfoxides	S=O stretch	~1050
Sulfones	S=O stretch	~1300 and ~1150
Sulfonates	S=O stretch	~1350 and ~11750
	S-O stretch	1000–750
Phosphines	P-H stretch	2320–2270
	PH bend	1090–810
Phosphine oxides	P=O	1210–1140

Source: Infrared Spectroscopy: IR Absorptions for Representative Functional Groups, http://www.chemistry.ccsu.edu/glagovich/teaching/316/ir/table.html.

B.8 Dissociation Temperature (T_{diss}) and Activation Energy (E_{diss}) of Some Typical Chemical Bonds

	Chemical Bond	T_{diss} in °C	E_{diss} in kJ/mol
1	$--CF_2--CF_2--$	500	400
2	$--Si--O--Si--$	500	400
3	$--CH_2--Ph$	420	380
4	$--CH_2--CH_2--$	400	320
5	$--CH_2--CH_2--C=C--$	390	300
6	$--C--S--S--C--$	380	300
7	CH_2--O--	345	330
8	$CH_2--CH--CO$	330	280
9	$--C--S_x--C--$	160	120

B.9 Heat Resistance of Some Engineering Elastomers

Heat Resistant up to	Elastomer
100°C	AU/EU, NR(IR), OT, SBR, PNR
125°C	CR, NBR, X-NBR
150°C	CO, ECO, EP(D)M, EVM, CM, CSM, (X)-IIR, H-NBR
175°C	ACM, EAM, PNF
200°C	FVMQ
225°C	MVQ
250°C	FKM

B.10 Brittleness Temperature of Some Engineering Elastomers

Low Temperature Flexibility (°C)	Rubber
–75	Q
–55	NR, IR, BR, CR, SBR, (X)IIR, EP(D)M, CM, CSM, FVMQ, PNF
–40	ECO, NBR, EP(D)M, CSM, FKM, AU, EU
–25	ACM, NBR, OT, FKM
–10	ACM, CO, FKM, TM, NBR

B.11 Typical Rubber Product Groups

Tires	Agricultural, aircraft, bicycle, motorcycle, scooter, car, earthmover, industrial tractor, perambulator, trolley, truck, bus
Belting	Fan, driving, elevator, flat transmission, conveyor, toothed, vee
Hose	Fueling, garden, brake, radiator, hydraulic, pneumatic, vacuum, chemical resistant, ventilation ducting, fire, sandblast, steam
Footwear	Wellingtons, canvas sports, sponge slippers, industrial, safety, heels, soles, stick-on soles, other components
Cables	Industrial covering, sheathing, insulating mats, plugs, sockets, clips, bushes
Mountings	Engine, machinery, instruments, shock absorbers, rail pads
Cellular	Pillows, mattresses, upholstery, carpet underlay, packaging, insulation
Inflatable	Beds, bags, cushions, meteorology balloons, dracones, dinghies
Seals	O-Rings, rectangular section, rotary shafts, gland rings, grommets, window, inflatable, fruit preserving jar rings, bellows, diaphragms, valves, magnetic rubbers
Sports	Bladders, handle grips, masks and fins for underwater, balls for golf, tennis, squash
Dipped	Balloons, golf ball sacks, teats, pen sacks
Flooring	Tiles, mats, sheetings
Linings	Tank, pipes
Adhesives	Cements, sealants
Sundries[a]	Hot water bottles, shower equipment, finger stalls, erasers, rubber bands, rubber stamps, toys
Proofing[a]	Ground sheets, hospital sheeting, tarpaulin
Rollers[a]	Wringer, typewriter, and business machines
Extrusions[a]	Glazing strips, channeling, welting, tubing, cord

[a] These groups are often collectively known as "General Rubber Goods."

Appendix C: Conversion Factors

Convert from	To	Multiply by
Atmosphere (760 mm Hg)	Pascal (Pa)	$1.013\,25 \times 10^{6}$
Board foot	Cubic meter (m^3)	$2.359\,737 \times 10^{-3}$
Btu	Joule (J)	$1.055\,056 \times 10^{3}$
Btu/h	Watt (W)	$2.930\,711 \times 10^{-1}$
Btu.In./s.ft^2.°F (k, thermal conductivity)	Watt/meter-kelvin	$5.192\,204 \times 10^{2}$
Calorie	Joule (J)	4.186 800[a]
Centipoises	Pascal second (Pa.s)	$1.000\,000^{a} \times 10^{-3}$
Centistrokes	$Meter^2$/second (m^2/s)	$1.000\,000^{a} \times 10^{-6}$
Circular mill	Square meter (m^2)	$5.067\,075 \times 10^{-10}$
Degree Fahrenheit	Degree Celsius	t°C = (t°F–32)/1.8
Foot	Meter (m)	$3.048\,000^{a} \times 10^{-1}$
Ft^2	Square meter (m^2)	$9.290\,304^{a} \times 10^{-2}$
Ft^3	Cubic meter (m^3)	$2.831\,685 \times 10^{-2}$
Foot-pound-force	Joule (J)	1.355 818
Foot-pound-force/min	Watt (W)	$2.259\,697 \times 10^{-2}$
Ft/s^2	Meter/$second^2$ (m/s^2)	$3.048\,000^{a} \times 10^{-1}$
Gallon (U.S. liquid)	Cubic meter (m^3)	$3.785\,412 \times 10^{-3}$
Horsepower (electric)	Watt (W)	$7.460\,000^{a} \times 10^{2}$
Inch	Meter (m)	$2.540\,000^{a}\ 10^{-2}$
$Inch^2$	Square meter (m^2)	$6.451\,600^{a} \times 10^{-4}$
$Inch^3$	Cubic meter (m^3)	$1.638\,706 \times 10^{-5}$
Inch of mercury (60°F)	Pascal (Pa)	$3.376\,85 \times 10^{3}$
Inch of water (60°F)	Pascal (Pa)	$2.488\,4 \times 10^{2}$
Kgf/cm^2	Pascal (Pa)	$9.806\,650^{a} \times 10^{4}$
Klp (1000 lbf)	Newton (N)	$4.448\,222 \times 10^{3}$
Klp/In^2 (ksl)	Pascal (Pa)	$6.894\,757 \times 10^{6}$
Ounce (U.S. fluid)	Cubic meter (m^3)	$2.957\,353 \times 10^{-6}$
Ounce-force	Newton (N)	$2.780\,139 \times 10^{-1}$
Ounce (avoirdupois)	Kilogram (kg)	$2.834\,952 \times 10^{-2}$
Ounce (avoirdupois)/ft^2	Kg/m^2	$3.051\,517 \times 10^{-1}$
Ounce (avoirdupois)/yd^2	Kg/m^2	$3.390\,575 \times 10^{-2}$
Oz (avoirdupois)/gal (U.S. liquid)	Kg/m^3	7.489 152
Pint (U.S. liquid)	Cubic meter (m^3)	$4.731\,765 \times 10^{-4}$
Pound-force (lbf)	Newton (N)	4.448 222
Pound (lb avoirdupois)	Kilogram (kg)	$4.535\,924 \times 10^{-1}$
Lbf/in^2 (psi)	Pascal (Pa)	$6.894\,757 \times 10^{3}$
Lb/$in.^3$	Kg/m^3	$2.767\,990 \times 10^{4}$
Lb/ft^3	Kg/m^3	$1.601\,846 \times 10$
Quart (U.S. liquid)	Cubic meter (m^3)	$9.463\,529 \times 10^{-4}$
Ton (short, 2000 lb)	Kilogram (kg)	$9.071\,847 \times 10^{2}$
Torr 9 mm Hg. (0°C)	Pascal (Pa)	$1.333\,22 \times 10^{2}$
Watt-hour	Joule (J)	$3.600\,000^{a} \times 10^{3}$
Yard	Meter (m)	$9.144\,000^{a} \times 10^{-1}$
$Yard^2$	M^2	$8.361\,274 \times 10^{-1}$
$Yard^3$	M^3	$7.645\,549 \times 10^{-1}$

[a] Exact.

Appendix D: Nuclear Magnetic Resonance (NMR) Spectroscopy

In order to determine the chemical structure unambiguously, NMR spectroscopy is often employed. Depending on the intrinsic nuclear spin value, when an atomic nucleus is subjected to any external magnetic field, it tends to align along with the field (α state, or lower energy state) or against the applied field (β state, or higher energy state). The difference between these two energy states is dependent on the strength of the external magnetic field and also on the chemical environment of the concerned nucleus. Considering the proton NMR, the various protons behave differently due to the shielding effect exerted on the nucleus of the proton by the surrounding electrons. For instance, if the same proton is attached to any electronegative group, there will be a de-shielding effect observed. This shielding and de-shielding effect will give rise to various peaks in the NMR spectra, which is normally expressed in terms of chemical shift (δ) value in ppm. Normally, the chemical shift value is expressed as a ratio between the "shift down field from TMS in Hz" to "spectrometry frequency in MHz". Here TMS (Tetra methyl silane) mentioned is the standard material used in proton NMR against which the chemical shift values are determined. A similar phenomenon is also true for ^{13}C-NMR. Owing to the lower proportion of availability of the ^{13}C nuclei, these NMR spectra are often associated with weak signals. The other factor responsible for weak signal intensity is the low gyromagnetic ratio of ^{13}C compared to proton. The essential point of difference between proton NMR and ^{13}C-NMR is that for the former, the intensity and splitting of the NMR signal gives an idea about the number of protons of the concerned atom as well as the number of protons on its adjacent atom; however, this fails to hold in case of ^{13}C-NMR spectroscopy. The situation is generally more complex for ^{13}C-NMR as the resonance due to the ^{13}C atom is split by the adjacent hydrogen atom. It is also highly unlikely that two 13-Carbons will reside side by side. Thus, to improve the scenario, a decoupling approach is applied, wherein a strong radio frequency with a broad range is applied to excite all the interfering neighboring H-nuclei and nullifying the coupling effect exerted onto the ^{13}C nuclei. This results in observation of a single NMR peak without any split in the case of the ^{13}C-NMR spectroscopy. The chemical shift values are shown in Tables D1 and D2, and a few representative spectra are given in Figures D.1–D.3.

TABLE D.1

Typical Chemical Shift Values Observed in ^{1}H-NMR

Type of Proton	Approximate Chemical Shift (δ)	Type of Proton	Approximate Chemical Shift (δ)
Alkane (-CH_3)	0.9	-C=C- (with CH_3 on C)	1.7
Alkane (-CH_2-)	1.3	Ph-H	7.2
Alkane (-CH-)	1.4	Ph-CH_3	2.3
-C(=O)-CH_3	2.1	R-CHO	9-10
-C=C-H	2.5	R-COOH	10-12
R-CH_2-X	3-4	R-OH	about 2–5
-C=C- (with H on C)	5-6	Ar-OH	about 4–7
		R-NH_2	about 1.5-4

TABLE D.2

Typical Chemical Shift Values Observed in ^{13}C-NMR

Type of Carbon	Approximate Chemical Shift (δ)	Type of Carbon	Approximate Chemical Shift (δ)
Carbonyl carbon atom	150-200	Saturated carbon with adjacent oxygen atom	50-100
Unsaturated carbon atom (both aromatic and aliphatic)	100-150	Saturated carbon atom	0-50

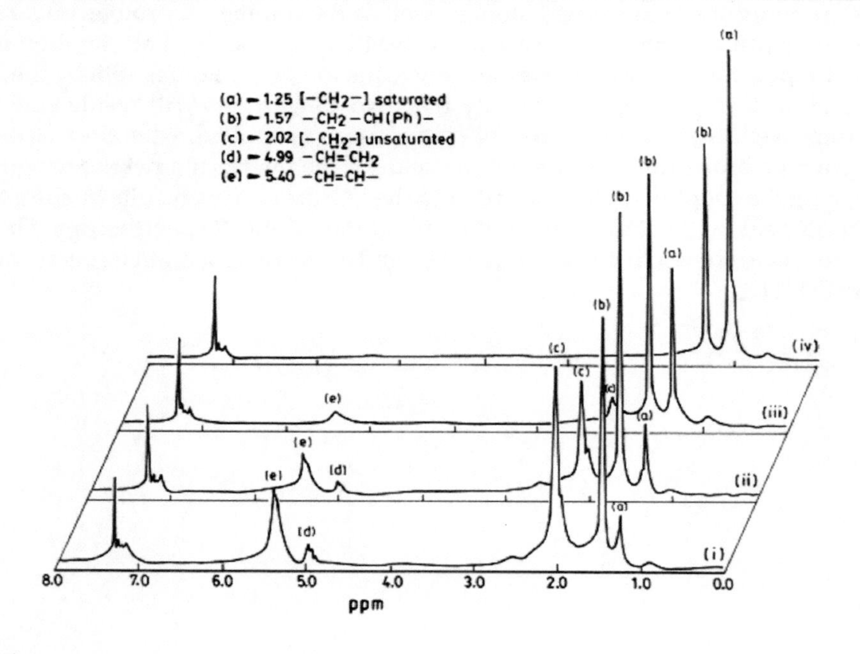

FIGURE D.1

1H-NMR spectra of (i) SBR and (ii,iii,iv) HSBR with 48, 62, and 92% hydrogenation, respectively. (De Sarkar, De and Bhowmick, *J. Appl. Polym. Sci.*, 1999, 71, 1581.)

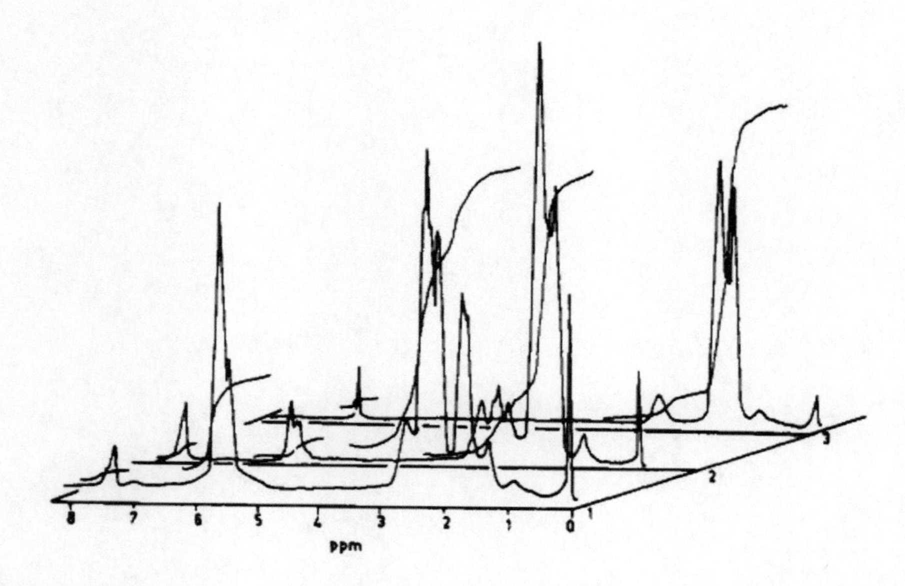

FIGURE D.2
^{1}H NMR spectra of (1) NBR and (2, 3) HNBR with 71% and 99.96% hydrogenation, respectively. (Bhattacharya, Bhowmick and Avasthi. *Ind. Eng. Chem. Res.* 1991, 30, 1086.)

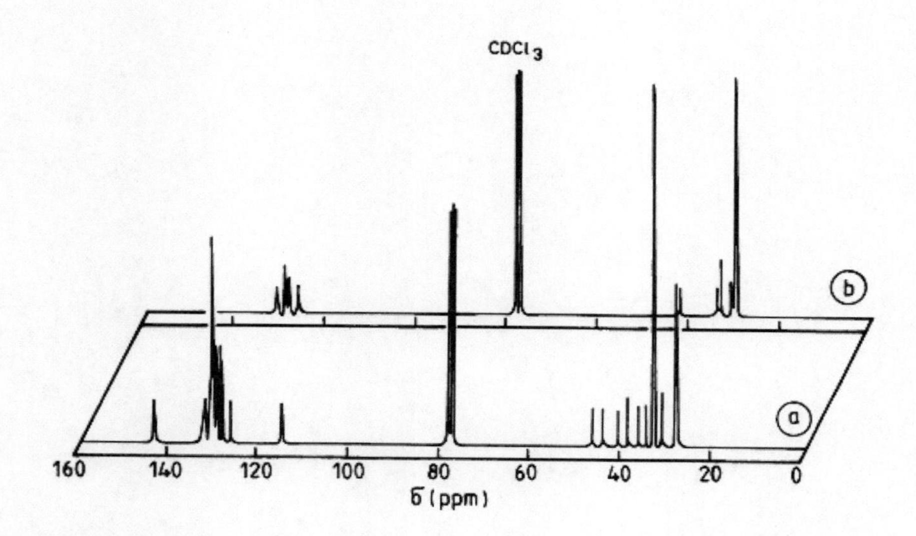

FIGURE D.3
^{13}CNMR spectra for (a) XSBR containing 1.2% methacrylic acid and (b) the same with 80% saturation. (De Sarkar, De and Bhowmick, *Polymer*, 2000, 41, 907.)

Glossary

Accelerator: A chemical that speeds the vulcanization of rubber used in the compounds to reduce curing time.

Activator: A rubber compound chemical used to help initiate the vulcanization process.

Alignment: Angles of the tire and suspension axes relative to each other and the ground: caster, camber, toe. Also, the adjustment of components to bring them into a predetermined position.

Alpha-Numeric: A load-based tire sizing system containing the load capacity, type of tire construction, aspect ratio, and the rim diameter in inches.

Antioxidant: A chemical that when added to a rubber compound prevents surface oxidation; used in tire tread and sidewall compounds to prevent weather checking and cracking.

Aramid: A fabric used in tire construction; a material that provides excellent high tensile strength to weight. The best known such aramid is Kevlar®.

Aspect Ratio: The dimensional relationship between the section height and section width; section height divided by section width.

Balance: A uniform mass distribution of a tire and wheel assembly about its axis of rotation.

Bale Rubber: The form in which natural rubber is shipped to tire manufacturers.

Banbury Mixer: An enclosed machine used for mixing the ingredients for rubber compounding. Commonly referred to as a Banbury.

Bead: The area of the mounted tire which seats against the wheel.

Bead Seat: The position where the tire rests and seals on the inside of the rim.

Belt: An assembly of fabrics and/or wire used to reinforce a tire's tread area. In radial tires, it also constrains the outside diameter against inflation pressure and centrifugal force.

Belt Edge Cushion: An extrusion of rubber placed under the edges of a belt used in radial tires to improve durability.

Belted Bias Tire: A bias tire with additional reinforcing belt(s) between the casing plies and the treads. Now, it is almost extinct. This tire was made mainly for the American market.

Bias Tire: A tire built with two or more casing plies that cross each other in the crown at an angle of 30 to 45 degrees to the tread centerline.

Bladder: A rubber bag used inside a tire during the molding and curing process; contains hot water or steam that presses the inside of the tire into the mold.

Blemish Tire: A tire with a cosmetic imperfection whose performance is unaffected.

Block: Part of a tire tread pattern created by lateral (side-to-side) grooves.

Breakaway: The point at which tire cornering traction is lost.

Breaking Torque: Torque applied by a brake to a tire/wheel assembly which slows or stops the vehicle.

Calendar: A machine consisting of two or more rollers which continuously rolls out a thin sheet of rubber compound. Also coats a fabric with a rubber compound as used in the plies of a tire.

Camber: The angle that the tire is leaning measured from true vertical. The inward or outward tilt of the wheel/tire at the top.

Camber Thrust: The cornering force developed by a tire due to its camber. A force in the same direction as the leaning of the tire.

Carbon Black: Very fine, specially structured particles of carbon; used in rubber compounds as a reinforcing filler.

Casing (Carcass): The structure of tire cords locked around wire beads.

Caster: The angle between the tire vertical and the steering pivot axis; the backward or forward tilt of the steering knuckle pivot points.

Center of Gravity: The center balance point of a vehicle: the single point where a car would be supported without tipping up or down.

Centerline: A plane dividing a tire, wheel, or vehicle into two symmetrical halves.

Centrifugal Force: The force that tends to throw a tire away from the center of rotation, same as lateral force.

Chafer: A finishing strip of calendered fabric used to protect the tire's bead area from the rim.

Chalk Test: A procedure to determine proper tire inflation pressure by chalking the outside shoulder at various points, running the tire, then checking to see how and where the chalk is worn away.

Chalking: A method for determining the right inflation pressure for a vehicle/tire/track combination. Tire shoulder blocks are chalked around the tire circumference, then inspected for the amount of chalk that is worn away after a test run.

Coefficient of Friction: The force required to slide an object, divided by the weight of the object; this indicates the difficulty in sliding one surface against another.

Compound: A mixture of rubber and chemicals; specially tailored to the needs of specific components.

Contact Patch: *See* Footprint.

Cord: Closely spaced parallel strands (warp) held together by light, widely spaced yarn (pick); used to form tire casings.

Cord Angle: Angle at which the tire cord crosses the centerline of the tread.

Cornering Coefficient: The cornering force generated by a tire at given conditions of slip angle, camber, inflation, etc., divided by the load.

Cornering Drag: A portion of a tire's cornering force which acts to slow down a car. It increases as slip angle increases.

Cornering Force: The force that turns a car around a corner. The opposite of lateral or centrifugal force.

Cornering Stability: An overall impression of a vehicle's stability during cornering; an indication of comparative ease with which the vehicle can be driven at its cornering limit.

Crown: The center part of the tread; the largest diameter in a rounded tire.

Crowned Road: A road design with a slope or pitch from its center to the curb or shoulder in order to facilitate water drainage.

Cure: To vulcanize; also time and temperature conditions used to vulcanize a tire.

Deflection: A turning or bending of the tire/wheel assembly, caused by the flexing of the wheel and resulting in reduced performance.

Design Rim: A rim with a specified width; used to measure tire dimensions.

Directional Stability: The ability of a car to travel in a straight line with a minimum of driver control.

Dog Tracking: A condition where the rear wheels do not follow the path of the front wheels.

Dot: A tire marking symbol that denotes that the tire meets the requirements of the American Department of Transportation.

Dual Compound Tread: A tire tread with two rubber compounds.

Durometer: A measure of the hardness of a rubber compound; its resistance to penetration of a spring-loaded blunt needle.

Dynamic Balance: Balance in motion. The balance of a wheel while it is rotating. A condition in which a tire and wheel assembly has weight distributed equally on both sides of the wheel's axis of rotation.

Eccentric Mounting: A condition in which a tire is unevenly mounted or cocked on the hub of a wheel.

E.C.E. Symbol: A tire performance certification based on regulations developed by the Economic Commission for Europe concerning physical dimensions, tire branding requirements, and high-speed endurance.

ETATO: European Tire and Rim Technical Organization.

Extrusion: The process of forcing rubber compounds through an aperture to give a profile or desired shape.

Fabric: An array of parallel cords used in tire manufacturing.

Fiberglass: A material used in belt construction consisting of fine spun glass coated with adhesive.

Filler: A rubber extrusion in the bead area of a tire; used to permit a smooth contour of casing plies around the bead and to the lower sidewall. Also used in enlarged form to stiffen the lower sidewall of the tire.

Flang and Height: The vertical height of the rim's side flange as measured from the top of the flange to the rim base.

Flipper: Cord that surrounds bead wire and bead filler; used to anchor the bead into the tire and to stiffen the lower sidewall.

Footprint: The area of a tire's tread under load in contact with the ground.

Fore-and-Aft Weight Transfer: A load factor where weight is transferred from the front tires to the rear tires during acceleration, and from the rear to the front tires during braking.

Green Tire: A tire that has not been vulcanized or cured.

Grooves: Circumferential channels between the tread ribs of a tire.

Grooving: A tread cutting process in which grooves of varying depths and angles are cut in braking, or lateral stability.

Heel: The rounded part of the bead which seats against the bead seat of the wheel.

Heel and Toe Wear: A type of wear peculiar to tread patterns with elements consisting of blocks or lugs where the trailing edge of the blocks wears more rapidly than the leading edge. When the leading edge of the flexible blocks comes into contact with the road surface, the blocks compress under load, and when coming off load as the compression is released, severe abrasion occurs to the trailing edge.

H-Rated: A speed rating category for tires which is used on vehicles with a top speed up to 130 mph.

Hydroplaning: Loss of traction at high speeds caused by a wedge of water that lifts a tire off the road surface.

Imbalance: A non-uniform distribution of mass in a tire and wheel assembly about its axis of rotation.

Independent Suspension: A suspension system in which the front or rear pair of wheels of a car is independently connected to the frame or underbody. In this system, deflection of the wheel on one side will not affect the wheel on the other side.

Inertia: The tendency of any mass at rest to stay motionless or any mass that is moving to remain moving in a straight line.

Innerliner: A thin layer of low permeability rubber on the inside of a tire; used to contain compressed air.

Kerb Weight: The total weight of a vehicle with no passengers and a full tank of fuel.

Kerfs: Small blades of metal on a tire mold which displace rubber; used to create minor grooves or slots in the tread. *See* Sipes.

Kevlar®: Brand name of E.I. DuPont de Nemours Co. for their aramid fibers.

Lateral Force: *See* Centrifugal Force.

Lateral Runout: Wobble or side-to-side motion of a rotating wheel or tire/wheel assembly.

Lateral Weight Transfer: A load factor in cornering where weight is transferred from the inside tires to the outside tires.

Lightness: Surface condition of a cured tire caused by the failure of the tire to contact the mold during vulcanization.

Load Range: The strength rating of a casing; determines maximum air pressure and load capacity.

Load Rating: The weight that a wheel is designed to support in normal service.

Loading: The amount of weight put on tires; increased load can increase cornering force.

Lower Sidewall: The part of the sidewall nearest the bead.

Lug Centric: The centering of a wheel by matching it up with the lug nuts, rather than by the center bore hole of the wheel; hub centric is the more accurate centering method.

MacPherson Strut: A front suspension assembly that combines the functions of the shock absorber, the upper steering pivot, and the wheel spindle in a single unit.

Match Mounting: A mounting procedure that matches the high point of a tire with the low point of its wheel. A dot or mark on the tire is matched with a dot, a sticker, or the valve hole on the wheel.

Metric Tire Size System: A tire sizing system using the cross section in millimeters, aspect ratio, speed category, tire construction, and the rim diameter in inches (e.g., 185/70SR13).

M+S, MIS, or M&S: A designation of a mud and snow tire.

Negative Camber: A condition where the top of the tire is leaning inward from the tire's vertical centerline as viewed from the top.

Numeric System: A tire sizing system using tire cross section width and rim diameter in inches (e.g., 7.35-14).

Offset: The distance from the centerline of the wheel to the mounting face of the wheel.

Off-the-Car Balancing: A procedure in which a tire and wheel assembly is balanced by a bubble or computerized electronic balancer while the assembly is off the vehicle; use of computerized electronic balancers is the best way to accurately measure dynamic balance.

Out-of-Round: A wheel or tire defect in which the wheel or tire is not round.

Overall Diameter: The maximum height of a tire when mounted on a wheel and inflated to rated pressure.

Overhead: A reinforcing ply or plies between the top belt ply and the tread; used to reinforce the tread/belt area.

Overinflation: The inflation of a tire above recommended pressure to achieve improved performance; negative byproducts are rough ride, bruise damage, and suspension system strain.

Oversteer: A cornering condition where rear tires operate at a greater slip angle than the front tires; the tendency of a car to turn more sharply than the driver intends while negotiating a turn.

Oxidation: Reaction of a material with oxygen usually resulting in degradation of the material.

Pitch: The length from a point on one tread block to the same point on the next tread block. Pitch is varied around a tire to minimize noise.

Plasticizer: A rubber compound chemical; used to make or keep rubber soft and flexible.

Plus 1/Plus 2 Concept: A concept to improve handling and performance through the mounting of tires with wider section widths and lower section heights to rims of 1, 2, and sometimes even 3 inches greater diameter.

Ply: A layer of rubber-coated fabric or wire making up the tire casing.

P-Metric System: A tire sizing system using the section width in millimeters, aspect ratio, type of tire construction, and rim diameter in inches (e.g., P225/70R15).

Polyester: A strong and lightweight synthetic cord material used in casing construction.

Polymer: A chemical compound made up of a large number of identical components linked together like a chain.

Positive Camber: A condition where the top of a tire is leaning outward from the tire's vertical centerline, as viewed from the top.

Post Cure Inflation: A manufacturing procedure of inflating a tire after it is released from a mold; required on thermoplastic materials like polyester to prevent shrinkage.

Power Oversteer: An oversteer condition caused by applying driving torque to the rear wheels during cornering.

Power Steering: A steering system that uses hydraulic pressure to reduce steering effort; the pressure powers pistons that are built into gear-box or rack-and-pinion steering mechanisms.

Pull: The tendency of a vehicle to veer to one side.

Pyramid Belt: A belt design in which the upper layer is narrower than the lower layer.

Pyrometer: A thermocouple device used for measuring tread temperatures in tires.

Radial Force Variation: The non-uniformity caused by changes in the spring rate around a tire; can cause vibration similar to the imbalance.

Radial Runout: The up-and-down motion of a rotating wheel or tire/wheel assembly: out-of-round.

Radial Tire: A tire built with casing plies that cross the crown at an angle of 90 degrees.

Rayon: A synthetic cord material used in casing and belt construction; provides high dynamic strength and good rubber adhesion.

Returnability: The ability of a vehicle to return to a straight-ahead attitude after removal of steering input.

Revolutions per Mile: The number of revolutions a tire makes in a mile at a given load, inflation, and speed.

Ribs: Parts of a tire tread pattern created by grooves that run circumferentially around the tire.

Rim Diameter: The diameter of the bead seat, not the diameter of the rim edge.

Rim Drop: The area of the wheel's rim having the smallest diameter.

Rim Flange: The outermost edge of a wheel's rim to which clip-on weights are attached.

Rim Width: The measurement inside of the rim flanges (i.e., from inside the flange on one side to inside the flange on the other side).

Road Wheel: A large diameter (typically 67") steer wheel capable of rotating at selected speeds; used to simulate road surface for tire testing. Also, a generic term describing wheels used on a vehicle.

Rolling Resistance: The force required to roll a loaded tire.

Rotation: The pattern of movement of tires to different positions on a vehicle to compensate for irregular or unequal tire wear.

Rubber-to-Void-Ratio: The ratio between the rubber area and the groove area in a tire footprint.

Runout Gauge: A device used to check and correct radial and lateral runout.

Section Height: The distance from the bottom of the bead to the top of the tread.

Section Width: The distance from sidewall to sidewall, exclusive of any raised lettering.

Self-Aligning Torque: The force that causes a tire/wheel assembly to return to its straight-ahead position after a turn.

Service Description (Load Index Speed Symbol): A speed rating system that describes the load capacity and high speed of a tire; includes numerical load indexes and alphabetical speed symbols.

Shock Absorber: A "damper" between the body or frame of the car and the suspension; used to cushion road bumps and bounces and keep the tires in contact with the road.

Sidewall: The side of a tire between the tread shoulder and the rim bead.

Sipes: Small cuts in the surface of a tread to improve traction.

Siping: The cutting of fine slits across the tread of a tire to improve traction in specialized competitive situations.

Spacer Ring: A metal ring used to increase the section width (and therefore the tread width) of a two-piece matrix so that tires of different dimensions and sizes may be cured in the same matrix. Spacers of different widths may be used.

Speed Rating System (Speed Category Markings): An alphabetical system describing a tire's capability to travel at established and predetermined speeds.

Sprung Weight: The total weight of a vehicle that is supported by the suspension system.

Squirm: The footprint distortion of a rolling tire usually hourglass in shape on a straight road and crescent shaped on curves.

S-Rated: A speed rating category for vehicles with a top speed up to 112 mph.

Stacked Belt: A belt design in which both layers are of equal width.

Standing Wave: Severe distortion of the tread and sidewalls of a tire when leaving the footprint at very high speeds; causes excessive heat buildup that can lead to high-speed tire failure.

Static Balance: Balance at rest. A condition in which a tire and wheel assembly has equal weight around the wheel's axis of rotation.

Static Loaded Radius: The measurement from the middle of the axle to the road surface; measured with the tire inflated to required pressure and carrying the rated load.

Steel Belt: A belt material used in radial tires. Its high stiffness provides good handling and low tread wear.

Sulfur: A chemical used in the vulcanizing process.

Swing Out: The tendency of the rear tires of a vehicle to break away during sudden steering maneuvers.

Synthetic Rubber: Rubber made from chemicals as a substitute for natural rubber; properties can be tailored for specific needs.

Tensile Strength: The maximum tensile force per cross-sectional area that a material can withstand before it breaks.

Tire Mixing: The installation of tires of different sizes and/or construction on a vehicle; a condition generally to be avoided. However, certain manufacturers recommend different tire sizes on the front and rear positions. Manufacturer's specifications should always be checked.

Toe (Bead Toe): The tapered inside edge of a tire bead.

Toe-In: A condition where the fronts of two tires on the same axle are closer together than at the rear.

Toe-Out: A condition where the fronts of two tires on the same axle are farther apart than at the rear.

Torque: The product of a force applied through a lever arm to produce a rotating or turning motion.

Torque Rating: The proper torque, expressed in foot-pounds, for tightening lug nuts of various diameters.

Torquing: The securing of the tire/wheel assembly to the vehicle by the tightening of the wheel's lug nuts to the studs of the vehicle's hub; in the case of speciality wheels, torquing should always be done with a manual torque wrench, containing an insert socket of plastic or Teflon®.

Track: The distance between the front tires on the front axle and the rear on the rear axle.

Tracking: The difference in distance between each of the rear wheels and the centerline of the vehicle.

Tractive Force: A frictional force generated by the tire against the road surface causing acceleration or deceleration.

Tramp: The action of a wheel in jumping, or hopping up and down.

Tread: The portion of a tire which contacts the road surface.

Tread Buffing: A process in which a portion of the tire tread is removed by buffing or grinding it down; similar to tread shaving.

Tread Depth: The distance from the tread surface to the bottom of the grooves.

Tread Radius: The radius of curvature of the tread arc across the tread.

Tread Shaving: The shaving of tread from a tire with a blade (usually to half of the original tread depth) to reduce tread squirm and tearing in racing applications.

Tread Wear Indicators: A raised area in the tread grooves which becomes level with the tread surface when the tire is worn to the legal wear outpoint in a tire's life.

Under-inflation: A condition where a tire is inflated below recommended pressure, resulting in sluggish response and greatly increased wear.

Undertread: The portion of the tread compound between the bottom of the tread grooves and the top of the uppermost ply belt.

Uniformity: A term describing the amount of radial and lateral force variation in a tire.

Unsprung Weight: The total weight of a vehicle not supported by the suspension system; tires and wheels, for example.

Upper Sidewall: The part of a tire's sidewall nearest the tread shoulder.

UTQGLBC: Uniform Tire Quality Grade Labeling; a performance measurement of a tire, based upon its test results in three categories—tread wear, traction, and temperature resistance.

Varied Pitch Ratio: Variations in angles and sizes of a tire's tread elements which reduce ride noise levels.

V-Rated: A speed category for vehicles with a top speed of more than 130 mph.

Vulcanization: The linking together, under heat and pressure, of rubber compound polymers which changes material from a sticky, putty-like substance to an elastic, bouncy substance.

Wheel Weights: Weights that are either clipped, taped, or self-adhered to the inside or outside of the wheel in order to balance the tire/wheel assembly.

Wires: High-tensile, brass-plated steel wires coated with a special adhesion-promoting compound which are used as tire reinforcement. Belts of radial tires and beads are two common uses.

Index

B

C

T